Patrick Moore's Practical Astronomy Series

Springer
London
Berlin
Heidelberg
New York
Barcelona
Hong Kong
Milan
Paris
Singapore
Tokyo

Other titles in this series

The Observational Amateur Astronomer
Patrick Moore (Ed.)

Telescopes and Techniques
Chris Kitchin

The Art and Science of CCD Astronomy
David Ratledge (Ed.)

The Observer's Year
Patrick Moore

Seeing Stars
Chris Kitchin and Robert W. Forrest

Photo-guide to the Constellations
Chris Kitchin

The Sun in Eclipse
Michael Maunder and Patrick Moore

Software and Data for Practical Astronomers
David Ratledge

Amateur Telescope Making
Stephen F. Tonkin

Observing Meteors, Comets, Supernovae
and other Transient Phenomena
Neil Bone

Astronomical Equipment for Amateurs
Martin Mobberley

Transit: When Planets Cross the Sun
Michael Maunder and Patrick Moore

Practical Astrophotography
Jeffrey R. Charles

Observing the Moon
Peter T. Wlasuk

Deep-Sky Observing
Steven R. Coe

AstroFAQs
Stephen F. Tonkin

The Deep-Sky Observer's Year
Grant Privett and Paul Parsons

Mike Inglis

With 60 Figures

Michael David Inglis, BSc, MSc, PhD, FRAS
Department of Astrophysics, Princeton University,
Princeton, New Jersey, USA

Cover illustrations: Inset photos by Michael Stecker
(mstecker@earthlink.net).

Patrick Moore's Practical Astronomy Series ISSN 1431-9756

ISBN 1-85233-630-7 Springer-Verlag London Berlin Heidelberg

British Library Cataloguing in Publication Data
Inglis, Mike
Field guide to the deep sky objects. – (Patrick Moore's
practical astronomy series)
1. Astronomy – Observations
I. Title
522
ISBN 1852336307

Library of Congress Cataloging-in-Publication Data
Inglis, Mike, 1954–
Field guide to the deep sky objects / Mike Inglis.
p. cm. – (Patrick Moore's practical astronomy series,
ISSN 1431–9756)
ISBN 1–85233–630–7 (alk. paper)
1. Astronomy – Observers' manuals. I. Title. II. Series.
QB64.I54 2001 00–058349
522–dc21

Printed in Singapore

Typeset by EXPO Holdings, Malaysia
Printed and bound by Kyodo Printing Co. (S'pore) Pte. Ltd., Singapore
58/3830-543210 Printed on acid-free paper SPIN 10732277

for
Karen Sara

Preface and Thanks

I was only knee-high to a tripod when, on a fateful day in April 1957, the first episode of the BBC television series, "The Sky at Night", presented by Patrick Moore, was broadcast on British TV. In the programme he mentioned that the comet Arend-Roland was visible to the naked eye, and my father decided that this was something I ought to see. I don't actually remember seeing the comet, but I do recall being carried out in my father's arms, under a pitch black sky in South Wales. From that moment on my life's purpose and direction were mapped out for me: I wanted to learn about the stars. So it is only right and proper that I begin this preface by thanking Patrick for providing the inspiration and desire (as he has also done for countless other people) to become, eventually, an astronomer.

The idea for a book of this type has been with me for many years, but it was only after meeting John Watson – of Springer-Verlag, the publisher – at the London Astrofest that I finally could begin work. John's knowledge of publishing, editing and indeed astronomy has been of incalculable worth during the writing of the book. It is easy to have a mind full of ideas, but to actually get these onto paper in a coherent manner so that they are understandable by everyone and not only those with astrophysics PhDs has been achieved only with the steady hand of John guiding me.

I began my journey into astronomy as an amateur, and to this day, it remains a very important part of my life. The number of amateur astronomers I have met is enormous, and some of these meetings have developed into deep friendships. I refer, of course, to those two erstwhile members of the South Bayfordbury Astronomical Society, Mike Hurrell and Don Tinkler. The observing sessions I have had with these characters are some of the most unique and unforgettable experiences of my life (I am still in therapy), and to this day I cannot recall them without breaking out into a smile, and more often than not, hysterical laughter. Their outlook on astronomy is enviable, as they just enjoy the subject for what it is, amazing and beautiful, a viewpoint that is often overlooked by focusing too much on the science of astrophysics and the obsession with the latest observing aids and equipment.

During my apprenticeship as a professional astronomer I have been fortunate enough to have met many wonderful people, not necessarily involved in astronomy, but rather as colleagues working at the same university. Between them, they managed to keep me sane, by listening to me complain about the injustice of life in general, buying me beer (an excellent source of contentment) and just being there. They remain to this day my dearest friends, and are indeed part of my extended family. Thank you, Bill and Pippa Worthington, Stuart and Mandy Young, Peter Harris and Andy Tye.

As a student and later as an educator and researcher, I was fortunate to meet people who had the same passion for astronomy as I. They too had decided to take the same path, and learn as much about the subject as possible. Learning about astronomy with a group of similar-minded individuals made the process so much more enjoyable, and to be able to teach students who were also fired with enthusiasm has been, and still is, a pleasure and privilege. Therefore I would like to thank my fellow travellers on this journey: Heather Reeder, Danny Gleeson, Chris Packham and Roger O'Brien. As an educator, I was also able to meet many people whose desire to learn about the universe and then pass this knowledge on to others matched my own. They made the sometimes bureaucratic nature of academia bearable: Jim Collett, Dave Axon, Chris McConville, Stuart Palmer and Chris Warwick.

However, all that I have achieved as an astronomer would have been for naught if it weren't for the astronomers at the University of Hertfordshire. Their knowledge of the way the universe works, and how to impart this knowledge to a student, is astounding. In my opinion they are the finest educators I have ever met. That they had the patience to teach me is something I still wonder about, as I remember only too well the number of mistakes and errors I made. But they took it all in good spirit and with humour, and to be associated with them is something that I look upon with humility. I would like to take this opportunity to acknowledge publicly the debt I owe these people: Iain Nicolson, Jim Hough, Alan McCall and Lou Marsh.

It is also important that I single out and mention Professor Chris Kitchin, who is director of the University of Hertfordshire Observatory at Bayfordbury, and Bob Forrest who is its principal technical officer. Their depth of knowledge about all matters astronomical, theoretical and observational is without parallel, and to be allowed to use the amazing variety of superb telescopes at the observatory was definitely a privilege afforded to few. Chris's knowledge of astronomy and astrophysics is very wide-ranging and impressive, and he has been my mentor in all that I have done. If it hadn't been for Chris, I would have never found out how much I like to teach astronomy, and I have learnt from him many of the techniques which make teaching a pleasure and not a chore. Bob is an astute observer and I learnt most of my observational skills at his side. He is also a patient and long-suffering man (especially when I am involved), as on many occasions when I could not get the telescopes to work (the Dec control just came off in my hand Bob, honest!), Bob would wander over, twiddle a control so that the object sprang magically into view, then walk away shaking his head and sighing deeply. Suffice to say that I owe both these astronomers a lot!

Writing a book is, strangely enough, just like observing. They are both, more often than not, solitary pastimes, and occur during the night hours. However, I have not been alone during either of these times, as I have had the music of the American musician Steve Roach as a companion. His tone poems are wonderful and inspiring, and it can truly be said that he has created the music of the spheres. Finally, I have to thank those people who are close to me, and whose constant, day-to-day presence makes life tolerable. Firstly I want to thank my partner in the journey of life, Karen, who smile makes everyday a happier time, and whose compassion and understanding when, sometimes, I complain bitterly about life, is a source of strength and comfort – thank you, Cariad. I also want to say a few words about my brother Bob. It is true to say that without his help, I would have never become an astronomer. He also enjoys a pint of beer as much as I do, which is an excellent measure of someone's character. (Can I buy you a pint Bob, and we'll call everything quits?) Finally, I want to thank my Mam and Dad. My father, who bought me my first astronomical telescope, always encouraged me to seek out my dream of becoming an astronomer, and the memory I have of showing him the rings of Saturn in the telescope, and his subsequent true and real joy, is something I will treasure forever. My mother, who has been with me all the way, with unconditional love, food parcels and clean socks. She has also been there for me, during those times when not everything was going according to plan.

To everyone I have mentioned, thank you all!

Dr Mike Inglis
St. Albans, Herts, UK
Lawrenceville, New Jersey, USA
January 2001

Acknowledgements

I would like to thank the following people and organisations for their help and permission to quote their work and for the use of the data and software they provided:

Michael Hurrell and Donald Tinkler of the South Bayfordbury Astronomical Society, England, for use of their observing notes.

Richard Dibon-Smith, of Toronto, Canada, for allowing me to quote freely the data from his books, *STARLIST 2000.0* and *The Flamsteed Collection*, and for the use of several of his computer programs.

The publishers of *The SKY Level IV* astronomical software, Colorado, USA, for permission to publish the star maps.

The European Space Organisation, for permission to use the Hipparcos and Tycho catalogues.

Jean Schneider, of the Observatoire de Paris, France, for permission to use the material from his website, *The Extra Solar Planets Encyclopedia*.

Gary Walker, of the American Association of Variable Star Observers, for information on the many types of variable star.

Cheryl Gundy, of the Space Telescope Science Institute, Baltimore, USA, for supplying astrophysical data on many of the objects discussed.

Dr Chris Packham, of the University of Florida, for information relating to galaxies, particularly active galaxies.

Mark Hendy, for the superb and professional work that he did editing the book, especially when text seemed to be disappearing via the electronic highway!

The Smithsonian Astrophysical Observatory, for providing data on many of the stars and star clusters.

In developing a book of this type, which presents a considerable amount of detail, it is nearly impossible to avoid errors. If any arise, I apologise for the oversight, and I would be more than happy to hear from you. Also, if you feel that I have omitted a star or object of particular interest, again, please feel free to contact me at: fg2dso@hotmail.com. I can't promise to reply to all e-mails, but I will certainly read them.

Contents

Chapter 1 Introduction

It's a beautiful, crisp, clear night, the seeing is perfect, light pollution is minimal, and you're equipped with high-quality binoculars or a telescope. Eager to pursue your passion for astronomy, you have decided that tonight you'll look at a whole range of interesting objects – beginning with a beautiful double star, perhaps a triple star system. Then once you're fully dark-adapted you'll try for a glimpse of a few compact star clusters and maybe a couple of spiral galaxies.

Easier said than done!

Even planning well ahead, you have to find out what you are able to see at this time of year. Putting together an evening's observing programme means that you have to know which double stars, galaxies and clusters are the best ones to observe. Star atlases – on paper or on the computer – give you the references and the positions, but won't indicate how easy your selected objects are to see. Other books tell you which objects are well placed to observe on a given night, but won't tell you which objects of a particular category – galaxies, say – are in good positions.

If it's a spur-of-the-moment decision, you may not have time to plan your observing in advance. In that case, when you get outside, not only will you have to carry your binoculars (or telescope) but also several books, catalogues and manuals. You'll need all these to discover which objects you are going to look at and where they are, and you'll probably spend an inordinate amount of time trying to find the positions and descriptions of the stars and galaxies you wish to see from several different books.

Then, unless you own a computer-controlled telescope, you've got to locate the objects in the heavens. It's quite likely that at this point, cold and frustrated, you will have lost your initial excitement, and what should have been a glorious time scanning the sky has instead become a trying one, leaving you with seriously dampened enthusiasm.

Amateur astronomy shouldn't be like that.

How much easier it would be if you had a guide to the night sky which not only provided you with a catalogue of each type of object you wanted to see, but also arranged those objects in such a way as to make it easy to locate them by their classification, ease of observing, and the date you observe. How much more useful than a list of objects made according to the constellation they appear in!

I couldn't find such a book, so I set out to write one myself. Here it is. I hope you find it as useful as I have.

My purpose is twofold: first to provide an introduction to the thousands upon thousands of objects within reach of the amateur astronomer, and second to help develop a lifelong interest in what is without a doubt the most beautiful, exciting and thought-provoking field of science.[1]

The notion of providing a means of finding an object by the type of object it is, and by the date you are observing it on (rather than by its position in the sky and what constellation it lies in) came partly from my own experiences as an amateur astronomer, when I spent many nights trying to find objects I had read about or seen pictures of, but partly also through my work as a professional astronomer teaching astronomy to undergraduates, where the time taken in locating a particular class of star, cluster, nebula or galaxy was often longer than the time I had been allocated to teach for!

As well as listing objects according to their type, I have also grouped them into three broad sub-categories according to how easy it is to see them – "easy", "moderate" and "difficult".

This helps observers to improve their skills in viewing different classes of object. For a given night, you can just work through the list in order. Start with the *easy* objects, then just work your way down to difficult. The best way to get better at observational astronomy is to look at progressively harder objects.

Most of the *easy* objects are very simple to find, and can be seen in binoculars and small telescopes, and perhaps even the naked eye. Those classified as *moderate* objects may be a bit fainter and smaller, needing a little more skill to observe properly. The *difficult* objects could be faint and located in sparsely populated regions of the sky and/or may need larger-aperture telescopes.

Locating and observing these objects in the order I have set them down should also help you hone your observing skills.

1.1 The Type of Objects

The number of objects that could be observed throughout a year, which could include interesting stars, clusters, nebulae and galaxies, is immense. Equipped with even a small pair of binoculars, you could easily spend a lifetime looking at (or for) these objects. Because this book is limited in size by simple economics, and cannot run to several volumes, this means a certain amount of judicious pruning has had to take place. I've therefore limited the number and type of objects to those which can be observed by most amateurs equipped with small (up to 10 inch, 250 mm) telescopes and binoculars. The categories are:

1. Stars (bright, coloured, double, triple, quadruple, multiple).
2. The spectral sequence.
3. Star clusters.
4. Globular clusters.
5. Stellar associations and streams.
6. Nebulae (emission, dark, reflection, planetary).
7. Supernova remnants.

---

[1]There are hardly any photographs in the book. That was a deliberate decision. The reason is that more often than not the photos give a completely false impression as to what can be seen through a telescope. I am seeking to give a true and accurate description of what can be observed. What you can actually see depends on your skills as an observer, and of course upon the kind of instrument you have available to observe *through*.

8. Galaxies (spiral, elliptical, irregular), galaxy clusters.
9. Several miscellaneous but interesting objects.

In some instances (like supernova remnants) there may only be a few objects described, even though many exist; they have been excluded simply because they are too difficult. (Readers who own large-aperture telescopes will find all of the listed objects easily; there are many other catalogues and handbooks that will satisfy a craving for smaller and fainter objects!)

I have tried, whenever possible, to include some objects that may not be so familiar to the amateur astronomer, along with some that are often described in specialist books and magazine articles, but usually get left out of general books like this one.

1.2 Observing the Night Sky

You may think that a section on observing the night sky is redundant. What is there to observing but just going outside with a pair of binoculars and getting on with it? Well, there are in fact several important topics that not only explain what limits the amount and type of observing that can be done, but also help you to understand how to achieve more when you go outside, and thus become a better observer.

Although I expect that most of you are very familiar with the use of a telescope, it wouldn't harm even experienced amateurs to read the following sections. Give it a shot – there may be a few things you don't know!

1.3 Binoculars and Telescopes

Once upon a time, all astronomy books would recommend that the first piece of equipment you ought to buy was be a pair of binoculars, because this would allow you to get a feel for the night sky and so introduce you to the constellations and objects that lie therein. For many years this advice was correct. But now, with the advent of small telescopes, of scarcely more than 80 mm (3 inch) aperture or thereabouts, equipped with superb optics, equatorial mounts, electric drives and computer databases of several thousand objects, this advice may no longer be entirely appropriate.

Furthermore, astronomy is now such a glamorous science, with incredible images from telescopes regularly seen in daily newspapers and on television and the Internet, that today's amateur often wants to go outside and find these objects immediately, and not bother with the slow and steady method of learning patiently over several months the shapes and contents of the constellations. And why not? There's plenty of room in the world for every sort of amateur astronomer.

It may be that the deciding factor on choosing optical equipment is now the cost. If money is no object, then by all means buy a mid-aperture telescope – a 200 mm (8 inch) reflector, say, or a 75 mm (3 inch) refractor. However, in every case, the quality of the optics must not be compromised. Whether you are buying binoculars or a telescope, always try to make sure that the optical system is of the highest quality. But how do you find out if what you are buying is of good quality, as there are, unfortunately, many pieces of astronomical equipment now available which are, to be blunt, far from appropriate? The answer is to make sure that you buy any and all of the equipment from a reputable supplier. Advertisements for these companies can often be seen in astronomy magazines, and in my experience, the people who run them are usually amateur astronomers themselves and will give you the best advice available, particularly if the company is large and well known.

Personally, I prefer to have both a pair of binoculars *and* a small-aperture telescope. The binoculars are used for scanning the night sky, to locate bright objects, and to sweep across the star fields of the Milky Way. The telescope is used for more detailed work on individual objects. My advice is find the piece of equipment you are comfortable with, and the best way to do this is to find your local astronomical society and attend their meetings.[2]

They are usually great fun, and their members will have both binoculars and telescopes; more often than not, one member at least will have a large-aperture telescope. They will be only too happy to give you good advice and help in your decision, and allow you the opportunity to use the different pieces of equipment.

Whatever you decide, whether it be a pair of good binoculars or a small telescope, or even a large computer-driven telescope, don't just rush out and buy it. Try, wherever possible, to use a similar piece of equipment before you buy one. This way, if you find that you cannot justify the purchase of a telescope because you feel you are not ready to progress to that size of aperture just yet, then all well and good: you not only will have saved yourself a considerable sum of money, but also will probably have changed what may have been just a passing interest in astronomy into a lifelong passion.

1.4 Magnification

Most of us have heard a story like this. Someone decides to "take up astronomy" and spends a lot of money on buying a telescope. It has superb optics, computer control, and comes with three eyepieces. A bright supernova has been reported as visible in the faint galaxy M33 in Triangulum, and he rushes out into the evening, sets up the telescope and using the highest magnification tells the computer to GOTO M33. Nothing is visible. Not a glimmer. The telescope ends up back in the shop, along with its irate owner.

Well, maybe that's just a *little* bit of an exaggeration. Most of you are already familiar with the night sky and have had some experience observing with binoculars or telescopes, so that the topic of magnification is not new to you. But who among us has not been so eager to try out some new piece of equipment, or view an exciting object, that we rush out and try to observe with an inappropriate magnification, only to be disappointed when the image doesn't live up to our expectations?[3] The purpose of this section is to explain how magnification works, and when best to use a certain magnification for a particular object, or when circumstances warrant it.

The topic of magnification can be a confusing one for newcomers to amateur astronomy. What is the best time to use high magnification, or low? Why doesn't higher magnification split close doubles? Why do some extended objects seem to get fainter at higher magnification, while others seem to get brighter with low magnification? This section will help clarify these and other points.

A full description of the physical process of magnification is beyond the scope of this book, but a few details are appropriate.[4] The magnification of a telescope (or binoculars) is given by a simple formula: the focal length of the objective lens (or mirror), f_o, divided by the focal length of the eyepiece, f_e. This is usually written as $M = f_o/f_e$.

---

[2]A list of useful addresses can be found in Appendix 2.
[3]I am guilty of this!
[4]A list of several books that deal with the subject in more detail can be found in Appendix 2.

For example, a telescope with a mirror of focal length 200 cm, used with an eyepiece of focal length 5 cm, gives a magnification of $200/5 = 40$. Thus the magnification of the telescope using this eyepiece is 40 times, sometimes written as 40×. In this way, a selection of eyepieces of differing focal lengths provides several different magnifications.

It usually comes as a surprise to newcomers to learn that there is a minimum magnification that can be usefully used. This is sometimes quoted as $M \geqslant 1.7D$, where D is the diameter of the telescope in cm.[5] What this really means is that there is a minimum magnification at which all of the light from the telescope passes into your eye. If you use a lower magnification some of the light is wasted as it is spread out over an area larger than the pupil of your eye.

As an example, if you have a telescope of diameter 20 cm, then the minimum magnification is $1.7 \times 20 = 34$. Thus an eyepiece as in the example above, which gives a magnification of 40×, would be appropriate.

It comes as less of a surprise that there is also a limit to the *highest* useful magnification that can be employed. In the past, advertisements for telescopes would quote ridiculously high values for magnifications, often several hundred times. While it is true to say that these magnifications are in theory possible, in practice they are, in a word, useless. Although there is no hard-and-fast rule for the limit of highest magnification, a good rule of thumb is that the highest power, on average, should be from about $10D$ to $20D$, where D is the diameter of the primary mirror (or lens) in centimetres. For a 20 cm telescope this would result in magnification from 200× to 400×. Such high magnifications are however, rarely used, as they suffer from the following drawbacks:

- A smaller field of view.[6]
- A decrease in brightness of extended or non-stellar objects.
- An exaggeration of atmospheric defects.
- An exaggeration of any tremors or defects in the mount or drive system.

Usually, most amateurs have a minimum of three eyepieces that provide a good range of magnifications. These are:

1. Low power, $2D$ to $3D$ – this shows the largest amount of sky.
2. Medium power, $5D$ to $8D$ – used for more general observations.
3. High power, $10D$ to $20D$ – useful for double star work.

The usual approach is always to observe with the lowest-power eyepiece, and then move on to other higher powers provided the conditions are right. On some very rare nights when the observing conditions are perfect you will be able to use very high powers, and the amount of detail that reveals itself will be staggering. But such nights are few and far between, and you will be the best judge of what eyepieces to use. It all comes down to experience, and gradually over time you will acquire a knowledge of the characteristics not only of the objects you observe but also of the limitations and expectations of your telescope and eyepieces.

Most people who are not familiar with observing would expect an object like a nebula to be brighter when viewed through a telescope than when seen with the naked eye. It's usually a surprise that this is not the case.[7] Basically, because of light losses and other effects, the brightness of a nebula – or any other extended deep-sky object – are fainter

---

[5]Other sources quote a value of $1.3D$.

[6]See Section 1.7.

[7]Note that this discussion does not apply to point sources, i.e. stars!

when seen through the telescope than if viewed by the naked eye! But what a telescope does is increase the apparent size of a nebula from an inappreciable to appreciable extent. And the background sky appears darker through a telescope than when seen with the naked eye. Too high a magnification, however, spreads the light out to such an extent as to make any detailed observation suspect.

Finally in this section on magnification, I should mention that there is a minimum magnification that is needed, if you are to *resolve* all the detail that your telescope is capable of achieving. Although the section on resolution comes later,[8] I think it is appropriate to mention this aspect here.

It's probably easiest to discuss this point by using an example. Take for instance a close double star. You know from the telescope's handbook (or you may have calculated for yourself) that the resolution of your telescope has a certain value. You see in the section on double stars in this book that your telescope is capable of splitting these stars. However, when you observe them, instead of seeing two separate and distinct stars, you see instead just one star, or maybe an elongated blur. This may be because the magnification is too low; although the double star should be resolvable by the objective lens (or mirror), the two components will not be observed as individual stars unless a high magnification is used in order to bring them above the resolving threshold of the human eye (about 2 to 3 minutes of arc).

Ignoring for the moment atmospheric effects and other considerations which limit resolution, this magnification is given a value anywhere from 10 to 16 times the telescope's aperture in centimetres. A ballpark number is 13 times the aperture in centimetres. Thus, to split very close double stars and to resolve detail close to your resolution limit, you need not only superb optics, good weather and so on, but also, on occasion, a high magnification.

Remember, however, that you can never increase the resolution by increasing magnification ad infinitum. As you will see in the next section, the resolution of your binoculars or telescope is constrained both by the size of the primary mirror or lens and by the physical nature of light.

1.5 Resolution

The topic of resolution is extremely theoretical, and a full description of the theory would be better suited to an undergraduate textbook in astrophysics.[9] Not surprisingly, it is also confusing for many amateur astronomers (and even a few professional astronomers), as more often than not, there are few books specifically written for the amateur which describe the Rayleigh resolution, the Dawes limit, the resolving power, the Airy disc, and so on. To that end I do not bother to explain where the theory and formulas come from, but just write them down without explanation as to their derivation.

Let's begin our foray into the area of telescopic resolution by starting with some simple theory. You might expect that stars which appear as incredibly small points of light to the naked eye because of their immense distance from us would, when magnified, still appear as small points of light – but observation tells us otherwise. The image of a star, when at the focus of a telescope, appears as a finite – although very small – disc of light. This is called the *Airy disc*. In fact, the Airy disc represents about 83 per cent of the star's light.

---

[8]See Section 1.5 – the Rayleigh criterion.

[9]For those with a theoretical mind, a list of astrophysics texts is in Appendix 2.

The remaining 17 per cent can often be seen as faint diffraction rings around the star's image.

This is the first counter-intuitive result: no matter how big a telescope you have, how perfect the optics, or how high a magnification you use, not all of the star's light goes into making the central image. This is a consequence of the wave nature of light.

The normally accepted definition of the theoretical resolution of a telescope is given by what is called the *Rayleigh criterion*, denoted by α and given by the formula $\alpha = 1.22\lambda/D$, where λ is the wavelength of light and D is the diameter in metres of the lens or mirror. However, the unit of measurement for this definition is the radian – one that strikes terror into most people! A more user-friendly formula is $\alpha = 1.22\lambda/D \times 206{,}265$, which gives an answer in arc seconds.[10] Even this can be simplified by assuming that the wavelength of light, λ, is about 550 nanometres, a perfectly acceptable value when using the telescope for optical observations. Thus the simplest formula is $r = 13.8/D$, where r is the angular resolution in arc seconds and D is the diameter in centimetres of the objective lens or mirror.

There's another definition of the highest resolution of a telescope to be found in the literature, and this is the *Dawes limit*. This one isn't derived from any theory, but is an empirical criterion, the result of a series of observations made with telescopes of various apertures. The resolution in arc seconds for the Dawes criterion is given by the formula $r_D = 11.6/D$, where D is the diameter of the objective lens or mirror in centimetres.

Both these resolution criteria are useful in that they can give a useful measure for the capabilities of your telescope, but in practice the performance of the telescope may be different from both the Rayleigh and Dawes criteria. The reasons for this are:

1. The visual acuity of the observer.
2. Both criteria only apply strictly for objects which are both of the same brightness. The bigger the difference in brightness, the greater is the discrepancy between what is expected and what is actually seen.[11]
3. The criteria have been calculated for a light wavelength of 550 nm, and thus it follows that a pair of bluish stars can be resolved at a smaller separation than a pair of reddish stars.
4. The type of telescope you use has can also change the resolution. A reflecting telescope has about 5 per cent greater resolution than a refractor of the same aperture, owing to diffraction effects at the support and flat in a Newtonian telescope.
5. Atmospheric turbulence, or scintillation, usually always stops you from achieving the expected resolutions.

It may seem to you that no matter what is the calculated or expected resolution of your own telescope, you will never achieve it, or even know what it is actually capable of resolving! Take heart, though – there *is* a way to discover what your telescope is actually capable of, namely by observing a number of double stars whose separations are known. By undertaking this series of observations, you will be able to determine the performance of your telescope under various conditions, and thus determine the resolution. Of course, the above list of conditions will have to be taken into account, but at least you will know how your telescope behaves, and thus will know what objects cannot be seen, and, more importantly, those which can be observed.

---

[10]206,265 is the number of arc seconds in one radian: $360 \times 60 \times 60$ divided by 2π.

[11]You may be forgiven in thinking that fainter stars would be easier to resolve than brighter ones, but this is not borne out from experience. In addition, bright double stars are found to be more difficult to resolve as they tend to "dazzle", and reduce the performance of the eye.

1.6 The Limiting Magnitude

Having decided what magnifications to use with your telescope, and the resolution you can expect to get, we now turn our attention to the topic of *limiting magnitude*, m_L, or *light grasp*. This determines what is the faintest object you can detect with a given telescope.

Once again, different books give different explanations and formulas, thus confusing the issue, and we should not forget that several factors similar to those I listed in the previous section will determine what you can see. And then there is the issue of whether continuous visibility is needed, or whether a fleeting glance can be considered as detection – an important point for the variable-star observer, who needs to make a definite magnitude determination,[12] as opposed to a glancing determination of magnitude.

The important point here is that the bigger the aperture, the greater the amount of light collected, and thus the fainter the objects detected. For example, a telescope with a 5 cm aperture has half the light grasp of a 7.5 cm one, which in turn has just over half the light grasp of a 10 cm telescope. In fact, the theoretical value for light grasp is given by the simple formula D^2/P^2, where P is the diameter of the eye's pupil.

In order to determine the limiting magnitude, a series of observations were carried out several years ago on the faint stars in the Pleiades star cluster, with telescopes of different apertures, and an empirical formula determined. This is $m_L = 7.71 + 5 \log D$, and represents the expected performance of typical observers under normal conditions.

Sometimes, when conditions permit, it may be possible to improve your visual limiting magnitude by using a higher magnification, because this reduces the total amount of light from the sky background.

Again, it is worth stressing that factors such as observing skill, atmospheric conditions, the magnification used and even the physiological structure of an observer's eyes (we're all different!) can and does influence the limiting magnitude observed, and any figures quoted are approximate.[13,14]

Table 1.1 gives the Rayleigh resolution, Dawes resolution and limiting magnitude for those telescope apertures most commonly used by amateurs.

Table 1.1. Rayleigh resolution, Dawes resolution and limiting magnitude

D (cm)	5.0	6.0	7.5	10.0	12.5	15.0	17.5	20.0	22.5	25.0
r (arcsec)	2.77	2.3	1.85	1.38	1.11	0.92	0.79	0.69	0.65	0.55
r_D (arcsec)	2.32	1.9	1.54	1.16	0.93	0.77	0.66	0.58	0.52	0.46
mL	11.2	11.5	12.1	12.7	13.2	13.6	13.9	14.2	14.5	14.7

[12] Magnitudes are discussed in Section 1.20.

[13,14] Many popular astronomy books will tell you that the faintest, or limiting magnitude, for the naked eye is around the 6th magnitude. This may well be true for those of us who live in an urban location. But the truth of the matter is that from exceptionally dark sites with a complete absence of light pollution, magnitudes as faint as 8 can be seen. This will come as a surprise to many amateurs. Furthermore, when eyes are fully dark-adapted, the technique of averted vision will allow you to see with the naked eye up to three magnitudes fainter! But before you rush outside to test these claims, remember that to see really faint objects, either with the naked eye, or telescopically, several factors mentioned in the text will need to be taken into consideration, with light pollution as the biggest evil.

1.7 Field of View

The field of view of binoculars and telescopes is an important topic. Field of view defines how much of the sky you see through your equipment. There are two definitions to be mentioned: the *apparent field of view*, α, and the *true field of view*, θ.

When a telescope is pointed at the sky and you place your eye at the eyepiece, you can see a circular area of light. The angular diameter of this disc of light is the *apparent field of view* of the eyepiece. Different eyepieces give different apparent fields of view. The apparent field can vary from 30° to 80°, depending on the type of eyepiece you use. The *true field of view*, θ, is the angular diameter of the area of the sky whose image is within the apparent field.

Dividing the apparent field of view of the eyepiece α by the magnification m resulting from the eyepiece gives the true field of view θ; thus $\theta \approx \alpha/m$. This equation shows that when the magnification increases, the size of the amount of sky visible – the true field – decreases. This is why it is so important to centre any object you view in eyepiece before switching to higher magnifications, especially with faint and small objects such as, say, planetary nebulae. If you don't do this and the object is off-centre, switching to a higher magnification will result in your losing the object, and a frustrating time can ensue as you try to find it again. Always centre objects in the eyepiece initially.[15]

To determine the field of view of your telescope with a particular eyepiece is very easy. Locate a star that lies on or very close to celestial equator – α Aquari and δ Orionis are good examples – and set up the telescope so that the star will pass through the centre of the field of view. Now, using any fine controls you may have, adjust the telescope in order to position the star at the extreme edge of the field of view. Turn off any motor drives and measure the time t it takes for the star to drift across the field of view. This should take several minutes and seconds, depending on the eyepiece used. Then multiply this time by 15, to determine the apparent field diameter of the eyepiece, conveniently also in minutes and seconds – but minutes and seconds of arc, rather than of time.

If you have to use a star which does not lie on or close to the celestial equator, then the formula $15t \cos \delta$, where δ is the declination of the star, can be used to find the apparent field diameter of the eyepiece in minutes and seconds of arc.

All of the previous sections have dealt with parameters that are defined by the optics of the system you are using. Let's now look at a factor that is beyond the control of most of us – the atmosphere.[16]

1.8 Effects of the Atmosphere

No matter what sort of equipment you have, whether it is a pair of binoculars, a small telescope, or a large-aperture "light-bucket", all equipped with superb optics, and a rock-solid mount, there is one element that can reduce us all to equals, and this is the atmosphere we

[15]With some observations it may be necessary to off-centre the object in order to locate fainter and more elusive structure within an object. This is good and proper observing technique, but centre it first, then move off.

[16]At the time of writing, equipment is coming onto the market which can, albeit in a small way, deal with some of the effects of the atmosphere using active optics.

live in. Not that I'm against it, of course (we have to breathe), but it really is the bane of astronomers the world over! There are three things to discuss: transparency, seeing and light pollution.[17]

1.9 Transparency

The term *transparency* is used to define the clarity of the atmosphere, an important factor when taking long-exposure photographs or measurements. It is dependent on the altitude of your observing site and there are several components which contribute to the transparency: clouds, fog, mist, smoke and particles suspended in the air. Such is the effect of these components that even under what may be considered perfect conditions, a star always appears nearly three magnitudes dimmer at the horizon than it would at the zenith.

Other factors can also affect transparency. Living near built-up areas[18] and even aurorae in the upper atmosphere can dim the stars.

1.10 Seeing

Seeing is something everyone is familiar with if they trouble to look, for the twinkling of the stars is dependent upon the condition of the atmosphere – whether the air is steady or turbulent. Seeing affects the quality of the telescopic image in that it can cause the image in the telescope to dance about, or deform the image, or even do both simultaneously.

The twinkling, or scintillation, comes about when the temperature of the air changes, altering the air's refractive index and hence causing the image to flicker. Surely everyone has noticed the scintillation that Sirius exhibits when it is close to the horizon on cold winter nights? It twinkles in all the colours of the rainbow as different wavelengths are dispersed by the atmospheric turbulence!

Seeing can be also be divided into two components, sometimes referred to as "high" and "low" seeing. High seeing is due to air currents found at altitudes of a 1000 metres and more, and causes the movement of an image in the field of view as mentioned above. Low seeing, as you might expect, depends on local conditions near ground level and also within a dome or telescope; for example, warm air trapped inside a dome or telescope tube causes the air to become unsteady.[19]

Telescopes with different apertures are affected in different ways. One with a large aperture "sees" a larger cross-section of turbulent air and so may produce a more deformed image than a smaller-aperture one.

---

[17]I have included light pollution under this heading as without an atmosphere, there wouldn't be any light pollution!

[18]This is covered in fuller detail on the section on light pollution.

[19]A surprising result is that when the transparency is poor, the seeing can be good. For example, if the there is a slight haze in the air, as on Autumn evenings, the atmosphere will be still, and this is often a good time for planetary observation. Similarly, when the transparency is good, and the stars appear as bright sharp points, the seeing can be at its worst. These nights however, are perfect for observations of nebulae and galaxies.

1.11 Light Pollution

It is a sad reflection on our times that nearly all amateur astronomers are familiar with, and complain of, light pollution. It has grown alarmingly over the past few decades, and although steps are being made to reduce it, it marches ever onward.

The cause of the problem is predominantly street lighting, shining up into the sky and being scattered and reflected throughout the atmosphere. This glow of diffuse light is instantly recognisable as the orange radiance seen over most of the horizon and extending quite high into the sky. From urban sites, it severely limits what can be seen, particularly of the fainter nebulae and galaxies.

The good news is that most of the brighter objects I talk about in this book can be seen from a town location.

There are two practical solutions to the problem of light pollution, (1) using filters to block it out,[20] and (2) making your observations from a dark location. The first may be of little use to the binocular observer, but should be a weapon in the arsenal of every telescope user, while the second may be impractical for casual observers or those of us with big telescopes.

However, observing from a dark sight is a truly awe-inspiring experience. I have observed from several locations throughout the world – usually, but not exclusively, at major professional observatories – and the experience of being unable to discern the constellations because of the plethora of stars visible was breathtaking – as was being able to see most of the Messier objects in Sagittarius with the naked eye!

If you ever have the opportunity to observe from a really dark location, then take it![21] It also makes sense to observe after it has rained, as rain removes some of the dust and larger cloud particles in the air, and reduce the scattering of the street lights.

The three atmospheric variables mentioned above – transparency, seeing and light pollution –all have an influence on what can be observed, and thus you should always record them. Any classification you use will always be of a subjective nature, and writing "poor" as a description of the conditions will not usually provide much information, at least for other people. Several observing scales have been introduced, and the one most widely used is the Antoniadi scale. It was originally intended for lunar and planetary work, but is just as applicable for other objects. You can write it down in "shorthand" like this:

I Perfect seeing, no movement whatsoever.
II Moments of calm lasting several seconds, with some slight undulation.
III Moderate seeing, accompanied with large air tremors.
IV Poor seeing, accompanied by constant air tremors.
V Bad seeing, preventing any worthwhile observing to be made.

That's enough about seeing. Now let's turn our attention to two matters that will allow you, when you have mastered them, to observe faint objects – *dark adaption* and *averted vision.*

---

[20]A discussion on light pollution filters can be found in Appendix 1.

[21]Note that the term "dark sky" refers to a sky that is clear, free of light pollution, and transparent – not one that is very black. In fact, contrary to popular belief, the more light that is seen from stellar objects, the brighter the sky will appear to be. From exceptionally "dark" sites, the light from the stars, Milky Way, zodiacal light, galaxies, and so on., will all combine to brighten the night sky. With such a "dark sky" objects with very faint magnitudes can be observed.

1.12 Dark Adaption and Averted Vision

The eye is a very complicated optical device. It uses a simple lens to focus light onto the retina, and changes focus by altering the geometry of the lens itself. The retina – the light-sensitive inside back surface of the eye – is composed of two sorts of photosensitive cell, *rods* and *cones*. These cells are packed closely together, stacked rather like the pile in a carpet. The cones are responsible for the perception of colour and for our excellent daytime colour vision; they also enable us to see fine detail. The rods, on the other hand, are sensitive to very low levels of illumination (and also to movement), but produce a low-resolution image and do not discriminate different colours. These different photosensitive cells are mixed together in the retina, but the concentration of each type is not even – there are many more rods toward the edge of the retina, and more cones near the eye's optical axis. Right in the centre of the retina is a small area called the *fovea centralis*, where there are no rods at all and the cones are packed extremely densely. This area is about 600 μm in diameter, and is the part of the retina that provides the highest resolution of detail; however, the absence of rods makes it relatively insensitive to light.

The eye has three mechanisms for regulating the amount of light reaching the retina. Short-term and fairly small variations are dealt with by the iris, which opens and closes to adjust the size of the eye's aperture. There is also a safety system (rather like that on the Hubble Space Telescope) that rapidly closes our eyelids if the light is too bright. The third mechanism is chemical in nature, and varies the sensitivity of the retina itself, like using fast or slow film in a camera. It is this chemical mechanism that causes the most dramatic change in the eye's sensitivity to light.

When you leave a brightly lit room and go outside and observe you can only see a few stars, but after a period of time, which can be as short as 10 minutes, it becomes apparent that more stars are becoming visible. In fact, after a period of about 15 minutes, the eye is six times more sensitive to low light levels than it was immediately after leaving a lit room. This is the process of *dark adaption*. If you spend even longer in the dark – at least 30 minutes – then the rod cells, located, you will remember, mostly nearer the edge of the retina, can become nearly *one thousand* times more sensitive.

That's why it really is very important to allow time for your eye to become adapted to the darkness before you start to observe. Try to make sure that no bright light can interfere with your observing (as this will destroy your dark adaption and you will have to wait another 30 minutes before you begin observing again). Such is the sensitivity of the eye that even bright-red light can affect your adaption, so a dull-red light should be used. Indeed, some extremely keen observers place thick black cloth over their heads (yes, it does look silly!) to totally exclude even the minutest trace of light, before beginning observations. It's worth thinking about keeping your eyes dark-adapted, and allowing time: it really does work, as faint objects will be much easier to locate.

A possible problem can arise for double-star observers because of the eye's response to differing levels of illumination. The cones, which are responsible for colour vision, peak at about 550 nanometres, that is, in the yellow–green region of the spectrum, while the rods peak at about 510 nanometres, the green region of the spectrum. This shift in sensitivity is called the *Purkinje effect* and may result in an observer underestimating the magnitude of a bright and hot star when comparing it with a cool, fainter one. This is also the reason why moonlight appears bluer than sunlight.

When you look directly at an object, you are using the fovea centralis, because it is in the optical centre of the retina, and probably some of the surrounding area of the retina in which there are relatively few rod cells. The human eye has evolved to provide us with high-resolution vision in daylight, along with low-resolution but sensitive night vision, particularly at the periphery – for daytime hunting and night-time avoidance of predators!

Observers can consciously make use of the sensitive rod cells to look at faint objects. This is the technique of using *averted vision*. It may go against all your experience, but don't look directly at the object. Instead, look to one side of it, shifting your gaze from one side of the object to the other. This works surprisingly well, and much more detail becomes apparent than you would expect.

It works best well with very faint objects of course, and should be the principal method of observation of such objects.[22]

1.13 What to Wear

This may seem a strange topic to include, but, as experienced observers know only too well, if you are not kept reasonably warm while you observe then in only a very short time you will be far too cold to even make any pretence of observation. Basically, common sense should prevail, and even if you don't actually wear any thick clothing to begin with, it should be available as the temperature drops. Nothing spoils a clear night's observing more than having cold hands and feet.

Those of us familiar with the climate of Northern Europe know that even in early and late summer, when the days may be warm and sometimes hot, the nights can be quite chilly. So to be sensible, take gloves and a hat (a very important piece of apparel), have a thick coat to hand, and wear thick shoes. Shoes with thick insulating soles – such as air-sprung sports shoes – are good. During winter nights, when the skies can be glorious, the temperature can easily fall way below zero, and warm clothing is a necessity. It also helps if you can keep out of any wind, as this will soon sap any heat from you. Several layers of loose clothing are preferable to one thick layer, as they trap pockets of air which are then warmed by your body heat.

My last piece of advice in this section is to know when it's time to stop observing! This is usually when your teeth are chattering, and your body is shaking from the cold.[23] This is the time to go indoors and have a hot drink, looking back in your mind over the incredible sights you have seen earlier that evening.

1.14 Taking Notes

This is a subject that, for some unknown reason, is often ignored by many amateur astronomers. If you are just casually observing, sweeping the sky at leisure, then recording your observations may indeed be superfluous, but once you start to search out and observe specific objects, like the stars, nebulae and galaxies found within this book, then note-taking should become second nature, if only to make a checklist of the items you have seen.

---

[22]There is something else that is applicable to observing, namely the *Troxler* phenomenon, which basically means that any image that remains on the same area of the retina for any amount of time will be ignored by the brain and seem to disappear. Thus, when observing, move your eyes about. The inverse is true as well, in that you should rest your eyes from time to time.

[23]Believe it or not, but many observers, myself included, have reached this stage on several occasions, not wanting to waste a minute of a clear night!

Basically, you should record the object viewed, the time of the observation,[24] the telescope used, with eyepiece and magnification, the seeing and transparency, and maybe a brief description of the object and a sketch. These notes can then, if needs be, be copied up into more formal notes later, preferably the next day. (Your initial notes should always be made at the telescope, so as to keep as accurate a record as possible.) Some observers use a separate book or notepad for each type of object observed. A pocket dictation machine can be a quick and easy substitute for the notebook (apart from the sketch!). Whatever method you use, keeping a record will help you keep track of your observations, and will help you to become a better observer.

1.15 The Science of Astronomy

This book was written for non-professional astronomers who want to observe the night sky, and doesn't deal in any great depth with the physics and mathematics of the subject – that is, astrophysics. There are plenty of books available that do that.

But it is inevitable that throughout this book, terms will arise which need defining, such as magnitude, light-years, right ascension and declination, the spectral sequence,[25] galaxy classification, and so on and so forth. I thought it would be instructive to explain these terms, even if you are already familiar with some of them.

1.16 Angular Distance

In some areas of astronomical observing it is useful to be able to estimate angular distances by eye alone.

From the horizon, to the point directly above your head – the zenith – is 90°. If you look due south, and scan the horizon going from south to west, continuing to the north, then east and back to south, you will have traversed 360°. The angular diameter of the Moon and also of the Sun is 0.5°. Other distances which may be of use are:

δ–ε Orionis	1.25°
α–γ Aquilae	2°
δ–ζ Orionis	3°
α–β Canis Majoris	4°
α–β Ursae Majoris	5°
α–δ Ursae Majoris	10°

Further approximate distances are:

The width of the nail of your index finger, at arm's length	1°.
The width of your clenched fist held at arm's length	8°.
The pan of your open hand held at arm's length	18°.

[24]A discussion of the time coordinates used in astronomy is in Section 1.17.

[25]The descriptions of the spectral sequence and galaxy classification are given in the appropriate chapters.

1.17 The Date and Time

We are all familiar with the passing of time – day and night, autumn following summer, family birthdays, and so on. But in astronomy accurate measurements rely on the accurate determination of the time of the observation. Not surprisingly, there are many time systems used, but we shall limit ourselves to only a few. When you are making notes at the telescope, there is a standard way to record the *date and time,* usually in units of increasing accuracy, for example:

2020 February 2d 2h 22m 2.4s

or

2020 February 2.204

A time system that is often seen, but little understood, is the *Julian date*, JD. It is the number of days that have elapsed since midday at Greenwich on 4713 BC January 1. The reason for such a strange date is that it doesn't rely on any secular calendar, and any astronomical events which we may wish to look at that may have happened in the past, for example, the past return of comets, have positive dates. To calculate a Julian date is straightforward but longwinded, and the interested reader is referred to any of the textbooks mentioned in the appendices.[26]

The *apparent solar day* is the interval between two successive transits of the Sun. It is a measure of the motion of the Sun across the sky. However, as the Earth's orbit is elliptical, this interval changes slightly throughout the year, and so the *mean solar day* is also used. This is the time interval between two successive transits of the mean Sun, which is thus equal to the mean value of the apparent solar day. The mean solar day is the time interval we are all used to, in day-to-day time-keeping. *Greenwich Mean Time,* GMT, is the mean solar time at the longitude of Greenwich, England, starting from midnight. It is also known (more correctly) as *Universal Time,* UT. All astronomical observations should be quoted in UT, even though this may need some recalculating if you don't live near the Greenwich meridian. For example, 7.00 UT in Greenwich would be 2.00 UT in New York.

1.18 Coordinate Astronomy

Throughout this book you will see that all the objects listed have definite coordinates, namely *right ascension,* RA, *and declination,* Dec. These allow you to locate, with the use of star maps, the position of any celestial body.[27]Because the vast majority of astronomical objects observed in the universe are so faint that to locate them by visual techniques is impossible,[28] a coordinate system – rather like a map reference – was devised. The RA and Dec (as they are usually referred to) are the celestial equivalents of longitude and latitude. The system may initially look complicated, but once mastered it is extremely useful, and indeed indispensable. Any book on basic astronomy will explain the details of the system,

---

[26]The modified Julian date, MJD, is the number of days that have elapsed since Greenwich Midnight 1989 June 7.75.

[27]There are several other coordinate systems in use in astronomy – horizontal, ecliptic, galactic, heliocentric – and descriptions of them can be found in most astronomy textbooks.

[28]Nearly all the objects in this book are visible to the naked eye, binoculars or small telescopes.

so here I just highlight the more important topics here. Basically, try to visualise the stars placed on a sphere, with the observer at its centre. This is the "celestial sphere". Now imagine the Earth's equator and north and south poles projected onto this celestial sphere. Call these the "celestial equator", and "celestial north and south poles".

The RA is measured east of the equivalent of the Greenwich meridian, and is described not in degrees but in hours, minutes and seconds, with the Greenwich meridian at 0^h. Declination is a measurement above and below the celestial equator: above (north of) the celestial equator it is preceded with a plus sign (+), and below (south of) it with a negative sign (–). The units of declination are degrees (°), minutes (′) and seconds (″). Using these coordinates, the position of any object may be located in the sky. Finding a given object in the sky can sometimes be a problem, even when you know the RA and Dec. There are several possible approaches. A detailed knowledge of the night sky is something that is be built up over time, and so eventually you will have an idea of where the object should be, but this background information isn't likely to be acquired overnight!

A popular technique is *star-hopping*. You begin by locating the position of the object on a star map, and then you try to see if there are any bright stars nearby that you can use as signposts to the fainter object, moving the telescope from one to another until you finally have your target in your field of view. This is actually quite a good way of finding double stars, nebulae and galaxies, which may reside in unfamiliar parts of the sky. Even this technique has its limitations, especially with the very faint objects and those that lie in parts of the sky where there are few bright stars. In such cases the next alternative is to use the *setting circles* which should be on the telescope. Many people find using setting circles difficult, yet spending a few minutes aligning them can allow you to locate many faint and hitherto unobserved objects. In order to use setting circles correctly, the telescope needs to be equatorially mounted and polar-aligned. A telescope that is equatorially mounted has its polar axis pointing to the north celestial pole (NCP), and its declination axis at right angles to it. This is illustrated in Fig. 1.1. The advantage of such a mounting is that once an object is located within the field of view, it is necessary only to move the telescope in right ascension in order to track it across the sky as the earth rotates. Most amateur astronomers use portable telescopes, and it is imperative that if you have one of these you

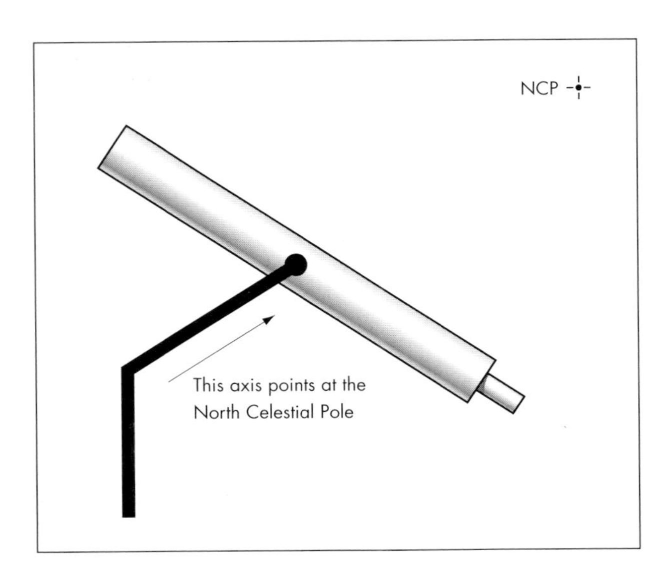

Figure 1.1. An equatorially mounted telescope with its polar axis pointing to the north celestial pole, NCP.

correctly align the polar axis so that it points towards the NCP. Different telescope manufacturers recommend different methods of doing this. Most involve at least a degree of trial and error.

For short-term visual observing, it is perfectly acceptable to try to align the axis on Polaris, the Pole Star. In most cases this suffices to track an object; however, after a few hours it will be necessary to realign the telescope. Some telescopes are equipped with a small polar alignment finder scope as part of the main telescope body or mounting, which allows you to set up the axis such that it is correctly positioned.

Having aligned the telescope, you can now begin to use the setting circles.

Do it like this. First find a bright star, for example Vega (α Lyrae), and centre this in the telescope. You need to know the RA and Dec of Vega; they are 18^h 35.9^m: 38° 47′. A book (including this one!), a chart or planetarium software will give you the figures you need.

Now align the setting circles so that the RA and Dec are at the coordinates of (in this case) Vega. The telescope having been aligned for Vega, it will now be set up for locating any other object.[29] It may seem a cumbersome procedure, but it will, I assure you, be a great advantage. Finally, there are computer-controlled telescopes. These have recently achieved affordable prices and great popularity. It is true to say that they have revolutionised amateur astronomy. Along with motor drives on both the RA and Dec axes, they also come equipped with a database of at least several thousand stars, nebulae, galaxies and the planets. All you have to do to find an object is key in its reference number (or RA and Dec coordinates), and the telescope will automatically slew to the desired position. These are fantastic telescopes because they allow you to spend much more time observing an object, instead of trying to find it! I can speak with some experience; having spent a large amount of time using telescopes *without* these facilities and then later with them, I can testify that I found and saw objects that previously I had only ever read about. They have opened up the field of deep-sky observing like no other product.

1.19 Distances in Astronomy

The best-known unit of astronomical distance is the *light year*. It is simply the distance that electromagnetic radiation travels, in a vacuum, in one year. As light travels at a speed of 3,000,000 km per second, the distance it travels in one year is 946,000,000,0000 km. It is often abbreviated to l.y. The next commonly-used distance unit is the *parsec*. This is the distance at which a star would have an annual parallax of 1 second of arc, hence the term *par*allax *sec*ond. The parallax is the angular measurement that is observed from the Earth when an object (a star or nebula) is seen from two different locations in the Earth's orbit, usually 6 months apart. It appears to shift its position with respect to more distant background stars, and an object that appears to shift by 1 second of arc will lie at a distance of 3.2616 l.y. Figure 1.2 will help; r is the distance, in this case one parsec, and P is the parallax, of one arc second.

[29]It is important to remember that the setting circles will be accurate only if the telescope is equatorially mounted. This means that the Dec axis points to the North (or South) celestial pole. Several telescopes which have an altazimuth mount have a "wedge" which allows you to tilt the telescope so that it is correctly mounted.

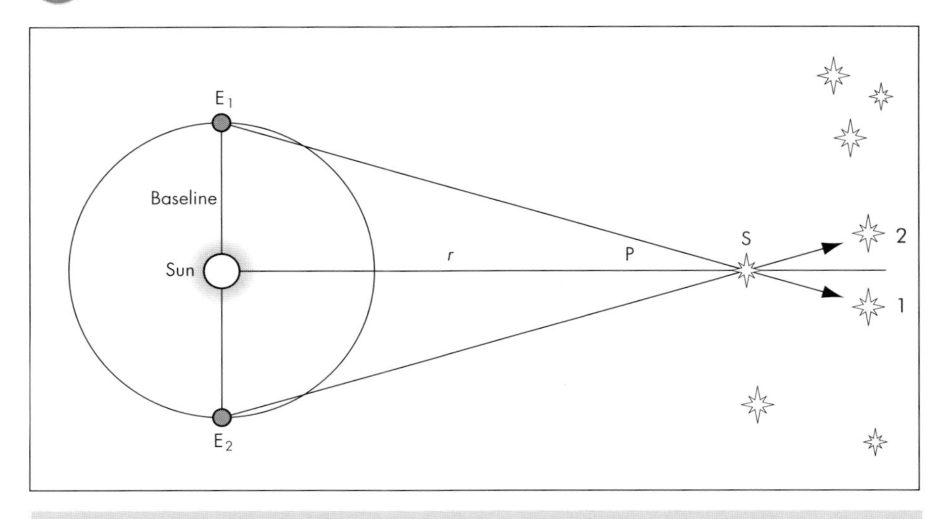

Figure 1.2. Diagram illustrating parallax: the star, S, when observed from Earth at E_1, is in the direction of the background star 1. Six months later, when star S is observed from Earth at E_2, S is in the direction of the background star 2. This apparent change in direction to the star results from its being observed at the two ends of the baseline, which is 2 AU long.

Another unit of distance sometimes used is the *astronomical unit,* AU, which is the mean distance between the Earth and Sun, and is 149,597,870 km. Note that one light year is nearly 63,200 AU.

1.20 The Magnitudes of Objects

The first thing that strikes even a casual observer is that the stars are of differing brightness. Some are faint, some are bright, and a few are very bright; this brightness is called the *magnitude* of a star.

The origins of this brightness system are historical, when all the stars seen with the naked eye were classified into one of six magnitudes, with the brightest being called a "star of the first magnitude", the faintest a "star of the sixth magnitude". Since then the magnitude scale has been extended to include negative numbers for the brightest stars, and decimal numbers used between magnitudes, along with a more precise measurement of the visual brightness of the stars. Sirius has a magnitude of −1.44, while Regulus has a magnitude of +1.36. Magnitude is usually abbreviated to m. Note that the brighter the star, the smaller the numerical value of its magnitude.

A difference between two objects of 1 magnitude means that the object is about 2.512 times brighter (or fainter) than the other. Thus a first-magnitude object (magnitude m = 1) is 2.512 times brighter than a second-magnitude object (m = 2). This definition means that a first-magnitude star is brighter than a sixth-magnitude star by the factor 2.512 raised to the power of 5. That is a hundredfold difference in brightness. The naked-eye limit of what you can see is about magnitude 6, in urban or suburban skies. Good observers report seeing stars as faint as magnitude 8 under exceptional conditions and locations. The magnitude brightness scale doesn't tell us whether a star is bright because it is close to us, or whether a star is faint because it's small or because it's distant. All that this classification tells us is the *apparent magnitude* of the object – that is, the brightness of an object as observed visually, with the naked eye or with a telescope. A more precise definition is the *absolute magnitude*, M, of an object. This is defined as the brightness an object would have at a distance of 10 parsecs from us. It's an arbitrary distance, deriving from the technique of distance determination known as parallax; nevertheless, it does quantify the brightness of objects in a more rigorous way. For example, Rigel has as an absolute magnitude of −6.7, and one of the faintest stars known, Van Biesbroeck's Star, has a value of +18.6.

Of course, the above all assumes that we are looking at objects in the visible part of the spectrum. It shouldn't come as any surprise to know that there are several further definitions of magnitude that rely on the brightness of an object when observed at a different wavelength, or waveband, the most common being the U, B and V wavebands, corresponding to the wavelengths 350, 410 and 550 nanometres respectively. There is also a magnitude system based on photographic plates: the *photographic magnitude*, m_{pg}, and the *photovisual magnitude*, m_{pv}. Finally, there is the *bolometric magnitude*, m_{BOL}, which is the measure of all the radiation emitted from the object.[30]

Where I say "magnitude" in the rest of this book, I mean the apparent magnitude.

1.21 Star Maps

The night sky can be a confusing place, even to a naked-eye observer, and through binoculars or a telescope even more so, with its myriad stars, and clusters, nebulae and galaxies. To make sense out of this apparent confusion, there is one item that should rank alongside a telescope and few decent eyepieces as an essential piece of equipment, and that's a star atlas.

Throughout this book you'll find several star maps, to help you to locate the constellations and the objects within them. The maps are very simple and are meant to be signposts to their *approximate location* of the objects in the sky. I cannot stress too highly the importance in getting a detailed star map, as this allows you to plot the location of the objects listed in this book and subsequently help you find them. An example of a good pocket-sized star atlas is *The Observer's Sky Atlas*, by E. Karkoschka; it is, incidentally, a perfect companion to this volume. Various other star atlases are available,[31] which show objects down to much fainter magnitudes and should be in the library of any serious observer. I should also mention a recent addition – within the last decade – to the many atlases that are available, and this is the Planetarium, or computerised star atlas. There is a plethora of computer programs which will plot a virtual night sky, as seen from anywhere in the world, and for whatever time you wish. They range in sophistication from very simple "star atlas" types, showing just the basic constellations and objects contained therein, to the more detailed "ephemeris" type which have stars and other celestial objects such as galaxies and nebulae, often down to a magnitude of 12 and even fainter. Several of them also possess the ability to guide and control various commercially -made telescopes.

---

[30]It is interesting to reflect that *all* magnitudes are in fact not a true representation of the brightness of an object, because every object will be dimmed by the presence of interstellar dust. All magnitude determinations therefore have to be corrected for the presence of dust lying between us and the object. It is dust that stops us from observing the centre of our Galaxy.

[31]These atlases are listed in Appendix 2

1.22 ... And Finally

Having read this first chapter, you should now have some idea of what's in store for you, and be at least theoretically familiar with the techniques that will allow you to become a competent observer. Before you rush outside, I'd just like to finish with a few words of prudence.

Astronomy, like any hobby or pastime (or lifelong devotion!), improves with time. The longer you spend observing, the better you will become. It won't be too long before you are familiar with most of the objects in this book, and know how to locate them without recourse to a star map. You will also, I hope, glean some background information as to the origin and structure of these fascinating and beautiful objects. Remember that success in seeing all the objects that you observe – or try to observe – will depend on many things: the seeing, the time of year, the instruments used and even your state of health! Don't be despondent if you can't find an object the first time round.[32] It may be beyond your capabilities, *at that particular time*, to see it. Just record the fact of your non-observation and move onto something new. You'll be able to go back to the elusive object another day, or month. It will still be there. Take your time. You don't need to rush through and observations. Try spending a long time on each object you observe. In the case of extended objects (nebulae, galaxies and star clusters) it is sometimes very instructive and fascinating (and often breathtaking) to let an object drift into the telescope's field of view. You'll be surprised at how much more detail you will seem to notice.

Finally, although I have tried to include all the famous objects, if I have omitted your favourite, then I apologise! To include everybody's favourite would be a nice idea, but an impossible task... Astronomy is a remarkable and exhilarating science, in which amateur astronomers can participate. We live in exciting times. Above all else – enjoy yourselves!

---

[32]It is often useful to be able to determine the night sky's observing conditions (light pollution, haze, cloud cover, transparency) before starting an observing session, so as to determine what type of objects will be visible and even allow you to decide whether observing is viable at all. A good way to do this is to use a familiar constellation, which should be observable every night of the year, and estimate what stars in the constellation are visible. If only the brighter stars are visible, then this would limit you to only bright stellar objects, while if the fainter stars in the constellation can be seen, then conditions may be ideal to seek out the more elusive, and faint objects. A favourite constellation used by many amateurs for just such a technique is Ursa Minor, the Little Bear. If, once outside, you can see θ UMi (mag 5.2) from an urban site, then the night is ideal for deep-sky observing. However, if θ UMi is not visible, then the sky conditions are not favourable for any serious deep-sky observing, but casual constellation observing may be possible. If the stars δ, ε and ζ UMi, located in the "handle" of the Little Bear, are not visible (magnitudes 4.3, 4.2 and 4.3 respectively) then do not bother observing at all, but go back indoors and read this book.

Most of the view of our galaxy is obscured by dust. The Hubble Space Telescope peered into the Sagittarius Star Cloud, a dust-free region, to provide this spectacular glimpse of a treasure chest full of stars. Some of these gems are among the oldest inhabitants of our galaxy. By studying the older stars that pack our Milky Way's hub scientists can learn more about the evolution of our galaxy

Many of the brighter stars in this image show vivid colours. A star's colour reveals its temperature one of its most "vital statistics". A star's temperature and the power of its radiation allow scientists to make conclusions about its age and mass. Most blue stars are young and hot, up to ten times hotter than our Sun. They consume their fuel much faster and live shorter lives than our Sun. Red stars come in two flavours: small stars and "red giants". Smaller red stars generally have a temperature about half that of our Sun, consuming their fuel slowly and thus living the longest. "Red giant" stars are at the end of their lives because they have exhausted their fuel. Although many "red giant" stars may have been ordinary stars like our Sun, as they die they swell up in size, become much cooler, and are much more luminous than they were during most of their stellar life

This image was obtained by combining three separate images taken through red, blue and green filters. This provides a true colour picture.

Hubble Heritage Team (AURA/STScI/NASA)

Chapter 2
Stars

This chapter deals with those objects which are easily the most familiar to us all, whether astronomers or not – the stars. Stars are the first things you notice as soon as you step out on a clear night, and are usually the initial target of any observing session, whether it be to align the telescope or judge the seeing conditions.

You may be forgiven in thinking that one star looks very like another, and to a casual observer this may well true, whereas an experienced amateur will know that the stars differ from each other in many fascinating ways and are a rich source for study. This chapter looks at the brightest and nearest stars, double, triple and even quadruple star groupings. I also present examples of the different spectral classification types of stars. There's even a section devoted to coloured stars!

I'm going to start this section by introducing some basic astrophysics, and it will be *basic*! From this point on, the information I give about the stars includes details such as their luminosity, their spectral classification, and so on, so I'll begin by explaining in detail what I mean. It is no use, and indeed pointless, if certain details about a star are given without an explanation of what is meant by the information.

2.1 The Light from a Star

When you go out in to the evening, the only apparent difference you notice between the stars is their brightness, or magnitude. Yet most amateur astronomers are aware that some stars are called white dwarfs, while others are red giants. Some are old stars, while there are a few that are relatively young. How were these things discovered? Furthermore, even to the untrained naked eye a few stars have a perceptible colour – Betelgeuse (α Orionis)[1]

[1]To explain how stars are named would take a book in itself. Suffice to say that many of the brighter stars were given Arabic names, for example, Betelgeuse, Rigel, and so on, for the simple reason that they were originally named by the ancient Arabian astronomers. The name in brackets signifies a star's listing in order of brightness in that particular constellation, so α Orionis should be brightest star in the constellation Orion, β Orionis (Rigel), the second brightest, and so on and so forth. When all the letters of the Greek alphabet are used up, numbers are then given to the stars, thus: 1 Orionis, 2, Orionis, and so on. The use of the Greek alphabet for referencing stars is called the Bayer classification system, while using a number is the Flamsteed system.

is most definitely red, Capella (α Aurigae) is yellow, and Vega (α Lyrae) is steely blue. What causes stars to have such different colours, and why aren't all the stars visibly multi-coloured?

Well, stars can be classified into various groups, and these groups relate to the stars' temperature, size and colour. In fact, the classification is so exact that in certain cases stars' masses and sizes can also be determined.

The history of stellar classification is a fascinating study in itself, but is not really important to us here. I will just explain the basic principles and how these are related to the stars that you observe. Before I go any further, however, it's probably worth discussing just what a star is!

A star is an immense ball of gas.

It is as simple, or as complex, as that, whichever way you wish to look at it. Owing to its very large mass, and its concomitant strong gravitational field, conditions in the centre of the ball of gas are such that the temperature can be about 10 *million* degrees Kelvin. At this temperature nuclear fusion occurs – and a star is born!

The gases composing the star are hydrogen, the most common element in the universe, along with some helium and then some other elements.[2] By and large, most stars are nearly all hydrogen, with just a few percent helium, and very small amounts of everything else. As the star ages, it uses up more and more hydrogen in order to keep the nuclear reactions going. A by-product of this reaction is helium. Thus, as time passes, the amount of hydrogen decreases and the helium increases. If conditions are right (these include a higher temperature and a large mass) then the helium itself will start to undergo nuclear fusion at the core of the star. After a long time, this in turn will produce, as a by-product of the reaction, the element carbon, and again, if conditions are suitable, this too will start to begin nuclear fusion and produce more energy. Note that each step requires a higher temperature to begin the nuclear reactions, and if a star does not have the conditions necessary to provide this high temperature then further reactions will not occur. So you will realise that the "burning" of hydrogen and helium is the power source for nearly all the stars you can see, and that the mass of the star determines how the reaction will proceed.

To determine the classification of star is a theoretically easy task, although it can be difficult in practice. What is needed is a spectroscope. This is an instrument that looks at the light from a star in a special way by utilising either a prism or a diffraction grating to analyse the light. You'll be aware that white light is in fact a mixture of many different colours, or wavelengths, so it's safe to assume that the light from a star is also a mixture of colours. Indeed it is, but usually with an added component. Using a spectroscope mounted at the eyepiece end of the telescope,[3] light from the star can be collected and photographed (these days with a CCD camera). The end result is something called a spectrum.

Basically, a spectrum is a map of the light coming from the star. It consists of all of the light from the star, spread out according to wavelength (colour) so that the different amounts of light at different wavelengths can be measured. Red stars have a lot of light at the red end of the spectrum, while blue stars have a correspondingly larger amount at the blue end. However, the important point to make is that in addition to this light, there will be a series of dark lines superimposed upon this rainbow-like array of colours. These are

---

[2]Astronomers designate as metals every element other than hydrogen and helium. It's odd, I agree, but don't worry about it – just accept it.

[3]Some spectroscopes place the prism or grating in front of the telescope, and thus the light from *every* star in the field of view is analysed simultaneously. This is called an *objective spectroscope*. The drawback is the considerable loss of detail, i.e., information, about the stars, but does allow initial measurements to be made.

called *absorption lines*, and are formed in the atmosphere of the star. In a few rare cases, there are also bright lines, called *emission lines*. These lines, although comparatively rare in stars, are very important in nebulae.[4] The origins of the absorption lines are due to the differing amounts of elements in the cooler atmosphere of the stars. (Recall that I mentioned earlier that in addition to hydrogen and helium, there are also the other elements, or metals, present, although in minute quantities.)

The factor that determines whether an absorption line will arise is the temperature of the atmosphere of a star. A hot star will have different absorption lines from a cool star – the classification of a star is determined by examining its spectrum and measuring aspects of the absorption lines. The observational classification of a star is determined primarily by the temperature of the atmosphere and not the core temperature. The structure of the absorption lines themselves can also be examined, and this gives further information on pressure, rotation and even whether a companion star is present.

2.2 Stellar Classification

Having seen how stars are distinguished by their spectra (and thus temperature), let's now think about the spectral type. For historical reasons a star's classification is designated by a capital letter thus, in order of *decreasing* temperature:

O B A F G K M R N S

Thus the sequence goes from hot blue stars types O and A to cool red stars K and M. In addition there are rare and hot stars called *Wolf–Rayet* stars, class WC and WN, exploding stars Q, and peculiar stars, P. The star types R, N and S actually overlap class M, and so R and N have been reclassified as C-type stars, the C standing for Carbon stars. Furthermore, the spectral types themselves are divided into ten spectral classes beginning with 0, 1, 2, 3 and so on up to 9. A class A1 star is thus hotter than a class A8 star, which in turn is hotter than a class F0 star. Further prefixes and suffixes can be used to illustrate additional features:

a star with emission lines (also called f in some O-type stars)	e
metallic lines	m
peculiar spectrum	p
a variable spectrum	v
a star with a blue or shift in the line (for example P-Cygni stars)	q

And so forth. For historical reasons, the spectra of the hotter star types O, A and B are sometimes referred to as *early-type* stars, while the cooler ones, K, M, C and S, are *later-type*. Also, F and G stars are *intermediate-type* stars.

Finally, a star can also be additionally classified by its *luminosity*, which is related to the star's intrinsic brightness, with the following system:

[4]See Chapter 4 for a discussion about the processes which make nebulae shine.

Supergiants[5]	I
Bright giants	II
Giants	III
Subgiants	IV
Dwarfs	V
Subdwarfs	VI
White dwarfs	VII

It's evident that astronomers use a complex and very confusing system! In fact several classes of spectral type are no longer in use, and the luminosity classification is also open to confusion. It will not surprise you to know that there is even disagreement among astronomers as to whether, for example, a star labelled F9 should be reclassified as G0 ! Nevertheless, it is the system generally used, and so will be adhered to here. Examples of classification are:

α Boötes (Arcturus)	K2IIIp
β Orionis (Rigel)	B8Ia
α Aurigae (Capella)	G8 III
P Cygni	B1Iapeq
Sun	G2V

Table 2.1. Spectral classification

Spectral-type	Absorption lines	Temperature °K	Colour	Notes
O	ionised helium (HeII)	35,000 +	blue–white	massive, short lived
B	neutral helium first appearance of hydrogen	20,000	blue–white	massive and luminous
A	hydrogen lines singly ionised metals	10,000	white	up to 100 times more luminous than Sun
F	ionised calcium (CaII), weak hydrogen	7,000	yellow–white	
G	CaII prominent, very weak hydrogen	6,000	yellow	Sun is G-type
K	neutral metals, faint hydrogen, hydrocarbon bands	4,000 to 4,700	orange	
M	molecular bands, titanium oxide (TiO)	2,500 to 3,000	red	most prolific stars in Galaxy

[5]These can be further sub classified into Ia and Ib, with Ia the brighter.

I conclude my section on spectral classification with what the spectral-type actually *refers* to.[6] You will recall that the classification was based on the detection of absorption lines, which in turn depend on the temperature of the star's atmosphere. Thus, the classification relies on the detection of certain elements in a star, giving rise to a temperature determination for that star. The classification can be summarised best by Table 2.1.

Having now briefly discussed the various stellar parameters and classifications, let's begin our exploration of the night sky.

2.3 The Brightest Stars

This section will list and detail the twenty brightest stars in the sky, as seen with the naked eye. Some stars, for instance Alpha Centauri, will be visible only from the southern hemisphere, while others can be observed only from a northerly location. The stars are listed in Table 2.2, going from the brightest to the faintest, but, as you will see from the detailed descriptions, the stars (and all other objects within the book) are catalogued in such a way

Table 2.2. The twenty brightest stars in the sky

Star	Apparent magnitude, m	Constellation
1 Sirius	-1.44_v	Canis Major
2 Canopus	-0.62_v	Carina
3 Alpha Centauri	-0.28	Centaurus
4 Arcturus	-0.05_v	Boötes
5 Vega	0.03_v	Lyra
6 Capella	0.08_v	Auriga
7 Rigel	0.18_v	Orion
8 Procyon	0.40	Canis Minor
9 Achernar	0.45_v	Eradinus
10 Betelgeuse	0.45_v	Orion
11 Hadar	0.61_v	Centaurus
12 Altair	0.76_v	Aquila
13 Acrux	0.77	Crux
14 Aldebaran	0.87	Taurus
15 Spica	0.98_v	Virgo
16 Antares	1.05_v	Scorpius
17 Pollux	1.16	Gemini
18 Formalhaut	1.16	Piscis Austrinus
19 Becrux	1.25_v	Crux
20 Deneb	1.25	Cygnus

[6]It usual for only the classes O, A, B, F, G, K, M, to be listed. The other classes are used and defined as and when they are needed.

as to allow you to observe objects at different times of the year, from January to December. All these stars in this section are easily visible with the naked eye, even from the most heavily light-polluted areas.

Throughout the rest of the book I'll use the following nomenclature: first is the common or popular name for the object (if it has one), followed by its scientific designation. The next item is its position in right ascension and declination, and the final term is the date of transit at midnight (at Greenwich) – this is the date when the object is at its highest in the sky, and so will be the best time to observe.[7]

The next line will then present information pertinent to the type of object. Thus, if it is a star, it will give its magnitudes, stellar classification and distance. If a double star is being described, then information about both components will be given, if a star cluster, its size, and so on. The positions quoted are for epoch 2000.0 and the source of data is the Hipparcos Catalogue. Some objects will be *circumpolar*, that is, they never set below the horizon, and thus are observable on every night of the year (weather permitting). (However, this is a double-edged sword because it also means that at certain times of the year the object, be it a star, nebula or galaxy, will be so close to the horizon that it would be a waste of time trying to observe it.) Use your own judgement to decide. Circumpolar[8] objects will be indicated by the symbol ©, which will be placed after the ease-of-observation designation. Note that the data is presented in a way which allows you to observe a class of object, be it a star, nebula or galaxy. A full description will be given for the month an object transits at midnight, whilst for other months, both before and after the transit date, just the physical data will be given, along with a reference to the transit date.

January

Sirius	α canis Majoris	$06^h\ 45.1^m$	$-16°\ 43'$	January 20
−1.44v[9]m	1.45M	A1 V	8.60 l.y.	
Also known as the Dog Star, this is the brightest star in the night sky. It is the closest bright star visible from a latitude of 40°N, with a parallax of 0.3792″ When observed from northerly latitudes, it is justly famous for the exotic range of colours it exhibits owing to the effects of the atmosphere. It also has a close companion star known as the Pup, which is a white dwarf star, the first ever to be discovered. Sirius is a dazzling site in a telescope.				

[7]Any object can of course be observed earlier or later than this date. Remember that a star or any astronomical object (except the Moon!) rises about 4 minutes earlier each night, nearly $^1/_2$ hour each week, and thus about 2 hours a month. To observe any object earlier than its transit date, you will have to get up in (or stay up to) the early hours of the morning. To observe a star later than the transit date will mean looking for the object earlier in the evening. As an example, Sirius transits on January 1 at midnight, but will transit on December 1 at about 2.00 a.m., and at around 4.00 a.m. on November 1. Similarly, it will transit on February 1 at approximately 10.00 p.m., and on March 1 at about 8.00 p.m.

[8]The circumpolar objects will be those that are visible from around a latitude of 40° north.

[9]Denotes that the star, and thus the magnitude, is a variable.

Procyon	α canis Minoris	07^h 39.3^m	–56° 13′	January 15
0.40m	2.68M	F5 IV	11.41 l.y.	

The eighth-brightest star in the sky, notable for the fact that it has, like nearby Sirius, a companion star which is a white dwarf. However, unlike Sirius, the dwarf star is not easily visible in small amateur telescopes, having a magnitude of 10.8 and a mean separation of 5 arcseconds.

Pollux	β Gem	07^h 45.3^m	+28° 02′	January 16
1.16 m	1.09 M	K0 IIIvar	33.72 l.y.	

The seventeenth-brightest star is the brighter star of the two in Gemini, the other being of course Castor. It is also, however, the less interesting. It has a ruddier colour than its brother, and thus is the bigger star.

Acrux	α Crucis	12^h 26.6^m	–63° 06′	March 29
0.72 m	–4.19 M	B0.5 IV + B1 V	321 l.y.	

See March.

Aldebaran	α Tauri	04^h 35.9^m	+16° 31′	November 29
–0.87 m	–0.63 M	K5 III	65.11 l.y.	

See November.

Rigel	β Orionis	05^h 14.5^m	–08° 12′	December 9
–0.18$_v$m	–6.69 M	B8 Iac	773 l.y.	

See December.

Capella	α Aurigae	05^h 16.7^m	+46° 00′	December 10
0.08$_v$m	–0.48 M	G5 IIIe	42 l.y.	

See December.

Betelgeuse	α Orionis	05^h 55.2^m	+07° 24′	December 20
0.45$_v$m	–5.14 M	M2 Iab	427 l.y.	

See December.

Canopus	α Carinae	06^h 24^m	–52° 42′	December 27
–0.62$_v$m	–5.53 M	F0 Ib	313 l.y.	

See December.

February

Sirius	α Canis Majoris	06^h 45.1^m	–16° 43′	January 1
–1.44$_v$m	1.45 M	A1 V	8.60 l.y.	

See January.

Procyon	α Canis Minoris	07^h 39.3^m	+05° 13′	January 15
0.40 m	2.68 M	F5 IV	11.41 l.y.	

See January.

Pollux	β Gem	$07^h\ 45.3^m$	+28° 02′	January 16
1.16 m	1.09 M	K0 IIIvar	33.72 l.y.	
See January.				

Acrux	α Crucis	$12^h\ 26.6^m$	−63° 06′	March 29
0.72 m	−4.19 M	B0.5 IV + B1 V	321 l.y.	
See March.				

Aldebaran	α Tauri	$04^h\ 35.9^m$	+16° 31′	November 29
0.87 m	−0.63 M	K5 III	65.11 l.y.	
See November.				

Rigel	β Orionis	$05^h\ 14.5^m$	−08° 12′	December 9
$0.08_v m$	−0.48 M	G5 IIIe	42 l.y.	
See December.				

Capella	α Aurigae	$05^h\ 16.7^m$	+46° 00′	December 10
$0.08_v m$	−0.48 M	G5 IIIe	42 l.y.	
See December.				

Betelgeuse	α Orionis	$05^h\ 55.2^m$	+07° 24′	December 20
$0.45_v m$	−5.14 M	M2 Iab	427 l.y.	
See December.				

Canopus	α Carinae	$06^h\ 24^m$	−52° 42′	December 27
$-0.62_v m$	−5.53 M	F0 Ib	313 l.y.	
See December.				

March

Acrux	α Crucis	$12^h\ 26.6^m$	−63° 06′	March 29
0.72[10]m	−4.19 M	B0.5 IV + B1 V	321 l.y.	
The thirteenth-brightest star in the sky, it is a double star, components about $4^1/_2''$ apart. Both stars are around the same magnitude, 1.4 for α^1 and 1.9 for α^2. The colours of the stars are white and blue–white respectively.				

Becrux	β Crucis	$12^h\ 47.7^m$	−59° 41′	April 3
$1.25_v m$	−3.92 M	B0.5 III	352 l.y.	
See April.				

Spica	α Virginis	$13^h\ 25.2^m$	−11° 10′	April 13
$0.98_v m$	−3.55 M	B1 V	262 l.y.	
See April.				

[10]This is the value for the combined magnitudes of the double star system .

Hadar	β Centauri	14^h 03.8^m	–60° 22′	April 22
0.58$_v$m	–5.45 M	B1 III	525 l.y.	
See April.				

Arcturus	α Boötis	14^h 15.6	+19° 11′	April 25
–0.16$_v$m	–0.10 M	K2 IIIp	36.7 l.y.	
See April.				

Sirius	α Canis Majoris	06^h 45.1^m	–16° 43′	January 1
–1.44$_v$m	1.45 M	A1 V	8.60 l.y.	
See January.				

Procyon	α Canis Minoris	07^h 39.3^m	+05° 13′	January 15
0.40 m	2.68 M	F5 IV	11.41 l.y.	
See January.				

Pollux	β Gem	07^h 45.3^m	+28° 02′	January 16
1.16 m	1.09 M	K0 IIIvar	33.72 l.y.	
See January.				

Rigel	β Orionis	05^h 14.5^m	–08° 12′	December 9
–0.18$_v$m	–6.69 M	B8 Iac	773 l.y.	
See December.				

Capella	α Aurigae	05^h 16.7^m	+46° 00′	December 10
0.08$_v$m	–0.48 M	G5 IIIe	42 l.y.	
See December.				

Betelgeuse	α Orionis	05^h 55.2^m	+07° 24′	December 20
0.45$_v$m	–5.14 M	M2 Iab	427 l.y.	
See December.				

Canopus	α Carinae	06^h 24^m	–52° 42′	December 27
–0.62$_v$m	–5.53 M	F0 Ib	313 l.y.	
See December.				

April

Becrux	β Crucis	12^h 47.7^m	–59° 41′	April 3
1.25$_v$m	–3.92 M	B0.5 III	352 l.y.	
The nineteenth-brightest and penultimate star in our list lies too far south for northern observers. It occurs in the same field as the Jewel Box cluster and is a pulsating variable with a very small change in brightness.				

Spica	α Virginis	$13^h 25.2^m$	−11° 10′	April 13
$0.98_v m$	−3.55 M	B1 V	262 l.y.	

The fifteenth-brightest star, and a fascinating one. It is a large spectroscopic binary with the companion star lying very close to it and thus eclipsing it slightly. Spica is also a pulsating variable star, though the variability and the pulsations are not visible with amateur equipment.

Hadar	β Centauri	$14^h 03.8^m$	−60° 22′	April 22
$0.58_v m$	−5.45 M	B1 III	525 l.y.	

The eleventh-brightest star in the sky, and unknown to northern observers because of its low latitude, lying as it does only $4^1/_2$° from α Centauri. It has a luminosity that is an astonishing 10,000 times that of the Sun. A definitely white star, it has a companion of magnitude 4.1, but is a difficult double to split as the companion is only 1.28 arcseconds from the primary.

Arcturus	α Boötis	$14^h 15.6^m$	+19° 11′	April 25
$-0.16_v m$	−0.10 M	K2 IIIp	36.7 l.y.	

The fourth-brightest star in the sky, and the brightest star north of the celestial equator. It has a lovely orange colour. Notable for its peculiar motion through space, Arcturus, unlike most stars, is not travelling in the plane of the Milky Way, but is instead circling the Galactic centre in an orbit which is highly inclined. Calculations predict that it will swoop past the Solar System in several thousand years' time, moving towards the constellation Virgo. Some astronomers believe that in as little as half a million years Arcturus will have disappeared from naked-eye visibility. At present it is about 100 times more luminous than the Sun.

Acrux	α Crucis	$12^h 26.6^m$	−63° 06′	March 29
0.72 m	−4.19 M	B0.5 IV + B1 V	321 l.y.	

See March.

Rigel Kentaurus	α Centauri	$14^h 39.6^m$	−60° 50′	May 1
−0.20 m	4.07 M	G2 V + K1 V	4.39 l.y.	

See May.

Antares	α Scorpii	$16^h 29.4^m$	−26° 26′	May 29
$1.06_v m$	−5.28 M	M1 Ib + B2.5 V	604 l.y.	

See May.

Procyon	α Canis Minoris	$07^h 39.3^m$	+05° 13′	January 15
0.40 m	2.68 M	F5 IV	11.41 l.y.	

See January.

Pollux	β Gem	$07^h 45.3^m$	+28° 02′	January 16
1.16 m	1.09 M	K0 IIIvar	33.72 l.y.	

See January.

Capella	α Aurigae	$05^h 16.7^m$	+46° 00′	December 10
$0.08_v m$	−0.48 M	G5 IIIe	42 l.y.	

See December.

Canopus	α Carinae	$06^h 24^m$	–52° 42′	December 27
$-0.62_{v}m$	–5.53 M	F0 Ib	313 l.y.	

See December.

May

Rigel Kentaurus	α Centauri	$14^h 39.6^m$	–60° 50′	May 1
$-0.20^{10}m$	4.07 M	G2 V + K1 V	4.39 l.y.	

The third-brightest star in the sky, this is in fact part of a triple system, with the two brightest components contributing most of the light. The system contains the closest star to the Sun, Proxima Centauri. The group also has a very large proper motion (its apparent motion in relation to the background). Unfortunately, it is too far south to be seen by any northern observer. Some observers have claimed that the star is visible in the daylight with any aperture.

Antares	α Scorpii	$16^h 29.4^m$	–26° 26′	May 29
$1.06_{v}m$	–5.28 M	M1 Ib + B2.5 V	604 l.y.	

The sixteenth-brightest star in the sky, this is a red giant, with a luminosity 6000 times that of the Sun, and a diameter hundreds of times bigger than the Sun's. But what makes this star especially worthy is the vivid colour contrast that is seen between it and its companion star. The star is often described as vivid green when seen with the red of Antares. The companion has a magnitude of 5.4, with a PA of 273°, lying 2.6″ away.

Becrux	β Crucis	$12^h 47.7^m$	–59° 41′	April 3
$1.25_{v}m$	–3.92 M	B0.5 III	352 l.y.	

See April.

Spica	α Virginis	$13^h 25.2^m$	–11° 10′	April 13
$0.98_{v}m$	–3.55 M	B1 V	262 l.y.	

See April.

Hadar	β Centauri	$14^h 03.8^m$	–60° 22′	April 22
$0.58_{v}m$	–5.45 M	B1 III	525 l.y.	

See April.

Arcturus	α Boötis	$14^h 15.6^m$	+19° 11′	April 25
$-0.16_{v}m$	–0.10 M	K2 IIIp	36.7 l.y.	

See April.

Vega	α Lyrae	$18^h 36.9^m$	+38° 47′	July 1
$0.03_{v}m$	0.58 M	A0 V	25.3 l.y.	

See July.

Altair	α Aquilae	$19^h 50.8^m$	+08° 52′	July 19
$0.76_{v}m$	2.20 M	A7 IV–V	16.77 l.y.	

See July.

Acrux	α Crucis	$12^h 26.6^m$	–63° 06′	March 29
0.72 m	–4.19 M	B0.5 IV + B1 V	321 l.y.	
See March.				

June

Vega	α Lyrae	$18^h 36.9^m$	+38° 47′	July 1
$0.03_v m$	0.58 M	A0 V	25.3 l.y.	
See July.				

Altair	α Aquilae	$19^h 50.8^m$	+08° 52′	July 19
$0.76_v m$	2.20 M	A7 IV–V	16.77 l.y.	
See July.				

Rigel Kentaurus	α Centauri	$14^h 39.6^m$	–60° 50′	May 1
–0.20 m	4.07 M	G2 V + K1 V	4.39 l.y.	
See May.				

Antares	α Scorpii	$16^h 29.4^m$	–26° 26′	May 29
$1.06_v m$	–5.28 M	M1 Ib + B2.5 V	604 l.y.	
See May.				

Becrux	β Crucis	$12^h 47.7^m$	–59° 41′	April 3
$1.25_v m$	–3.92 M	B0.5 III	352 l.y.	
See April.				

Spica	α Virginis	$13^h 25.2^m$	–11° 10′	April 13
$0.98_v m$	–3.55 M	B1 V	262 l.y.	
See April.				

Hadar	β Centauri	$14^h 03.8^m$	–60° 22′	April 22
$0.58_v m$	–5.45 M	B1 III	525 l.y.	
See April.				

Arcturus	α Boötis	$14^h 15.6^m$	+19° 11′	April 25
$-0.16_v m$	–0.10 M	K2 IIIp	36.7 l.y.	
See April.				

Deneb	α Cygni	$20^h 41.4^m$	+45° 17′	August 1
$1.25_v m$	–8.73 M	A2 Ia	3228 l.y.	
See August.				

July

Vega	α Lyrae	$18^h\ 36.9^m$	+38° 47′	July 1
$0.03_v m$	0.58 M	A0 V	25.3 l.y.	

The fifth-brightest star, familiar to northern observers, located high in the summer sky. Although similar to Sirius in composition and size, it is three times as distant, and thus appears fainter. Often described as having a steely-blue colour, it was one of the first stars observed to have a disc of dust surrounding it – a possible proto-solar system in formation. Vega was the Pole Star some 12,000 years ago, and will be again in a further 12,000 years.

Altair	α Aquilae	$19^h\ 50.8^m$	+08° 52′	July 19
$0.76_v m$	2.20 M	A7 IV – V	16.77 l.y.	

The twelfth-brightest star, this has the honour of being the fastest-spinning of the bright stars, completing one revolution in approximately $6^1/_2$ hours. Such a high speed deforms the star into what is called a flattened ellipsoid, and it is believed that because of this amazing property the star may have an equatorial diameter twice that of its polar diameter. The star's colour has been reported as completely white, although some observers see a hint of yellow.

Deneb	α Cygni	$20^h\ 41.4^m$	+45° 17′	August 1
$1.25_v m$	–8.73 M	A2 Ia	3228 l.y.	

See August.

Rigel Kentaurus	α Centauri	$14^h\ 39.6^m$	–60° 50′	May 1
–0.20 m	4.07 M	G2 V + K1 V	4.39 l.y.	

See May.

Antares	α Scorpii	$16^h\ 29.4^m$	–26° 26′	May 29
$1.06_v m$	–5.28 M	M1 Ib + B2.5 V	604 l.y.	

See May.

Becrux	β Crucis	$12^h\ 47.7^m$	–59° 41′	April 3
$1.25_v m$	–3.92 M	B0.5 III	352 l.y.	

See April.

Spica	α Virginis	$13^h\ 25.2^m$	–11° 10′	April 13
$0.98_v m$	–3.55 M	B1 V	262 l.y.	

See April.

Hadar	β Centauri	$14^h\ 03.8^m$	–60° 22′	April 22
$0.58_v m$	–5.45 M	B1 III	525 l.y.	

See April.

Arcturus	α Boötis	$14^h\ 15.6^m$	+19° 11′	April 25
$-0.16_v m$	–0.10 M	K2 IIIp	36.7 l.y.	

See April.

August

Deneb	α Cygni	$20^h\ 41.4^m$	+45° 17′	August 1
$1.25_v m$	–8.73 M	A2 Ia	3228 l.y.	

The twentieth and final star is very familiar to observers in the northern hemisphere. This pale-blue supergiant has recently been recognised as the prototype of a class of non-radially pulsating variable star. Although the magnitude change is very small, the time scale is from days to weeks. It is believed that the luminosity of Deneb is some 60,000 times that of the Sun, with a diameter 60 times greater.

Formalhaut	α Pisces Austrini	$22^h\ 57.6^m$	–29° 37′	September 5
1.17 m	1.74 M	A3 V	25.07 l.y.	

See September.

Vega	α Lyrae	$18^h\ 36.9^m$	+38° 47′	July 1
$0.03_v m$	0.58 M	A0 V	25.3 l.y.	

See July.

Altair	α Aquilae	$19^h\ 50.8^m$	+08° 52′	July 19
$0.76_v m$	2.20 M	A7 IV–V	16.77 l.y.	

See July.

Antares	α Scorpii	$16^h\ 29.4^m$	–26° 26′	May 29
$1.06_v m$	–5.28 M	M1 Ib + B2.5 V	604 l.y.	

See May.

September

Formalhaut	α Pisces Austrini	$22^h\ 57.6^m$	–29° 37′	September 5
1.17 m	1.74 M	A3 V	25.07 l.y.	

The eighteenth-brightest star is a white one, which often appears reddish to northern observers owing to the effect of the atmosphere. It lies in a barren area of the sky, and is remarkable only for the fact that a star close to it, which is not bound gravitationally, yet lies at the same distance from Earth, is moving through space in a manner and direction similar to Formalhaut's. It has been suggested that the two stars are remnants of a star cluster or star association which has long since dispersed. The star is an orange 6.5-magnitude object about 2° south of Formalhaut.

Deneb	α Cygni	$20^h\ 41.4^m$	+45° 17′	August 1
$1.25_v m$	–8.73 M	A2 Ia	3228 l.y.	

See August.

Archenar	α Eridani	$01^h\ 37.7^m$	–57° 14′	October 15
$0.45_v m$	–2.77 M	B3 Vpe	144 l.y.	

See October.

Vega	α Lyrae	$18^h\ 36.9^m$	+38° 47′	July 1
0.03$_v$m	0.58 M	A0 V	25.3 l.y.	
See July.				

Altair	α Aquilae	$19^h\ 50.8^m$	+08° 52′	July 19
0.76$_v$m	2.20 M	A7 IV–V	16.77 l.y.	
See July.				

October

Archenar	α Eridani	$01^h\ 37.7^m$	–57° 14′	October 15
0.45$_v$m	–2.77 M	B3 Vpe	144 l.y.	

The ninth-brightest star in the sky lies too far south for northern observers, at the southernmost end of the constellation. Among the bright stars it is one of the very few which has the designation "p" in its stellar classification, indicating that it is a "peculiar" star.

Formalhaut	α Pisces Austrini	$22^h\ 57.6^m$	–29° 37′	September 5
1.17 m	1.74 M	A3 V	25.07 l.y.	
See September.				

Aldebaran	α Tauri	$04^h\ 35.9^m$	+16° 31′	November 29
0.87 m	–0.63 M	K5 III	65.11 l.y.	
See November.				

Deneb	α Cygni	$20^h\ 41.4^m$	+45° 17′	August 1
1.25$_v$m	–8.73 M	A2 Ia	3228 l.y.	
See August.				

November

Aldebaran	α Tauri	$04^h\ 35.9^m$	+16° 31′	November 29
0.87 m	–0.63 M	K5 III	65.11 l.y.	

The fourteenth-brightest star, apparently located in the star cluster the Hyades. However, it is not physically in the cluster, lying as it does twice as close as the cluster members. This pale-orange star is around 120 times more luminous than the Sun. It is also a double star, but a very difficult one to separate owing to the extreme faintness of the companion. The companion star, a red dwarf star, magnitude 13.4, lies at a PA of 34° at a distance of 121.7″.

Rigel	β Orionis	$05^h\ 14.5^m$	–08° 12′	December 9
–0.18$_v$m	–6.69 M	B8 Iac	773 l.y.	
See December.				

Capella	α Aurigae	05^h 16.7^m	+46° 00′	December 10
$0.08_v m$	–0.48 M	G5 IIIe	42 l.y.	

See December.

Betelgeuse	α Orionis	05^h 55.2^m	+07° 24′	December 20
$0.45_v m$	–5.14 M	M2 Iab	427 l.y.	

See December.

Canopus	α Carinae	06^h 24^m	–52° 42′	December 27
$-0.62_v m$	–5.53 M	F0 Ib	313 l.y.	

See December.

Archenar	α Eridani	01^h 37.7^m	–57° 14′	October 15
$0.45_v m$	–2.77 M	B3 Vpe3	144 l.y.	

See October.

Formalhaut	α Pisces Austrini	22^h 57.6^m	–29° 37′	September 5
1.17 m	1.74 M	A3 V	25.07 l.y.	

See September.

December

Rigel	β Orionis	05^h 14.5^m	–08° 12′	December 9
$-0.18_v m$	–6.69 M	B8 Iac	773 l.y.	

The seventh-brightest star in the sky, Rigel is in fact brighter than α Orionis. This supergiant star is one of the most luminous stars in our part of the galaxy, almost 560,000 times more luminous than our Sun but at a greater distance than any other nearby bright star. Often described as a bluish star, it is a truly tremendous star, with about 50 times the mass of the Sun, and around 50 times the diameter. It has a close bluish companion at a PA of 202°, apparent magnitude 6.8, at a distance of 9 arcseconds, which should be visible with a 15 cm telescope, or one even smaller under excellent observing conditions.

Capella	α Aurigae	05^h 16.7^m	+46° 00′	December 10
$0.08_v m$	–0.48 M	G5 IIIe	42 l.y.	

The sixth-brightest star in the sky. High in the sky in winter, it has a definite yellow colour, reminiscent of the Sun's own hue. It is in fact a spectroscopic double, and is thus not split in a telescope; however, it has a fainter 10th-magnitude star about 12 arc seconds to the south-east, at a PA of 137°. This is a red dwarf star, which in turn is itself a double (only visible in larger telescopes). Thus, Capella is in fact a quadruple system.

Betelgeuse	α Orionis	$05^h\ 55.2^m$	+07° 24′	December 20
$0.45_v m$	–5.14 M	M2 Iab	427 l.y.	

The tenth-brightest star in the sky, and a favourite among observers, this orange–red star is a giant variable, with an irregular period. Recent observations by the Hubble Space Telescope have shown that it has features on its surface that are similar to Sunspots, but much larger, covering perhaps a tenth of the surface. It also has a companion star, which may be responsible for the non-spherical shape it exhibits. Although a giant star, it has a very low density and a mass only 20 times greater than the Sun's, which together mean that the density is in fact about 0.000000005 that of the Sun. A lovely sight in a telescope of any aperture; subtle colour changes have been reported as the star goes through its variability cycle.

Canopus	α Carinae	$06^h\ 24^m$	–52° 42′	December 27
$-0.62_v m$	–5.53 M	F0 Ib	313 l.y.	

The second-brightest star in the sky, although its position makes it very difficult to observe for northern latitudes. An intrinsically brilliant star, it is some 30 times larger than the Sun, and over 1000 times more luminous. Its parallax is 0.0104″.

Aldebaran	α Tauri	$04^h\ 35.9^m$	+16° 31′	November 29
0.87 m	–0.63 M	K5 III	65.11 l.y.	

See November.

Sirius	α Canis Majoris	$06^h\ 45.1^m$	–16° 43′	January 1
$-1.44_v m$	1.45 M	A1 V	8.60 l.y.	

See January.

Procyon	α Canis Minoris	$07^h\ 39.3^m$	+05° 13′	January 15
0.40 m	2.68 M	F5 IV	11.41 l.y.	

See January.

Pollux	β Gem	$07^h\ 45.3^m$	+28° 02′	January 16
1.16 m	1.09 M	K0 IIIvar	33.72 l.y.	

See January.

2.4 The Nearest Stars

Let's now look at the nearest stars to us. The layout will be similar to that above, and in addition, several of the stars will be the same as those presented above. In such a case, the information will not be duplicated, but the reader will de directed to the relevant section. Also, the nearest stars to us will not necessarily be the brightest, and so some of the stars listed will be very faint, and will need correspondingly larger aperture telescopes in order to be visible. The data was provided by the Hipparcos catalogue up to 11th magnitude.

Table 2.3. The 20 nearest stars in the sky

Star	Distance, l.y.	Constellation
1 Sun	—	—
2 Proxima Centauri	4.22	Centaurus
3 Alpha Centauri A[12]	4.39	Centaurus
4 Barnard's Star	5.94	Ophiuchus
5 Wolf 359	7.8	Leo
6 Lalande 21185	8.31	Ursa Major
7 Sirius A[12]	8.60	Canis Major
8 UV Ceti A[12]	8.7	Cetus
9 Ross 154	9.69	Sagittarius
10 Ross 248	10.3	Andromeda
11 Epsilon Eridani	10.49	Eradinus
12 HD 217987	10.73	Piscis Austrinus
13 Ross 128	10.89	Virgo
14 L 789-6 A[12]	11.2	Aquarius
15 61 Cygni A	11.35	Cygnus
16 Procyon A[12]	11.42	Canis Minoris
17 61 Cygni B	11.43	Cygnus
18 HD173740	11.47	Draco
19 HD173739	11.64	Draco
20 GX Andromadae[12]	11.64	Andromeda

I have introduced an additional observing parameter, which gives an indication of the ease of observability – *easy, moderate* or *difficult*.[11] *Easy* objects are those within naked eye limit, or just beyond it, and so will be relatively easy to locate. *Moderate* objects include those which are beyond naked eye visibility, or may be hard to detect from an urban location, thus needing a somewhat more careful approach to find and observe. Whilst *difficult* objects are those requiring very dark skies, or may lie in a sparse area of the sky, and will defiantly need a telescope of moderate to large aperture, say 12 cm and greater. With this definition, you will not only be able to go out on any clear evening and find many different types of objects, but over time, will improve your observing skills as you locate the fainter and thus more difficult objects.

[11] This parameter is used throughout the book, but will be modified to take into account the different types of objects observed.

[12] This signifies that the star is part of a multiple star system and the distance quoted is for component A.

January

The Sun				January–December
–26.78m	4.82 m	G2 V		
Our closest star, and the object without which no life would have evolved on Earth. Visible every day, throughout the year, unless you happen to live in the UK.				

Sirius A	α Canis Majoris	06^h 45.1^m	–16° 43′	January 1
–1.44 m + 8.4 m 4	1.45 M 11.34 M	A1 V DA2	*8.60 l.y.*	easy
The sixth-closest star is also the brightest star. The companion star, a *white dwarf*, can be seen under excellent conditions and when its distance from the primary is at its greatest. For further details see Section 2.3 – The Brightest Stars.				

Procyon A, B	α Canis Minoris	07^h 39.3^m	+05° 13′	January 15
–0.40 m 10.7 m 13.0 m	2.68 M	F5 IV DF	*11.41 l.y.*	easy
The fifteenth nearest star is a very easy object to observe. See Section 2.3 for further information.				

Epsilon Eridani	HD 22049[13]	03^h 32.9^m	–09° 27′	November 13
3.72 m	6.18 M	K2 V	*10.49 l.y.*	easy
See November.				

Lalande 21185	HD 95735	11^h 03.3^m	+35° 58′	March 8
7.49 m	10.46 M	M2 V	*8.31 l.y.*	moderate
See March.				

Wolf 359	CN Leo	10^h 56.5^m	+07° 01′	March 6
13.46_vm	16.57 M	M6.5 Ve	*7.8 l.y.*	difficult
See March.				

Ross 128	FI Vir	11^h 47.6^m	+00° 48′	March 19
11.12_vm	13.50 M	M4.5 V	*10.89 l.y.*	difficult
See March.				

February

Sirius A	*α Canis Majoris*	*06^h 45.1^m*	*–16° 43′*	*January 1*
–1.44 m + 8.4 m 4	1.45 M 11.34 M	A1 V DA2	*8.60 l.y.*	easy
See January.				

Procyon A, B	α Canis Minoris	07^h 39.3^m	+05° 13′	January 15
–0.40 m 10.7 m 13.0 m	2.68 M	F5 IV DF	*11.41 l.y.*	easy
See January.				

[13]This reference indicates that the star is the 22,049th object in the Henry Draper catalogue.

Epsilon Eridani	HD 22049	03^h 32.9^m	–09° 27′	November 13
3.72 m	6.18 M	K2 V	*10.49 l.y.*	easy

See November.

Lalande 21185	HD 95735	11^h 03.3^m	+35° 58′	March 8
7.49 m	10.46 M	M2 V	*8.31 l.y.*	moderate

See March.

Wolf 359	CN Leo	10^h 56.5^m	+07° 01′	March 6
13.46$_v$m	16.57 M	M6.5 Ve	*7.8 l.y.*	difficult

See March.

Ross 128	FI Vir	11^h 47.6^m	+00° 48′	March 19
11.12$_v$m	13.50 M	M4.5 V	*10.89 l.y.*	difficult

See March.

March

Sirius A	α Canis Majoris	06^h 45.1^m	–16° 43′	January 1
–1.44 m + 8.4 m 4	1.45 M 11.34 M	A1 V DA2	*8.60 l.y.*	easy

See January.

Procyon A, B	α Canis Minoris	07^h 39.3^m	+05° 13′	January 15
–0.40 m 10.7 m 13.0 m	2.68 M	F5 IV DF	*11.41 l.y.*	easy

See January.

Epsilon Eridani	HD 22049	03^h 32.9^m	–09° 27′	November 13
3.72 m	6.18 M	K2 V	*10.49 l.y.*	easy

See November.

Lalande 21185	HD 95735	11^h 03.3^m	+35° 58′	March 8
7.49 m	10.46 M	M2 V	*8.31 l.y.*	moderate

The fifth-closest star is once again a red dwarf star, and has the eighth-largest known proper motion of 4.84 arc seconds per year. Measurements indicate that it may have an unseen companion of very low mass.

Wolf 359	CN Leo	10^h 56.5^m	+07° 01′	March 6
13.46$_v$m	16.57 M	M6.5 Ve	*7.8 l.y.*	difficult

The fourth-closest star, a *red dwarf*, is an extremely faint object and thus difficult to observe. It is one of the least luminous stars that can be seen. Like Barnard's Star, it too is a flare star, with a proper motion of 4.7 arc seconds per year.

Ross 128	FI Vir	11^h 47.6^m	+00° 48′	March 19
11.12$_v$m	13.50 M	M4.5 V	*10.89 l.y.*	difficult

The twelfth-nearest star is once again a red dwarf, difficult to observe.

Proxima Centauri	V645 Cen	$14^h 29.7^m$	–62° 41′	April 29
11.01_vm	15.45 M	M5 Ve	*4.22 l.y.*	difficult

See April.

April

Rigel Kentaurus	α Cen	$14^h 39.6^m$	–60° 50′	May 1
–0.01 m +1.35 m	4.34 M + 5.70 M	G2 V + K1 V	*4.39 l.y.*	easy

See May.

Sirius A	α Canis Majoris	$06^h 45.1^m$	–16° 43′	January 1
–1.44 m +.8.4 m	41.45 M 11.34 M	A1 V DA2	*8.60 l.y.*	easy

See January.

Procyon A, B	α Canis Minoris	$07^h 39.3^m$	+05° 13′	January 15
0.40 m 10.7 m	2.68 M 13.0 M	F5 IV DF	*11.41 l.y.*	easy

See January.

Lalande 21185	HD 95735	$11^h 03.3^m$	+35° 58′	March 8
7.49 m	10.46 M	M2 V	*8.31 l.y.*	moderate

See March.

Proxima Centauri	V645 Cen	$14^h 29.7^m$	–62° 41′	April 29
11.01_vm	15.45 M	M5 Ve	*4.22 l.y.*	difficult

The second-closest star to the Earth, but the closest star to the solar system. It is a very faint red dwarf star and also a flare star, with frequent bursts having a maximum amplitude of around one magnitude.

Wolf 359	CN Leo	$10^h 56.5^m$	+07° 01′	March 6
13.46_vm	16.57 M	M6.5 Ve	*7.8 l.y.*	difficult

See March.

Ross 128	FI Vir	$11^h 47.6^m$	+00° 48′	March 19
11.12_vm	13.50 M	M4.5 V	*10.89 l.y.*	difficult

See March.

May

Rigel Kentaurus	α Cen	$14^h 39.6^m$	–60° 50′	May 1
–0.01 m +1.35 m	4.34 M + 5.70 M	G2 V + K1 V	*4.39 l.y.*	easy

The second closest star to us. For more details see Section 2.3.

Barnard's Star	21185	$17^h 57.8^m$	+4° 38′	June 21
9.54 m	13.24 M	M3.8 V	*5.94 l.y.*	moderate

See June.

Lalande 21185	HD 95735	$11^h 03.3^m$	+35° 58′	March 8
7.49 m	10.46 M	M2 V	*8.31 l.y.*	moderate
See March.				

Wolf 359	CN Leo	$10^h 56.5^m$	+07° 01′	March 6
13.46_vm	16.57 M	M6.5 Ve	*7.8 l.y.*	difficult
See March.				

Ross 128	FI Vir	$11^h 47.6^m$	+00° 48′	March 19
11.12_vm	13.50 M	M4.5 V	*10.89 l.y.*	difficult
See March.				

Proxima Centauri	V645 C5en	$14^h 29.7^m$	–62° 41′	April 29
11.01_vm	15.45 M	M5 Ve	*4.22 l.y.*	difficult
See April.				

June

Rigel Kentaurus	α Cen	$14^h 39.6^m$	–60° 50′	May 1
–0.01 m +1.35 m	4.34 M + 5.70 M	G2 V + K1 V	*4.39 l.y.*	easy
See May.				

Barnard's Star	21185	$17^h 57.8^m$	+4° 38′	June 21
9.54 m	13.24 M	M3.8 V	*5.94 l.y.*	moderate
The third-closest star is a red dwarf. It also has the largest proper motion of any star: 0.4 arc seconds per year. Thus it would take about 150 years for the star to move the distance equivalent to the Moon's diameter across the sky.				

HD 173739	Σ2398[14]	$18^h 42.7^m$	+59° 38′	July 2
8.94 m	11.18 M	K5	*11.64 l.y.*	moderate ©
See July.				

HD 173740	Σ2398	$18^h 42.7^m$	+59° 37′	July 2
9.70 m	11.97 M	K5	*11.47 l.y.*	difficult ©
See July.				

Ross 154	V1216 Sgr	$18^h 49.8^m$	–23° 50′	July 4
10.37 m	13.00 M	M3.5 Ve	*9.69 l.y.*	difficult
See July.				

Proxima Centauri	V645 Cen	$14^h 29.7^m$	–62° 41′	April 29
11.01_vm	15.45 M	M5 Ve	*4.22 l.y.*	difficult
See April.				

[14]The Σ signifies that the star is 2398th object in the F. G. Wilhelm Struve 1827 catalogue.

July

61 Cyg A	V 1803 Cyg	$21^h 06.9^m$	+38° 45′	August 8
5.20_vm	*7.49 M*	K5 V	*11.35 l.y.*	easy

See August.

Rigel Kentaurus	*α* Cen	$14^h 39.6^m$	–60° 50′	May 1
–0.01 m +1.35 m	4.34 M + 5.70 M	G2 V + K1 V	*4.39 l.y.*	easy

See May.

HD 173739	Σ2398	$18^h 42.7^m$	+59° 38′	July 2
8.94 m	11.18 M	K5	*11.64 l.y.*	moderate ©

The eighteenth-closest star is the brighter but more distant half of the double star closest to the solar system.

61 Cyg B	HD 201092	$21^h 06.9^m$	+38° 44′	August 8
6.05_vm	8.33 M	K7 V	*11.42 l.y.*	moderate

See August.

HD 173740	Σ2398	$18^h 42.7^m$	+59° 37′	July 2
9.70 m	11.97 M	K5	*11.47 l.y.*	difficult ©

The seventeenth-closest star is the fainter but nearer half of one of the closest double stars to the solar system. Both stars are red dwarfs.

Ross 154	V1216 Sgr	$18^h 49.8^m$	–23° 50′	July 4
10.37 m	13.00 M	M3.5 Ve	*9.69 l.y.*	difficult

The eighth-closest star is, like so many of its peers, a red dwarf star. It is also a difficult object to observe owing to its faintness.

August

61 Cyg A	V 1803 Cyg	$21^h 06.9^m$	+38° 45′	August 8
5.20_vm	*7.49 M*	K5 V	*11.35 l.y.*	easy

The fourteenth-nearest star is a famous double, the stars separated by 30.3 arc seconds at a PA of 150°. Both stars are dwarfs, and have a nice orange colour. This was the first star to have its distance measured successfully by F. W. Bessel in 1838 using the technique of parallax.

61 Cyg B	HD 201092	$21^h 06.9^m$	+38° 44′	August 8
6.05_vm	8.33 M	K7 V	*11.42 l.y.*	moderate

The sixteenth-nearest star is the fainter of the two stars in the 61 Cygni double star system.

HD 173739	Σ2398	$18^h 42.7^m$	+59° 38′	July 2
8.94 m	11.18 M	K5	*11.64 l.y.*	moderate ©

See July.

Lacille 9352	HD 217987	$23^h 05.5^m$	–35° 52′	September 7
7.35 m	9.76 M	M2/M3 V	*10.73 l.y.*	moderate
See September.				

GX And	Grb 34	$00^h 18.2^m$	+44° 01′	September 25
8.09_vm	10.33 M	M1 V	*11.64 l.y.*	moderate
See September.				

L789–6		$22^h 38.5^m$	–15° 19′	August 31
12.32 m	14.63 M	M5 Ve	*11.2 l.y.*	difficult
The thirteenth-nearest star is a red dwarf star, difficult to observe.				

HD 173740	Σ2398	$18^h 42.7^m$	+59° 37′	July 2
9.70 m	11.97 M	K5	*11.47 l.y.*	difficult ©
See July.				

Ross 154	V1216 Sgr	$18^h 49.8^m$	–23° 50′	July 4
10.37 m	13.00 M	M3.5 Ve	*9.69 l.y.*	difficult
See July.				

Ross 248		$23^h 41.6^m$	+44° 10′	September 16
12.27 m	14.77 M	M5.5 Ve1	*0.3 l.y.*	difficult
See September.				

September

61 Cyg A	V 1803 Cyg	$21^h 06.9^m$	+38° 45′	August 8
5.20_vm	7.49 M	K5 V	*11.35 l.y.*	easy
See August.				

Lacille 9352	HD 217987	$23^h 05.5^m$	–35° 52′	September 7
7.35 m	9.76 M	M2/M3 V	*10.73 l.y.*	moderate
The eleventh-nearest star is a red dwarf, with the fourth-fastest proper motion of any known star. It traverses a distance of nearly 7 arc seconds a year, and thus would take about 1000 years to cover the angular distance of the full Moon, which is half a degree.				

GX And	Grb 34	$00^h 18.2^m$	+44° 01′	September 25
8.09_vm	10.33 M	M1 V	*11.64 l.y.*	moderate
The nineteenth-closest star to the solar system, and the twentieth-closest to Earth, this is half of a noted red dwarf binary system. The primary is in itself a spectroscopic binary.				

61 Cyg B	HD 201092	$21^h 06.9^m$	+38° 44′	August 8
6.05_vm	8.33 M	K7 V	*11.42 l.y.*	moderate
See August.				

HD 173739	Σ2398	18^h 42.7^m	+59° 38′	July 2
8.94 m	11.18 M	K5	*11.64 l.y.*	moderate ©
See July.				

Ross 248		23^h 41.6^m	+44° 10′	September 16
12.27 m	*14.77 M*	M5.5 Ve1	*0.3 l.y.*	difficult
The ninth-closest star to the solar system is a red dwarf star, and a difficult object to observe.				

UV Ceti		01^h 38.8^m	–17° 57′	October 16
12.56$_v$m 12 96$_v$m	15.42 M 15.81 M	M5.5 Ve M5.5 Ve	*8.7 l.y.*	difficult
See October.				

October

Epsilon Eridani	HD 22049	03^h 32.9^m	–09° 27′	November 13
3.72 m	6.18 M	K2 V	*10.49 l.y.*	easy
See November.				

61 Cyg A	V 1803 Cyg	21^h 06.9^m	+38° 45′	August 8
5.20$_v$m	7.49 M	K5 V	*11.35 l.y.*	easy
See August.				

Lacille 9352	HD 217987	23^h 05.5^m	–35° 52′	September 7
7.35 m	9.76 M	M2/M3 V	*10.73 l.y.*	moderate
See September.				

GX And	Grb 34	00^h 18.2^m	+44° 01′	September 25
8.09$_v$m	10.33 M	M1 V	*11.64 l.y.*	moderate
See September.				

61 Cyg B	HD 201092	21^h 06.9^m	+38° 44′	August 8
6.05$_v$m	8.33 M	K7 V	*11.42 l.y.*	moderate
See August.				

UV Ceti		01^h 38.8^m	–17° 57′	October 16
12.56$_v$m 12 96$_v$m	15.42 M 15.81 M	M5.5 Ve M5.5 Ve	*8.7 l.y.*	difficult
The seventh-closest star is a red dwarf system and is a very difficult but not impossible object to observe. The UV prefix indicates that the two components are flare stars, and the fainter is referred to in older texts as "Luyten's Flare Star", after its discoverer, W.J. Luyten, who first observed it in 1949.				

Ross 248		23^h 41.6^m	+44° 10′	September 16
12.27 m	*14.77 M*	M5.5 Ve1	*0.3 l.y.*	difficult
See September.				

November

Epsilon Eridani	HD 22049	03^h 32.9^m	−09° 27′	November 13
3.72 m	6.18 M	K2 V	*10.49 l.y.*	easy
The tenth-closest star is a naked-eye object, which some observers describe as having a yellow colour, while others say it is more orange. Recent observations indicate that there may be an unseen companion star with an extremely small mass, approximately 0.048 that of the Sun.				

Lacille 9352	HD 217987	23^h 05.5^m	−35° 52′	September 7
7.35 m	9.76 M	M2/M3 V	*10.73 l.y.*	moderate
See September.				

GX And	Grb 34	00^h 18.2^m	+44° 01′	September 25
8.09$_v$m	10.33 M	M1 V	*11.64 l.y.*	moderate
See September.				

UV Ceti		01^h 38.8^m	−17° 57′	October 16
12.56$_v$m 12 96$_v$m	15.42 M	M5.5 Ve M5.5 Ve	*8.7 l.y.*	difficult
See October.				

Ross 248		23^h 41.6^m	+44° 10′	September 16
12.27 m	14.77 M	M5.5 Ve1	*0.3 l.y.*	difficult
See September.				

December

Epsilon Eridani	HD 22049	03^h 32.9^m	−09° 27′	November 13
3.72 m	6.18 M	K2 V	*10.49 l.y.*	easy
See November.				

Sirius A	α Canis Majoris	06^h 45.1^m	−16° 43′	January 1
−1.44 m +.8.4 m	41.45 M 11.34 M	A1 V DA2	*8.60 l.y.*	easy
See January.				

Procyon A, B	α Canis Minoris	07^h 39.3^m	+05° 13′	January 15
0.40 m 10.7 m	2.68 M 13.0 M	F5 IV DF	*11.41 l.y.*	easy
See January.				

GX And	Grb 34	00^h 18.2^m	+44° 01′	September 25
8.09$_v$m	10.33 M	M1 V	*11.64 l.y.*	moderate
See September.				

UV Ceti		01^h 38.8^m	−17° 57′	October 16
12.56$_v$m 12 96$_v$m	15.42 M	M5.5 Ve M5.5 Ve	*8.7 l.y.*	difficult
See October.				

Ross 248		$23^h\ 41.6^m$	+44° 10′	September 16
12.27 m	14.77 M	M5.5 Ve1	*0.3 l.y.*	difficult
See September.				

2.5 The Spectral Sequence

This section will look at several examples of the spectral classification of stars. You will recall that I discussed this system in Sections 2.1 and 2.2, and have used the scheme in the previous two sections on bright stars and nearby stars. Even though most amateurs observe the stars without paying too much attention to their astrophysical classification, it is always a fascinating project to be able to search out and observe examples of the various classes. After all, it is a system that is used by all astronomers in the world, and to be able too understand, albeit at an introductory level, how the system is applied, will give you an added level of enjoyment to your observing sessions.

Several classes have already been introduced in describing the bright stars, and to reproduce them again serves no positive use, therefore I have tried to include several stars which may not be familiar to you, (but remember that all the stars you can see, either with the naked eye, or binoculars and/or telescopes are classified in this manner, and so there is no limit, literally, to the number of stars you can observe and classify). Also, not every class is represented, as some are not used, and some representative stars may be too faint and thus beyond the scope of small telescopes.[15] Finally, the stars have been listed as before, by date. In this way, you should be able to observe nearly several of the classes at any given time of the year[16].

January

θ Orionis C	θ Ori	$05^h\ 35.3^m$	−05° 23′	December 14
4.96 m	−5.04 M	*O6*	Orion	easy
See December.				

15 Monocerotis	HD47839	$06^h\ 40.9^m$	+09° 54′	December 31
4.66_vm	−2.3 M	*O7*	Monoceris	easy
See December.				

Plaskett's Star	HD47129	$06^h\ 37.4^m$	+06° 08′	December 30
6.05 m	−3.54 M	*O8*	Monoceros	easy
See December.				

[15]The values for the apparent and absolute magnitudes are taken from the Hipparchos catalogue, and in nearly every star listed the values differ from those previously published (pre-Hipparchos). It is an interesting exercise to compare the old and new values as sometimes there is a considerable difference.

[16]Note that the stars are listed in order of spectral classification.

Iota Orionis	ι Ori	$05^h 35.4^m$	–05° 55′	December 15
2.75 m	–5.30 M	*O9 III*	Orion	easy
See December.				

Murzim	β CMa	$06^h 22.7^m$	–17° 57′	December 27
1.98_vm	–3.96 M	*B1 II*	Canis Major	easy
See September.				

λ CMa	HD 45813	$06^h 28.2^m$	–32° 35′	December 28
4.47 m	–1.01 M	*B4 IV*	Canis Major	easy
See December.				

Aludra	η CMa	$07^h 24.1^m$	–29° 18′	January 11
2.45 m	7.51 M	*B5 I*	Canis Major	easy
A highly luminous supergiant, with an estimated luminosity 50,000 times that of the Sun.				

Regulas	α Leo	$10^h 08.3^m$	+11° 58′	February 19
1.36 m	–0.52 M	*B7 V*	Leo	easy
See February.				

Alhena	Gem	$06^h 37.7^m$	+16° 23′	December 30
1.93 m	–0.60 M	*A0 IV*	Gemini	easy
See December.				

Castor	α Gem	$07^h 34.6^m$	+31° 53′	January 14
1.43 m	0.94 M	*A1 V*	Gemini	easy
Part of the famous multiple star system, and fainter brother to Pollux. The visual magnitude stated is the result of combining the magnitudes of the two brighter components of the system, 1.9 and 2.9. [See also Sirius.][17]				

Beta Aurigae	β Aur	$05^h 59.5^m$	+44° 57′	December 21
1.90_vm	–0.10 M	*A2 V*	Auriga	easy
See December.				

2 Mon	HD 40536	$05^h 59.1^m$	–09° 33′	December 21
5.01 m	0.02 M	*A6*	Monoceros	easy
See December.				

Canopus	α Car	$06^h 23.9^m$	–52° 41′	December 27
–0.62 m	–5.53 M	*F0 I*	Carina	easy
See December.				

b Velorum	HD74180	$08^h 40.6^m$	–46° 39′	January 30
3.84 m	–6.12 M	*F3 I*	Vela	easy
This star is unremarkable except that its luminosity has been calculated to be that of 180,000 Suns!				

[17]Denotes a similar class star.

Polaris	α UMi	$02^h\ 31.8^m$	+89° 16′	October 29
1.97$_v$m	–3.64 M	*F7 I*	Ursa Minor	easy ©
See October.				

111 Tau	HD 35296	$05^h\ 24.4^m$	+17° 23′	December 12
5.00 m	4.17 M	*F8 V*	Taurus	easy
See December.				

Algeiba	γ^2Leo	$10^h\ 19.9^m$	+19° 50′	February 25
3.64 m	0.72 M	*G7 III*	Leo	easy
See February.				

β LMi	HD 90537	$10^h\ 27.8^m$	+36° 42¢	February 27
4.20 m	0.9 M	*G8 III*	Leo Minor	easy
See February.				

ν^2 CMa	HD 47205	$06^h\ 36.7^m$	–19° 15′	December 30
3.95 m	2.46 M	*K1 III*	Canis Major	easy
See December.				

γ^2 Vel	HD 68273	$08^h\ 09.5^m$	–47° 20′	January 23
1.99$_v$m	0.05 M	*WC 8*	Vela	easy
The brightest and closest of all Wolf–Rayet stars, which are believed to be precursors to the formation of planetary nebulae. Extremely luminous stars, Wolf–Rayets have luminosities that may reach 100,000 times that of the Sun, and temperatures in excess of 50,000 K. γ^2 Vel is an easy double, of colours white and greenish-white.				

February

15 Monocerotis	HD47839	$06^h\ 40.9^m$	+09° 54′	December 31
4.66$_v$m	–2.3 M	*O7*	Monoceris	easy
See December.				

Plaskett's Star	HD47129	$06^h\ 37.4^m$	+06° 08′	December 30
6.05 m	–3.54 M	*O8*	Monoceros	easy
See December.				

Aludra	η CMa	$07^h\ 24.1^m$	–29° 18′	January 11
2.45 m	7.51 M	*B5 I*	Canis Major	easy
See January.				

Regulas	α Leo	$10^h\ 08.3^m$	+11° 58′	February 19
1.36 m	–0.52 M	*B7 V*	Leo	easy
Alpha Leonis is the handle of the Lion's sickle. An easy double star with an 8th-magnitude companion 3′ away, colour orange–red. The companion is itself a double, but visible only in large instruments.				

Alhena	Gem	06^h 37.7^m	+16° 23′	December 30
1.93 m	–0.60 M	*A0 IV*	Gemini	easy

See December.

Castor	α Gem	07^h 34.6^m	+31° 53′	January 14
1.43 m	0.94 M	*A1 V*	Gemini	easy

See January.

Denebola	β Leo	11^h 49.1^m	+14° 34′	March 19
2.14$_v$m	1.92 M	*A3 V*	Leo	easy

See March.

Delta Leonis	δ Leo	11^h 14.1^m	+20° 31′	March 10
2.56 m	1.32 M	*A4 V*	Leo	easy

See March.

b Velorum	HD74180	08^h 40.6^m	–46° 39′	January 30
3.84 m	–6.12 M	*F3 I*	Vela	easy

See January.

Polaris	α UMi	02^h 31.8^m	+89° 16′	October 29
1.97$_v$m	–3.64 M	*F7 I*	Ursa Minor	easy ©

See October.

β Vir	HD 102870	11^h 50.7^m	+01° 46′	March 20
3.59 m	3.40 M	*F8 V*	Virgo	easy

See March.

Algeiba	γ^2Leo	10^h 19.9^m	+19° 50′	February 25
3.64 m	0.72 M	*G7 III*	Leo	easy

A famous double; most observers report orange–yellowish colours, but some see the G7 star as greenish.

β LMi	HD 90537	10^h 27.8^m	+36° 42′	February 27
4.20 m	0.9 M	*G8 III*	Leo Minor	easy

A constellation in which there is no star given the classification α, β LMi has the misfortune of not even being the brightest star in the constellation; that honour goes to 46 LMi.

ν^2 CMa	HD 47205	06^h 36.7^m	–19° 15′	December 30
3.95 m	2.46 M	*K1 III*	Canis Major	easy

See December.

Gacrux	γ^AA Crucis	12^h 31.2m	–57° 07′	March 30
1.59 m	–0.56 M	*M4 III*	Crux	easy

See March.

γ^2 Vel	HD 68273	08^h 09.5^m	–47° 20′	January 23
1.99$_v$m	0.05 M	*WC 8*	Vela	easy

See January.

March

Regulas	α Leo	$10^h\ 08.3^m$	+11° 58′	February 19
1.36 m	−0.52 M	*B7 V*	Leo	easy
See February.				

Gamma Centauri	γ Cen	$12^h\ 41.5^m$	−48° 58′	April 2
2.17 m	−0.6 M	*A1 IV*	Centaurus	easy
See April.				

Denebola	β Leo	$11^h\ 49.1^m$	+14° 34′	March 19
2.14_vm	1.92 M	*A3 V*	Leo	easy
Several companion stars are visible in a variety of instruments. The star has only recently been designated a variable.				

Delta Leonis	δ Leo	$11^h\ 14.1^m$	+20° 31′	March 10
2.56 m	1.32 M	*A4 V*	Leo	easy
Also called Zozma, it lies at a distance of 80 l.y., with a luminosity of 50 Suns.				

Polaris	α UMi	$02^h\ 31.8^m$	+89° 16′	October 29
1.97_vm	−3.64 M	*F7 I*	Ursa Minor	easy ©
See October.				

ν Vir	HD 102870	$11^h\ 50.7^m$	+01° 46′	March 20
3.59 m	3.40 M	*F8 V*	Virgo	easy
A close star at 34 l.y., only 3 times as luminous as the Sun.				

Algeiba	γ^2Leo	$10^h\ 19.9^m$	+19° 50′	February 25
3.64 m	0.72 M	*G7 III*	Leo	easy
See February.				

β LMi	HD 90537	$10^h\ 27.8^m$	+36° 42′	February 27
4.20 m	0.9 M	*G8 III*	Leo Minor	easy
See February.				

Gacrux	γ^AA Crucis	$12^h\ 31.2^m$	−57° 07′	March 30
1.59 m	−0.56 M	*M4 III*	Crux	easy
The top star of the Southern Cross, this is a giant star. γ^A and γ^B do not form a true binary as they are apparently moving in different directions.				

θ Apodis	HD 122250	$14^h\ 05.3^m$	−76° 48′	April 23
5.69_vm	−0.67 M	*M6.5 III*	Apus	easy
See April.				

April

Zubeneschamali	β Lib	$15^h 17.0^m$	–09° 23′	May 11
2.61 m	–0.84 M	*B8 V*	Libra	easy

See May.

Gamma Centauri	γ Cen	$12^h 41.5^m$	–48° 58′	April 2
2.17 m	–0.6 M	*A1 IV*	Centaurus	easy

A close binary star, with both members being almost identical. [See also Vega.]

Zeta Virginis	ζ Vir	$13^h 34.7^m$	–00° 36′	April 15
3.38 m	1.62 M	*A3 V*	Virgo	easy

A nice white star, also called Heze. Only 30 times as luminous as the Sun and lying at a distance of 92 l.y.

Delta Leonis	δ Leo	$11^h 14.1^m$	+20° 31′	March 10
2.56 m	1.32 M	*A4 V*	Leo	easy

See March.

Gamma Herculis	γ Her	$16^h 21.8^m$	+19° 09′	May 27
3.74 m	–0.15 M	*A9 III*	Hercules	easy

See May.

Zubenelgenubi	α^1 Lib	$14^h 50.7^m$	–15° 60′	May 4
5.15 m	3.28 M	*F4 IV*	Libra	easy

See May.

Polaris	α UMi	$02^h 31.8^m$	+89° 16′	October 29
1.97_vm	–3.64 M	*F7 I*	Ursa Minor	easy ©

See October.

β Vir	HD 102870	$11^h 50.7^m$	+01° 46′	March 20
3.59 m	3.40 M	*F8 V*	Virgo	easy

See March.

Kornephorus	β Her	$16^h 30.2^m$	+21° 29′	May 29
2.78 m	–0.50 M	*G8 III*	Hercules	easy

See May.

ν^1 Boö	HD 138481	$15^h 30.9^m$	+40° 50′	May 14
5.04 m	–2.10 M	*K5 III*	Boötes	easy

See May.

Antares	α Sco	$16^h 29.4^m$	–26° 26′	May 29
1.06_vm	–5.28 M	*M1 I*	Scorpio	easy

See May.

Gacrux	γ^AA Crucis	$12^h\ 31.2^m$	–57° 07′	March 30
1.59 m	–0.56 M	*M4 III*	Crux	easy

See March.

θ Apodis	HD 122250	$14^h\ 05.3^m$	–76° 48′	April 23
5.69_vm	–0.67 M	*M6.5 III*	Apus	easy

A reddish-tinted star which stands out nicely in contrast to its background of faint stars. This is a semi-regular variable with a period of 119 days and a range of 5th to nearly 8th magnitude. The titanium bands are now at their strongest.

May

Zubeneschamali	β Lib	$15^h\ 17.0^m$	–09° 23′	May 11
2.61 m	–0.84 M	*B8 V*	Libra	easy

A mysterious star for two reasons. Historical records state that it was much brighter than it is seen today, while observers of the past 100 years have declared that it is greenish or pale emerald in colour. Observe for yourself and decide if it is one of the rare green-coloured stars!

Eta Sagitai	ϵ Sgr	$18^h\ 24.2^m$	–34° 23′	June 27
1.79 m	–1.44 M	*B9.5 III*	Sagittarius	easy

See June.

Nu Draconis^1	ν^1Dra	$17^h\ 32.2^m$	+55° 11′	June 14
4.89 m	2.48 M	*Am*	Draco	easy ©

See June.

Gamma Centauri	γ Cen	$12^h\ 41.5^m$	–48° 58′	April 2
2.17 m	–0.6 M	*A1 IV*	Centaurus	easy

See April.

Zeta Virginis	ζ Vir	$13^h\ 34.7^m$	–00° 36′	April 15
3.38 m	1.62 M	*A3 V*	Virgo	easy

See April.

Ras Alhague	α Oph	$17^h\ 34.9^m$	+12° 34′	June 15
2.08 m	1.30 M	*A5 III*	Ophiucus	easy

See June.

Gamma Herculis	γ Her	$16^h\ 21.8^m$	+19° 09′	May 27
3.74 m	–0.15 M	*A9 III*	Hercules	easy

An optical double system, lying at a distance of 144 l.y., and with a luminosity of 46 Suns.

Zubenelgenubi	α^1 Lib	$14^h\ 50.7^m$	–15° 60′	May 4
5.15 m	3.28 M	*F4 IV*	Libra	easy

An easily resolvable double star, α^1 is also a spectroscopic binary. The colours are a nice faint yellow and pale blue.

Polaris	α UMi	$02^h\ 31.8^m$	+89° 16′	October 29
1.97_vm	–3.64 M	*F7 I*	Ursa Minor	easy ©

See October.

Kornephorus	β Her	$16^h\ 30.2^m$	+21° 29′	May 29
2.78 m	–0.50 M	*G8 III*	Hercules	easy

A spectroscopic binary star, it lies at a distance of 100 l.y., and is some 60 times as luminous as the Sun. [See also Capella.]

Ras Algethi	α^2 Her	$17^h\ 14.7^m$	+14° 23′	June 10
5.37 m	0.03 M	*G5 III*	Hercules	easy

See June.

ζ^2 Sco	HD 152334	$16^h\ 54.6^m$	–42° 22′	June 5
3.62 m	0.3 M	*K4 III*	Scorpius	easy

See June.

ν^1 Boö	HD 138481	$15^h\ 30.9^m$	+40° 50′	May 14
5.04 m	–2.10 M	*K5 III*	Boötes	easy

The star lies at a distance of 385 l.y. and has a luminosity of 104 Suns. [See also Aldebaran.]

Antares	α Sco	$16^h\ 29.4^m$	–26° 26′	May 29
1.06_vm	–5.28 M	*M1 I*	Scorpio	easy

A gloriously coloured star of fiery red (or, as some astronomers of the last century observed, saffron-rose), it contrasts nicely with its fainter green companion. A giant star measured to be some 600 times the diameter of our Sun. [See also Betelgeuse.]

θ Apodis	HD 122250	$14^h\ 05.3^m$	–76° 48′	April 23
5.69_vm	–0.67 M	*M6.5 III*	Apus	easy

See April.

June

Eta Sagitai	ϵ Sgr	$18^h\ 24.2^m$	–34° 23′	June 27
1.79 m	–1.44 M	*B9.5 III*	Sagittarius	easy

A brilliant orange star lying at a distance of 125 l.y. with a luminosity of 250 Suns.

Zubeneschamali	β Lib	$15^h\ 17.0^m$	–09° 23′	May 11
2.61 m	–0.84 M	*B8 V*	Libra	easy

See May.

Nu Draconis[1]	ν^1Dra	$17^h\ 32.2^m$	+55° 11′	June 14
4.89 m	2.48 M	*Am*	Draco	easy ©

A classic double star system visible in binoculars or small telescopes. Both stars are nearly identical in magnitude and stellar class, and have a lovely white colour. A true binary star system

Sarin	δ Her	$17^h\ 15.0^m$	+24° 50′	June 10
3.12 m	1.21 M	*A3 IV*	Hercules	easy

A good example of an optical double. What is astonishing about these stars is the range of colours ascribed to them. They have been called greenish and pale violet, green and ashy white, pale yellow and bluish-green, white and azure, and finally pale yellow and ruddy purple! Spectral class indicates that the stars should be yellow and orange; what colours do you see?

Ras Alhague	α Oph	$17^h\ 34.9^m$	+12° 34′	June 15
2.08 m	1.30 M	*A5 III*	Ophiucus	easy

An interesting star for several reasons. It shows the same motions through space as several other stars called the Ursa Major Group (see Chapter 3). It also shows interstellar absorption lines in its spectrum. Finally, measurements show an oscillation, or wobble, in its proper motion, which would indicate an unseen companion star. [See also β Triangulum.]

Gamma Herculis	γ Her	$16^h\ 21.8^m$	+19° 09′	May 27
3.74 m	–0.15 M	*A9 III*	Hercules	easy

See May.

Albaldah	π Sgr	$19^h\ 09.8^m$	–21° 01′	July 9
2.88 m	–2.77 M	*F3 III*	Sagittarius	easy

See July.

Zubenelgenubi	α^1 Lib	$14^h\ 50.7^m$	–15° 60′	May 4
5.15 m	3.28 M	*F4 IV*	Libra	easy

See May.

Polaris	α UMi	$02^h\ 31.8^m$	+89° 16′	October 29
1.97_vm	–3.64 M	*F7 I*	Ursa Minor	easy ©

See October.

Ras Algethi	α^2 Her	$17^h\ 14.7^m$	+14° 23′	June 10
3.03 m	–2.32 M	*M5 II*	Hercules	easy

As stated below, a beautiful double star, with colours of ruddy orange and blue–green. The spectral class refers to the primary of α^2 Her, which is a spectroscopic double, and thus visually inseparable with any telescope.

Kornephorus	β Her	$16^h\ 30.2^m$	+21° 29′	May 29
2.78 m	–0.50 M	*G8 III*	Hercules	easy

See May.

ζ^2 Sco	HD 152334	16^h 54.6^m	−42° 22′	June 5
3.62 m	0.3 M	*K4 III*	Scorpius	easy

The brighter of the two stars in this naked-eye optical double star system, the orange supergiant star contrasts nicely with its slightly fainter blue supergiant companion.

ν^1 Boö	HD 138481	15^h 30.9^m	+40° 50′	May 14
5.04 m	−2.10 M	*K5 III*	Boötes	easy

See May.

Antares	α Sco	16^h 29.4^m	−26° 26′	May 29
1.06_vm	−5.28 M	*M1 I*	Scorpio	easy

See May.

Ras Algethi	α^2 Her	17^h 14.7^m	+14° 23′	June 10
5.37 m	0.03 M	*G5 III*	Hercules	easy

A fine double-star system. The M5 semi-regular star is an orange supergiant, in contrast to its companion, a blue–green giant. However, it must be pointed out here that it can be resolved only with a telescope and not binoculars, as the two stars are less than 5″ apart. The changes in brightness are attributed to actual physical changes to the star, as it increases and then decreases in diameter.

July

Eta Sagitai	ϵ Sgr	18^h 24.2^m	−34° 23′	June 27
1.79 m	−1.44 M	*B9.5 III*	Sagittarius	easy

See June.

Nu Draconis^1	ν^1Dra	17^h 32.2^m	+55° 11′	June 14
4.89 m	2.48 M	*Am*	Draco	easy ©

See June.

Deneb	α Cyg	20^h 41.3^m	+45° 17′	August 1
1.25_vm	−8.73[18]M	*A2 I*	Cygnus	easy

See August.

Sarin	δ Her	17^h 15.0^m	+24° 50′	June 10
3.12 m	1.21 M	*A3 IV*	Hercules	easy

See June.

Ras Alhague	α Oph	17^h 34.9^m	+12° 34′	June 15
2.08 m	1.30 M	*A5 III*	Ophiucus	easy

See June.

[18]This value is in question. The data are awaiting reassessment.

Aldermin	α Cep	$21^h\ 18.6^m$	+62° 35′	August 11
2.45 m	1.58 M	*A7 IV*	Cepheus	easy ©

See August.

Albaldah	π Sgr	$19^h\ 09.8^m$	–21° 01′	July 9
2.88 m	–2.77 M	*F3 III*	Sagittarius	easy

A triple star system which, alas, is seen only in the largest amateur telescopes. [See also β^2 Sagittarii.]

Polaris	α UMi	$02^h\ 31.8^m$	+89° 16′	October 29
1.97_vm	–3.64 M	*F7 I*	Ursa Minor	easy ©

See October.

Sadal Suud	β Aqr	$21^h\ 31.6^m$	–05° 34′	August 14
2.90 m	–3.47 M	*G0 I*	Aquarius	easy

See August.

Sadal Melik	Aqr	$22^h\ 05.8^m$	–00° 19′	August 23
2.95 m	–3.88 M	*G2 I*	Aquarius	easy

See August.

Ras Algethi	α^2 Her	$17^h\ 14.7^m$	+14° 23′	June 10
5.37 m	0.03 M	*G5 III*	Hercules	easy

See June.

Gienah	ϵ Cyg	$20^h\ 46.2^m$	+33° 58′	August 2
2.48 m	0.76 M	*K0 III*	Cygnus	easy

See August.

Enif	ϵ Peg	$21^h\ 44.2^m$	+09° 52′	August 17
2.38_vm	–4.19 M	*K2 I*	Pegasus	easy

See August.

ζ^2 Sco	HD 152334	$16^h\ 54.6^m$	–42° 22′	June 5
3.62 m	0.3 M	*K4 III*	Scorpius	easy

See June.

Ras Algethi	α^2 Her	$17^h\ 14.7^m$	+14° 23′	June 10
5.37 m	0.03 M	*G5 III*	Hercules	easy

See June.

August

Algenib	γ Peg	$00^h\ 13.2^m$	+15° 11′	September 24
2.83_vm	–2.22 M	*B2 V*	Pegasus	easy

See September.

Eta Sagitai	ϵ Sgr	18^h 24.2^m	–34° 23′	June 27
1.79 m	–1.44 M	*B9.5 III*	Sagittarius	easy

See June.

Nu Draconis[1]	ν^1Dra	17^h 32.2^m	+55° 11′	June 14
4.89 m	2.48 M	*Am*	Draco	easy ©

See June.

Deneb	α Cyg	20^h 41.3^m	+45° 17′	August 1
1.25_vm	–8.73[19]M	*A2 I*	Cygnus	easy

The faintest star of the Summer Triangle (the others being Altair and Vega). A supergiant star with a definite pale-blue colour. The prototype of a class of pulsating variable star.

Sarin	δ Her	17^h 15.0^m	+24° 50′	June 10
3.12 m	1.21 M	*A3 IV*	Hercules	easy

See June.

Ras Alhague	α Oph	17^h 34.9^m	+12° 34′	June 15
2.08 m	1.30 M	*A5 III*	Ophiucus	easy

See June.

Aldermin	α Cep	21^h 18.6^m	+62° 35′	August 11
2.45 m	1.58 M	*A7 IV*	Cepheus	easy ©

This is a rapidly rotating star which results in the spectral lines becoming broad and less clear. It also has the dubious distinction of becoming the Pole Star in AD 7500 [See also Altair.]

Albaldah	π Sgr	19^h 09.8^m	–21° 01′	July 9
2.88 m	–2.77 M	*F3 III*	Sagittarius	easy

See July.

Polaris	α UMi	02^h 31.8^m	+89° 16′	October 29
1.97_vm	–3.64 M	*F7 I*	Ursa Minor	easy ©

See October.

Sadal Suud	β Aqr	21^h 31.6^m	–05° 34′	August 14
2.90 m	–3.47 M	*G0 I*	Aquarius	easy

A giant star, and a close twin to α Aqr. It lies at a distance of 990 l.y., and is 5000 times more luminous than the Sun.

Sadal Melik	Aqr	22^h 05.8^m	–00° 19′	August 23
2.95 m	–3.88 M	*G2 I*	Aquarius	easy

Although it has the same spectral class and surface temperature of the Sun, α Aqr is a giant star, whereas the Sun is a main sequence star. [See also Sun, Alpha Centauri A.]

---

[19]This value is in question. The data are awaiting reassessment.

Ras Algethi	α^2 Her	$17^h\ 14.7^m$	+14° 23′	June 10
5.37 m	0.03 M	*G5 III*	Hercules	easy
See June.				

Gienah	ϵ Cyg	$20^h\ 46.2^m$	+33° 58′	August 2
2.48 m	0.76 M	*K0 III*	Cygnus	easy
Marking the eastern arm of the Northern Cross, the star is a spectroscopic binary. In the K-class stars the metallic lines are now becoming more prominent than the hydrogen lines.				

Enif	ϵ Peg	$21^h\ 44.2^m$	+09° 52′	August 17
2.38_vm	–4.19 M	*K2 I*	Pegasus	easy
This star lies at a distance of 740 l.y. with a luminosity 7450 times that of the Sun. The two faint stars in the same field of view have been mistakenly classified as companions, but analysis has now shown them to be stars in the line of sight.				

ζ^2 Sco	HD 152334	$16^h\ 54.6^m$	–42° 22′	June 5
3.62 m	0.3 M	*K4 III*	Scorpius	easy
See June.				

Scheat	βPeg	$23^h\ 03.8^m$	+28° 045	September 6
2.44_vm	–1.49 M	*M2 II*	Pegasus	easy
See September.				

Ras Algethi	α^2 Her	$17^h\ 14.7^m$	+14° 23′	June 10
5.37 m	0.03 M	*G5 III*	Hercules	easy
See June.				

September

Gamma Cassiopeiae	γ Cas	$00^h\ 56.7^m$	+60° 43′	October 5
2.15_vm	–4.22 M	*B0 IV*	Cassiopeia	easy
See October.				

Algenib	γ Peg	$00^h\ 13.2^m$	+15° 11′	September 24
2.83_vm	–2.22 M	*B2 V*	Pegasus	easy
A member of the type β CMa (Canis Majoris) variable star. It is the south-eastern corner star of the famed square of Pegasus.				

Achernar	α Eri	$01^h\ 37.7^m$	–57° 14′	October 1
0.45_vm	–2.77 M	*B3 V*	Eradinus	easy
See October.				

Deneb	α Cyg	$20^h\ 41.3^m$	+45° 17′	August 1
1.25_vm	–8.7320M	*A2 I*	Cygnus	easy
See August.				

[20]This value is in question. The data are awaiting reassessment.

Aldermin	α Cep	$21^h\ 18.6^m$	+62° 35′	August 11
2.45 m	1.58 M	*A7 IV*	Cepheus	easy ©

See August.

Polaris	α UMi	$02^h\ 31.8^m$	+89° 16′	October 29
1.97_vm	–3.64 M	*F7 I*	Ursa Minor	easy ©

See October.

Sadal Suud	β Aqr	$21^h\ 31.6^m$	–05° 34′	August 14
2.90 m	–3.47 M	*G0 I*	Aquarius	easy

See August.

Sadal Melik	α Aqr	$22^h\ 05.8^m$	–00° 19′	August 23
2.95 m	–3.88 M	*G2 I*	Aquarius	easy

See August.

ξ^1 Cet	HD 15318	$02^h\ 12.0^m$	+08° 51′	October 24
4.36 m	–0.87 M	*G8 II*	Cetus	easy

See October.

β Cet	HD 4128	$00^h\ 43.6^m$	–17° 59′	October 2
2.04 m	–0.30 M	*G9.5 III*	Cetus	easy

See October.

Gienah	ϵ Cyg	$20^h\ 46.2^m$	+33° 58′	August 2
2.48 m	0.76 M	*K0 III*	Cygnus	easy

See August.

Hamal	α Ari	$02^h\ 07.2^m$	+23° 28′	October 23
2.01 m	0.48 M	*K2 III*	Aries	easy

See October.

Almach	γ^1 And	$02^h\ 03.9^m$	+42° 20′	October 22
2.33 m	–2.86 M	*K3 III*	Andromeda	easy

See October.

Mirach	β And	$01^h\ 09.7^m$	+35° 37′	October 8
2.07 m	–1.86 M	*M0 III*	Andromeda	easy

See October.

Scheat	β Peg	$23^h\ 03.8^m$	+28° 045	September 6
2.44_vm	–1.49 M	*M2 II*	Pegasus	easy

Marking the north-western corner of the Square of Pegasus, this is a red irregular variable star. It is noted for having been one of the first stars to have its diameter measured by the technique of interferometry, at 0.021″. Being variable, its size oscillates, to a maximum diameter of 160 Suns.

Mira	o Cet	02^h 19.3^m	−02° 59′	October 26
2.00$_v$m	−3.54 M	*M5*	Cetus	easy

See October.

Mira at minimum	o Cet	02^h 19.3^m	−02° 59′	October 26
10$_v$m	−0.5 M	*M9*	Cetus	difficult

See October.

October

Gamma Cassiopeiae	γ Cas	00^h 56.7^m	+60° 43′	October 5
2.15$_v$m	−4.22 M	*B0 IV*	Cassiopeia	easy

A peculiar star in that it has bright emission lines in its spectrum, indicating that it ejects material in periodic outbursts. The middle star of the familiar W-shape of Cassiopeia.

Algenib	γ Peg	00^h 13.2^m	+15° 11′	September 24
2.83$_v$m	−2.22 M	*B2 V*	Pegasus	easy

See September.

Achernar	α Eri	01^h 37.7^m	−57° 14′	October 1
0.45$_v$m	−2.77 M	*B3 V*	Eradinus	easy

A hot and blue star. It lies so far south that it can never be seen from the UK.

Electra	17 Tau	03^h 44.9^m	+24° 07′	November 16
3.72 m	−1.56 M	*B6 III*	Taurus	easy

See November.

Altas	27 Tau	03^h 49.2^m	−24° 03′	November 18
3.62 m	−1.72 M	*B8 III*	Taurus	easy

See November.

Deneb	α Cyg	20^h 41.3^m	+45° 17′	August 1
1.25$_v$m	−8.73[21]M	*A2 I*	Cygnus	easy

See August.

Aldermin	α Cep	21^h 18.6^m	+62° 35′	August 11
2.45 m	1.58 M	*A7 IV*	Cepheus	easy ©

See August.

Algenib	α Per	03^h 24.3^m	+49° 52′	November 11
1.79 m	−4.5 M	*F5 I*	Perseus	easy ©

See November.

[21]This value is in question. The data are awaiting reassessment.

Polaris	α UMi	$02^h\ 31.8^m$	+89° 16′	October 29
1.97_vm	−3.64 M	*F7 I*	Ursa Minor	easy ©

An interesting and famous star, even though it is only the 49th-brightest star in the sky. It is a *cepheid variable* type II (the W Virginis class); it will be closest to the celestial pole in AD 2102, and is a binary star (the companion reported as being pale bluish), being a good test for small telescopes.

Sadal Melik	α Aqr	$22^h\ 05.8^m$	−00° 19′	August 23
2.95 m	−3.88 M	*G2 I*	Aquarius	easy

See August.

ξ^1 **Cet**	HD 15318	$02^h\ 12.0^m$	+08° 51′	October 24
4.36 m	−0.87 M	*G8 II*	Cetus	easy

A star with an interesting background. Although about 550 times as luminous as the Sun, various measurements place it at 130, 175 and 640 l.y. distant!

β **Cet**	HD 4128	$00^h\ 43.6^m$	−17° 59′	October 2
2.04 m	−0.30 M	*G9.5 III*	Cetus	easy

The star lies at a distance of 60 l.y. with a luminosity of 42 Suns.

Hamal	α Ari	$02^h\ 07.2^m$	+23° 28′	October 23
2.01 m	0.48 M	*K2 III*	Aries	easy

This star lies at a distance of 63 l.y. with a luminosity 45 times that of the Sun. [See also Arcturus.]

Almach	γ^1 And	$02^h\ 03.9^m$	+42° 20′	October 22
2.33 m	−2.86 M	*K3 III*	Andromeda	easy

A famous binary star. The colours are gold and blue, although some observers see orange and greenish blue. Nevertheless, the fainter companion is hot enough to truly show a blue colour. It is also a binary in its own right, but not observable in amateur instruments.

Mirach	β And	$01^h\ 09.7^m$	+35° 37′	October 8
2.07 m	−1.86 M	*M0 III*	Andromeda	easy

With this stellar class, the bands of titanium oxide are strengthening. This red giant star is suspected of being slightly variable, like so many other stars of the same type. In the field of view is the galaxy NGC 404. At magnitude 12, a good test for large telescopes.

Menkar	α Cet	$03^h\ 02.3^m$	+0.4° 05′	November 6
2.54_vm	−1.61 M	*M2 III*	Cetus	easy

See November.

Eta Sagitai	η Per	$02^h\ 50.7^m$	+55° 54′	November 3
3.77 m	−4.28 M	*M3 I*	Perseus	easy ©

See November.

Mira	*ο* Cet	02^h 19.3^m	–02° 59′	October 26
2.00_vm	–3.54 M	*M5*	Cetus	easy

An important star, and maybe the first variable star ever observed. Written records certainly exist as far back as 1596. The prototype of the long-period pulsating variable, it varies from 3rd to 10th magnitude over a period of 332 days, and is an ideal star for the first-time variable star observer. See entry below.

Mira at minimum	*ο* Cet	02^h 19.3^m	–02° 59′	October 26
10_vm	–0.5 M	*M9*	Cetus	difficult

At minimum, the star is a deeper red colour, but of course fainter. It now has a lower temperature of 1900 K. The period, however, is subject to irregularities, as is its magnitude. It has been observed for maximum light to reach 1st magnitude – similar to Aldebaran!

November

***θ* Orionis C**	*θ* Ori	05^h 35.3^m	–05° 23′	December 14
4.96 m	–5.04 M	*O6*	Orion	easy

See December.

15 Monocerotis	HD47839	06^h 40.9^m	+09° 54′	December 31
4.66_vm	–2.3 M	*O7*	Monoceris	easy

See December.

Plaskett's Star	HD47129	06^h 37.4^m	+06° 08′	December 30
6.05 m	–3.54 M	O8	Monoceros	easy

See December.

Iota Orionis	*ι* Ori	05^h 35.4^m	–05° 55′	December 15
2.75 m	–5.30 M	*O9 III*	Orion	easy

See December.

Gamma Cassiopeiae	*γ* Cas	00^h 56.7^m	+60° 43′	October 5
2.15_vm	–4.22 M	*B0 IV*	Cassiopeia	easy

See October.

***λ* CMa**	HD 45813	06^h 28.2^m	–32° 35′	December 28
4.47 m	–1.01 M	*B4 IV*	Canis Major	easy

See December.

Electra	17 Tau	03^h 44.9^m	+24° 07′	November 16
3.72 m	–1.56 M	*B6 III*	Taurus	easy

Located within the Pleiades star cluster. A breathtaking and spectacular view when seen through binoculars, the cluster is a highlight of the night sky. [See also Taygeta (19 Tau) and Merope (23 Tau) in the Pleiades cluster.]

El Nath	β Tauri	$05^h 26.3^m$	+28° 36′	December 12
1.65 m	–1.37 M	*B7 III*	Taurus	easy

See December.

Atlas	27 Tau	$03^h 49.2^m$	–24° 03′	November 18
3.62 m	–1.72 M	*B8 III*	Taurus	easy

A lovely blue star. [See also Maia (20 Tau), Asterope (21 Tau) and Pleione (28 Tau) in the Pleiades.]

Alhena	γ Gem	$06^h 37.7^m$	+16° 23′	December 30
1.93 m	–0.60 M	*A0 IV*	Gemini	easy

See December.

Beta Aurigae	β Aur	$05^h 59.5^m$	+44° 57′	December 21
1.90_vm	–0.10 M	*A2 V*	Auriga	easy

See December.

2 Mon	HD 40536	$05^h 59.1^m$	–09° 33′	December 21
5.01 m	0.02 M	*A6*	Monoceros	easy

See December.

Algenib	α Per	$03^h 24.3^m$	+49° 52′	November 11
1.79 m	–4.5 M	*F5 I*	Perseus	easy ©

The star lies within Melotte 20, a loosely bound stellar association, also known as the Perseus OB–3, or Alpha Persei Association. About 75 stars with magnitudes down to 10 are contained within the group. All are stellar infants, only 50 million years old, lying 550 l.y. away. The metallic lines now increase through the F class, especially the H and K lines of ionised calcium. Has been described as having a pale yellow hue. [See also Procyon.]

Polaris	α UMi	$02^h 31.8^m$	+89° 16′	October 29
1.97_vm	–3.64 M	*F7 I*	Ursa Minor	easy ©

See October.

111 Tau	HD 35296	$05^h 24.4^m$	+17° 23′	December 12
5.00 m	4.17 M	*F8 V*	Taurus	easy

See December.

ξ^1 Cet	HD 15318	$02^h 12.0^m$	+08° 51′	October 24
4.36 m	–0.87 M	*G8 II*	Cetus	easy

See October.

β Cet	HD 4128	$00^h 43.6^m$	–17° 59′	October 2
2.04 m	–0.30 M	*G9.5 III*	Cetus	easy

See October.

ν^2 **CMa**	HD 47205	06^h 36.7^m	−19° 15′	December 30
3.95 m	2.46 M	*K1 III*	Canis Major	easy

See December.

Menkar	α Cet	03^h 02.3^m	+0.4° 05′	November 6
2.54_vm	−1.61 M	*M2 III*	Cetus	easy

An orange-coloured giant star, which contrasts nicely with a fainter blue star (93 Ceti nearly at due north), seen in the same field of view with small telescopes at low power.

Eta Sagitai	η Per	02^h 50.7^m	+55° 54′	November 3
3.77 m	−4.28 M	*M3 I*	Perseus	easy ©

The yellowish star in a easily resolved double star system. The colour contrasts nicely with its blue companion.

Mira	o Cet	02^h 19.3^m	−02° 59′	October 26
2.00_vm	−3.54 M	*M5*	Cetus	easy

See October.

Mira at minimum	o Cet	02^h 19.3^m	−02° 59′	October 26
10_vm	−0.5 M	*M9*	Cetus	difficult

See October.

December

θ **Orionis C**	θ Ori	05^h 35.3^m	−05° 23′	December 14
4.96 m	−5.04 M	*O6*	Orion	easy

A member of the famous Trapezium multiple star system in the Orion Nebula. Splitting the group is always a test for small telescopes. A fairly new star, maybe only several thousand years old, and as a consequence most of the star's light is emitted at ultraviolet wavelengths.

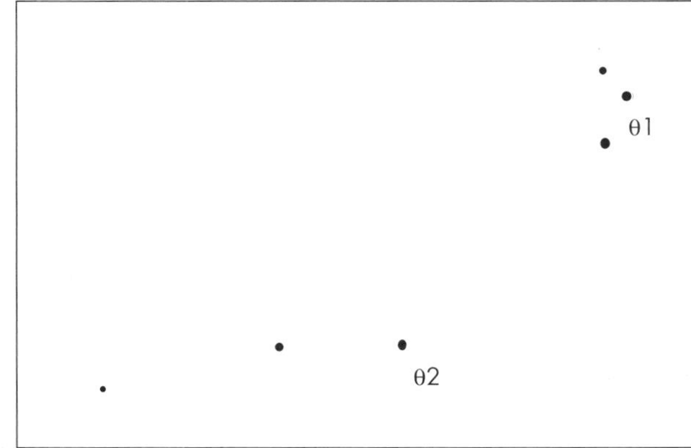

Figure 2.1. θ Orionis C.

15 Monocerotis	HD47839	06^h 40.9^m	+09° 54′	December 31
4.66_vm	–2.3 M	*O7*	Monoceris	easy

Both a visual binary and a variable star, it is located in the star cluster NGC 2264, which in turn is encased in a diffuse nebula. About 1° south is the famous Cone Nebula, visible only in the largest amateur telescopes under perfect conditions.

Plaskett's Star	HD47129	06^h 37.4^m	+06° 08′	December 30
6.05 m	–3.54 M	*O8*	Monoceros	easy

This is actually composed of two stars, a spectroscopic binary system, with an estimated mass of around 110 Suns, making it one of the most massive known.

Iota Orionis	ι Ori	05^h 35.4^m	–05° 55′	December 15
2.75 m	–5.30 M	*O9 III*	Orion	easy

The brightest star in the sword of Orion is in fact a fine triple star system, with reported colours of white, blue and red.

Murzim	β CMa	06^h 22.7^m	–17° 57′	December 27
1.98_vm	–3.96 M	*B1 II*	Canis Major	easy

This is the prototype of a class of variable star now classified as β Cepheid stars, which are pulsating variables. The magnitude variation is too small to be observed visually. [See also Spica and Beta Centauri in previous sections.]

λ CMa	HD 45813	06^h 28.2^m	–32° 35′	December 28
4.47 m	–1.01 M	*B4 IV*	Canis Major	easy

A nice bluish-white star.

Electra	17 Tau	03^h 44.9^m	+24° 07′	November 16
3.72 m	–1.56 M	*B6 III*	Taurus	easy

See November.

El Nath	β Tauri	05^h 26.3^m	+28° 36′	December 12
1.65 m	–1.37 M	*B7 III*	Taurus	easy

Located on the border of Auriga, it is sometimes mistakenly called γ Aurigae. It lies at a distance of 160 l.y.

Atlas	27 Tau	03^h 49.2^m	–24° 03′	November 18
3.62 m	–1.72 M	*B8 III*	Taurus	easy

See November.

Alhena	γ Gem	06^h 37.7^m	+16° 23′	December 30
1.93 m	–0.60 M	*A0 IV*	Gemini	easy

The star is relatively close at about 58 l.y., with a luminosity of 160 Suns.

Castor	α Gem	07^h 34.6^m	+31° 53′	January 14
1.43 m	0.94 M	*A1 V*	Gemini	easy

See January.

Beta Aurigae	β Aur	05^h 59.5^m	+44° 57′	December 21
1.90$_v$m	–0.10 M	*A2 V*	Auriga	easy

This is a good example of the Algol type of variable star, which is due to stars eclipsing each other. A spectral class A2 signifies that the hydrogen lines are now at their strongest.

2 Mon	HD 40536	05^h 59.1^m	–09° 33′	December 21
5.01 m	0.02 M	*A6*	Monoceros	easy

The star lies at a distance of over 1900 l.y., with a luminosity of 5000 Suns.

Canopus	α Car	06^h 23.9^m	–52° 41′	December 27
–0.62 m	–5.53 M	*F0 I*	Carina	easy

The second brightest star in the sky. Its colour is often reported as orange or yellow, as it is usually seen lying low down in the sky, and is thus apt to be affected by the atmosphere. Its true colour is white.

Algenib	α Per	03^h 24.3^m	+49° 52′	November 11
1.79 m	–4.5 M	*F5 I*	Perseus	easy ©

See November.

Polaris	α UMi	02^h 31.8^m	+89° 16′	October 29
1.97$_v$m	–3.64 M	*F7 I*	Ursa Minor	easy ©

See October.

111 Tau	HD 35296	05^h 24.4^m	+17° 23′	December 12
5.00 m	4.17 M	*F8 V*	Taurus	easy

A close star at 52 l.y., only 2 times as luminous as the Sun.

ν^2 CMa	HD 47205	06^h 36.7^m	–19° 15′	December 30
3.95 m	2.46 M	*K1 III*	Canis Major	easy

This star lies at a distance of 60 l.y. with a luminosity 7 times that of the Sun.

γ^2 Vel	HD 68273	08^h 09.5^m	–47° 20′	January 23
1.99$_v$m	0.05 M	*WC 8*	Vela	easy

See January.

2.6 Red Stars

This section will deal with the topic of the colours of stars. It may seem to a casual observer that the stars do not possess many bright colours, and only the brightest stars show any perceptible colour: Betelgeuse can be seen to be red, and Capella, yellow, whilst Vega is blue, and Aldebaran has an orange tint, but beyond that, most stars seem to be an overall white. To the naked eye, this is certainly the case, and it is only with some kind of optical equipment that the full range of star colour becomes apparent.

But what is meant by the colour of a star? A scientific description of a star's colour is one that is based on the stellar classification, which in turn is dependent upon the chemical composition and temperature of a star. In addition, a term commonly used by astronomers is the *colour index*. This is determined by observing a star through two filters, the *B* and the *V* filters, which correspond to wavelengths of 440 nm and 550 nm respectively, and measuring its brightness. Subtracting the two values obtained gives B – V, the colour index. Usually, a blue star will have a colour index that is negative, i.e., −0.3, orange-red stars could have a value greater than 0.0, and upwards to about 3.00 and greater for very red stars (M6 and greater).[22] But as this is an observationally based book, the scientific description will not generally apply. The most important factor which determines what the colour of a star is, is you – the observer! It is purely a matter of both physiological and psychological influences. What one observer describes as a blue star, another may describe as a white star, or one may see an orange star, whilst another observes the same star as being yellow. It may even be that you will observe a star to have different colour when using different telescopes or magnifications, and atmospheric conditions will certainly have a role to play. The important thing to remember is that whatever colour you observe a star to have, then that is the colour you should record.[23]

As mentioned above, red, yellow, orange and blue stars are fairly common, but are there stars which have, say, a purple tint, or blue, or violet, crimson, lemon, and the ever elusive green colour? The answer is, yes, but with the caution that it depends on how you describe the colour. A glance at the astronomy books from the last century and beginning of the twentieth century, will show you that star colour was a hot topic, and descriptions such as Amethyst (purple), Cinerous (wood-ash tint), Jacinth (pellucid orange), and Smalt (deep blue), to name but a few, were used frequently. Indeed, the British Astronomical Association even had a section devoted to star colours. But today, observing and cataloging star colour is just a pleasant past-time. Nevertheless, under good seeing conditions, with a dark sky, the keen-eyed observer, will be able to see the faint tinted colours from deepest red to steely blue, with all the colours in between.

It is worth noting that several distinctly coloured stars occur as part of a double star system. The reason for this may be that although the colour is difficult to see in an individual star, it may appear more intense when seen together with a contrasting colour. Thus, in the section on double and triple stars, there are descriptions of many beautifully coloured systems. For instance, in the following double star systems, the fainter of the two stars in η Cassiopeiae has a distinct purple tint, whilst in γ Andromadae and α Herculis, the fainter stars are most definitely green.

Many of the strongly coloured stars have already been described in the previous section, and thus will not be repeated here, and other stars will be described in the sections on double and triple stars. However, there is a star colour upon which most observers agree – the red stars, and to that end, the list below will catalogue the most famous and brightest of this type of star. All of them are classified as either N or R type stars, as well as a few C type stars. The N and R classification signifies that although the temperature may be of the same order as M type stars, these stars show different chemical compositions, whilst the C type stars are the carbon stars mentioned earlier. Some of these stars are intensely red, and have a deeper colour than even Betelgeuse and Arcturus!

---

[22]Note that in this section the magnitude quoted is the Hipparcos value, whilst the B–V value and the magnitude ranges have been taken from other sources.

[23]An interesting experiment is to observe a coloured star first through one eye, and then the other. You may be surprised by the result!

January

Hind's Crimson Star	R Leporis	$04^h 59.6^m$	–14° 48′	December 5
$7.71_v m$	B-V:3.4	C7		easy
See December.				

W Ori	HD 32736	$05^h 0.4^m$	+01° 11′	December 7
$6.3_v m$	B-V:3.33	N5		easy
See December.				

V Hydrae	Lalande 16	$10^h 51.6^m$	–21° 15′	March 5
$7.0_v m$	B-V:4.5	C9		easy
See March.				

X Cnc	HD 76221	$08^h 55.4^m$	+17° 14′	February 3
$6.12_v m$	B-V:2.97	C6		moderate
See February.				

February

V Hydrae	Lalande 16	$10^h 51.6^m$	–21° 15′	March 5
$7.0_v m$	B-V:4.5	C9		easy
See March.				

La Superba	U CVn	$12^h 45.1^m$	+45° 26′	April 2
$5.4_v m$	B-V:2.9	C7		easy
See April.				

RY Dra	HD 112559	$12^h 56.3^m$	+66° 00′	April 5
$6.4_v m$	B-V:3.3	C7		easy ©
See April.				

Hind's Crimson Star	R Leporis	$04^h 59.6^m$	–14° 48′	December 5
$7.71_v m$	B-V:3.4	C7		easy
See December.				

W Ori	HD 32736	$05^h 0.4^m$	+01° 11′	December 7
$6.3_v m$	B-V:3.33	N5		easy
See December.				

X Cnc	HD 76221	$08^h 55.4^m$	+17° 14′	February 3
$6.12_v m$	B-V:2.97	C6		moderate
An extremely orange star, this semi-regular variable star, classification SRB, has a period of 180 to 195 days, and has been observed to range in magnitude from 5.6 to 7.5.				

March

V Hydrae	Lalande 16	$10^h 51.6^m$	−21° 15′	March 5
7.0_vm	B-V:4.5	C9		easy

The star, another classic long-period variable, period about 533 days, varies in brightness between 6 and 12 m. It also has a second periodicity of 18 years. One of the rare *carbon stars* visible in amateur instruments, it has been described as coloured a "magnificent copper red". It is, however, difficult to observe owing to its large magnitude range.

La Superba	U CVn	$12^h 45.1^m$	+45° 26′	April 2
5.4_vm	B-V:2.9	C7		easy

See April.

RY Dra	HD 112559	$12^h 56.3^m$	+66° 00′	April 5
6.4_vm	B-V:3.3	C7		easy ©

See April.

Hind's Crimson Star	R Leporis	$04^h 59.6^m$	−14° 48′	December 5
7.71_vm	B-V:3.4	C7		easy

See December.

W Ori	HD 32736	$05^h 0.4^m$	+01° 11′	December 7
6.3_vm	B-V:3.33	N5		easy

See December.

X Cnc	HD 76221	$08^h 55.4^m$	+17° 14′	February 3
6.12_vm	B-V:2.97	C6		moderate

See February.

April

La Superba	U CVn	$12^h 45.1^m$	+45° 26′	April 2
5.4_vm	B-V:2.9	C7		easy

The colour of this star is best seen through binoculars or a small telescope. With a period of 159 days, and varying in magnitude between 4.9 and 6.0 m, this red giant has a diameter of 400 million kilometres.

RY Dra	HD 112559	$12^h 56.3^m$	+66° 00′	April 5
6.4_vm	B-V:3.3	C7		easy ©

A red giant variable with poorly understood periodicity, class SRB, with a period believed to be 200 days, and a magnitude range of 6.0–8.0 m, this star has a lovely red colour.

V Pav	HD 160435	$17^h 43.3^m$	−57° 43′	June 17
6.65_vm	B-V:2.45	C5		easy

See June.

V Hydrae	Lalande 16	$10^h 51.6^m$	–21° 15′	March 5
7.0_vm	B-V:4.5	C9		easy
See March.				

X Cnc	HD 76221	$08^h 55.4^m$	+17° 14′	February 3
6.12_vm	B-V:2.97	C6		moderate
See February.				

T Lyr		$18^h 32.3^m$	+37° 00′	June 29
8.5_vm	B-V:	C8		moderate
See June.				

X Cnc	HD 76221	$08^h 55.4^m$	+17° 14′	February 3
6.12_vm	B-V:2.97	C6		moderate
See February.				

May

La Superba	U CVn	$12^h 45.1^m$	+45° 26′	April 2
5.4_vm	B-V:2.9	C7		easy
See April.				

RY Dra	HD 112559	$12^h 56.3^m$	+66° 00′	April 5
6.4_vm	B-V:3.3	C7		easy ©
See April.				

V Pav	HD 160435	$17^h 43.3^m$	–57° 43′	June 17
6.65_vm	B-V:2.45	C5		easy
See June.				

V Aql	HD 177336	$19^h 04.4^m$	–05° 41′	July 8
7.5_vm	B-V:5.46	C5		easy
See July.				

RS Cyg	HD 192443	$20^h 13.3^m$	+38° 44′	July 25
8.1_vm	B-V:3.3	C5		easy
See July.				

V Hydrae	Lalande 16	$10^h 51.6^m$	–21° 15′	March 5
7.0_vm	B-V:4.5	C9		easy
See March.				

T Lyr		$18^h 32.3^m$	+37° 00′	June 29
8.5_vm	B-V:	C8		moderate
See June.				

June

V Pav	HD 160435	$17^h 43.3^m$	−57° 43′	June 17
6.65_vm	B-V:2.45	C5		easy

A red giant variable star, class SRB, varying in brightness from 6.3 to 8.2 m, over a period of 225.4 days. It also has a secondary period of about 3735 days. A glorious deep-red colour.

La Superba	U CVn	$12^h 45.1^m$	+45° 26′	April 2
5.4_vm	B-V:2.9	C7		easy

See April.

RY Dra	HD 112559	$12^h 56.3^m$	+66° 00′	April 5
6.4_vm	B-V:3.3	C7		easy ©

See April.

V Aql	HD 177336	$19^h 04.4^m$	−05° 41′	July 8
7.5_vm	B-V:5.46	C5		easy

See July.

RS Cyg	HD 192443	$20^h 13.3^m$	+38° 44′	July 25
8.1_vm	B-V:3.3	C5		easy

See July.

Garnet Star	μ Cep	$21^h 43.5^m$	+58° 47′	August 17
4.2_vm	B-V:2.3	M2		easy

See August.

T Lyr		$18^h 32.3^m$	+37° 00′	June 29
8.5_vm	B-V:	C8		moderate

An extremely red-coloured star, this is another with an irregular period; magnitude range 7.5 to 9.3.

S Cephei	HD 206362	$21^h 35.2^m$	+78° 37′	August 15
7.9_vm	B-V:2.7	C6		moderate/difficult ©

See August.

July

V Aql	HD 177336	$19^h 04.4^m$	−05° 41′	July 8
7.5_vm	B-V:5.46	C5		easy

A semi-regular variable star, with a period of about 350 days, varying in magnitude from 6.6 to 8.1 m. A very deep red in colour.

RS Cyg	HD 192443	$20^h 13.3^m$	+38° 44′	July 25
8.1_vm	B-V:3.3	C5		easy

A red giant star with a persistent periodicity, class SRA, it has a period of 417.39 days, with a magnitude range of 6.5 to 9.5 m. A strange star where the light curve can vary appreciably, with the maxima sometimes doubling. A deeply red-coloured star.

V Pav	HD 160435	$17^h 43.3^m$	–57° 43′	June 17
6.65_vm	B-V:2.45	C5		easy
See June.				

La Superba	U CVn	$12^h 45.1^m$	+45° 26′	April 2
5.4_vm	B-V:2.9	C7		easy
See April.				

RY Dra	HD 112559	$12^h 56.3^m$	+66° 00′	April 5
6.4_vm	B-V:3.3	C7		easy ©
See April.				

Garnet Star	μ Cep	$21^h 43.5^m$	+58° 47′	August 17
4.2_vm	B-V:2.3	M2		easy
See August.				

19 Piscium	TX Psc	$23^h 46.4^m$	+03° 29′	September 17
4.95_vm	B-V:2.5	C5		easy
See September.				

T Lyr		$18^h 32.3^m$	+37° 00′	June 29
8.5_vm	B-V:	C8		moderate
See June.				

S Cephei	HD 206362	$21^h 35.2^m$	+78° 37′	August 15
7.9_vm	B-V:2.7	C6		moderate/difficult ©
See August.				

August

Garnet Star	μ Cep	$21^h 43.5^m$	+58° 47′	August 17
4.2_vm	B-V:2.3	M2		easy
Located on the north-eastern edge of the nebulosity IC1396, the Garnet Star, named by William Herschel, is one of the reddest stars in the entire sky. It has a deep orange or red colour seen against a backdrop of faint white stars. It is a pulsating red giant star, with a period of about 730 days, varying from 3.4 to 5.1 m.				

V Aql	HD 177336	$19^h 04.4^m$	–05° 41′	July 8
7.5_vm	B-V:5.46	C5		easy
See July.				

RS Cyg	HD 192443	$20^h 13.3^m$	+38° 44′	July 25
8.1_vm	B-V:3.3	C5		easy
See July.				

V Pav	HD 160435	$17^h 43.3^m$	–57° 43′	June 17
6.65_vm	B-V:2.45	C5		easy
See June.				

19 Piscium	TX Psc	$23^h 46.4^m$	+03° 29′	September 17
4.95_vm	B-V:2.5	C5		easy
See September.				

R Scl	HD 8879	$01^h 26.9^m$	–32° 33′	October 13
5.79_vm	B-V:1.4	C6		easy
See October.				

T Lyr		$18^h 32.3^m$	+37° 00′	June 29
8.5_vm	B-V:	C8		moderate
See June.				

S Cephei	HD 206362	$21^h 35.2^m$	+78° 37′	August 15
7.9_vm	B-V:2.7	C6		moderate/difficult ©
A moderately difficult star to observe, owing to its magnitude range of between 7 and 12 magnitudes, it nevertheless has a very high colour index, making it one of the reddest stars in the sky if not *the* reddest.				

September

19 Piscium	TX Psc	$23^h 46.4^m$	+03° 29′	September 17
4.95_vm	B-V:2.5	C5		easy
A slow, irregular-period variable star. Classification LB, with a magnitude range of 4.8 to 5.2 m. The colour is an orange–red, best seen in small instruments.				

Garnet Star	μ Cep	$21^h 43.5^m$	+58° 47′	August 17
4.2_vm	B-V:2.3	M2		easy
See August.				

V Aql	HD 177336	$19^h 04.4^m$	–05° 41′	July 8
7.5_vm	B-V:5.46	C5		easy
See July.				

RS Cyg	HD 192443	$20^h 13.3^m$	+38° 44′	July 25
8.1_vm	B-V:3.3	C5		easy
See July.				

R Scl	HD 8879	$01^h 26.9^m$	–32° 33′	October 13
5.79_vm	B-V:1.4	C6		easy
See October.				

U Cam		$03^h 41.8^m$	+62° 39′	November 1
8.3_vm	B-V:4.9	N7		moderate ©
See November.				

S Cephei	HD 206362	$21^h 35.2^m$	+78° 37′	August 15
7.9_vm	B-V:2.7	C6		moderate/difficult ©
See August.				

October

R Scl	HD 8879	$01^h\ 26.9^m$	–32° 33′	October 13
5.79_vm	B-V:1.4	C6		easy

A semi-regular-period variable star, with a period of between 140 and 146 days, it varies in brightness from 5.0 to 6.5.

19 Piscium	TX Psc	$23^h\ 46.4^m$	+03° 29′	September 17
4.95_vm	B-V:2.5	C5		easy

See September.

Garnet Star	μ Cep	$21^h\ 43.5^m$	+58° 47′	August 17
4.2_vm	B-V:2.3	M2		easy

See August.

Hind's Crimson Star	R Leporis	$04^h\ 59.6^m$	–14° 48′	December 5
7.71_vm	B-V:3.4	C7		easy

See December.

W Ori	HD 32736	$05^h\ 0.4^m$	+01° 11′	December 7
6.3_vm	B-V:3.33	N5		easy

See December.

U Cam		$03^h\ 41.8^m$	+62° 39′	November 1
8.3_vm	B-V:4.9	N7		moderate ©

See November.

S Cephei	HD 206362	$21^h\ 35.2^m$	+78° 37′	August 15
7.9_vm	B-V:2.7	C6		moderate/difficult ©

See August.

November

U Cam		$03^h\ 41.8^m$	+62° 39′	November 16
8.3_vm	B-V:4.9	N7		moderate ©

A semi-regular variable star, period 412 days with a magnitude range of 7.7 to 9.5 m. It has a very deep-red colour.

R Scl	HD 8879	$01^h\ 26.9^m$	–32° 33′	October 13
5.79_vm	B-V:1.4	C6		easy

See October.

19 Piscium	TX Psc	$23^h\ 46.4^m$	+03° 29′	September 17
4.95_vm	B-V:2.5	C5		easy

See September.

Garnet Star	μ Cep	21h 43.5m	+58° 47′	August 17
4.2_vm	B-V:2.3	M2		easy
See August.				

Hind's Crimson Star	R Leporis	04h 59.6m	–14° 48′	December 5
7.71_vm	B-V:3.4	C7		easy
See December.				

W Ori	HD 32736	05h 0.4m	+01° 11′	December 7
6.3_vm	B-V:3.33	N5		easy
See December.				

X Cnc	HD 76221	08h 55.4m	+17° 14′	February 3
6.12_vm	B-V:2.97	C6		moderate
See February.				

December

Hind's Crimson Star	R Leporis	04h 59.6m	–14° 48′	December 5
7.71_vm	B-V:3.4	C7		easy
The star, a classic long-period variable, period about 432 days, varies in brightness between 6.0 and 9.7 m. At maximum brightness it displays the famous ruddy colour that gives it its name. Discovered in 1845 by J. R. Hind with a colour described as "intense smoky red". Some amateurs believe this to be the reddest star.				

W Ori	HD 32736	05h 0.4m	+01° 11′	December 7
6.3_vm	B-V:3.33	N5		easy
A red giant variable star, classification SRB, with a period of 212 days, although a secondary period of 2450 days is believed to occur. Varies in magnitude from 5.5 to 7.7 m. A deep-red star.				

U Cam		03h 41.8m	+62° 39′	November 16
8.3_vm	B-V:4.9	N7		moderate ©
See November.				

R Scl	HD 8879	01h 26.9m	–32° 33′	October 13
5.79_vm	B-V:1.4	C6		easy
See October.				

19 Piscium	TX Psc	23h 46.4m	+03° 29′	September 17
4.95_vm	B-V:2.5	C5		easy
See September.				

X Cnc	HD 76221	08h 55.4m	+17° 14′	February 3
6.12_vm	B-V:2.97	C6		moderate
See February.				

2.7 Double Stars

Having now started your observation of the night sky by looking at single stars, you are now able to use the skills so far developed to study and observe objects which not only present a wonderful array of colours, but also allow precise measurements to be made, double star systems. The study of double stars is one that has a great pedigree. It was, and in fact still is in some respects, an area of astronomy where the observer can make useful detailed observations. In addition, many double stars present a glittering range of colours.

Double stars are stars, that although appear to be just one single star, will on observation with either binoculars or telescopes resolve itself into two stars. Indeed, some apparently single stars turn out to be several stars! Many appear as double stars due to them lying in the same line of sight as seen from the Earth, and this can only be determined by measuring the spectra of the stars and calculating their red (or blue) shifts. Such stars are called *optical doubles*. It may well be that the two stars are separated in space by a vast distance. Some, however are actually gravitationally bound and may orbit around each other, over a period of days, or even years. Nevertheless, whatever type of double star it actually is, they still present a marvelous challenge and delight to the amateur.

The classification of some double stars is quite complex, and lies really in the realms of astrophysics. For instance, many cannot be resolved by even the largest telescopes, and are called *spectroscopic binaries*, the double component only being fully understood when the spectra is analyzed. Others are *eclipsing binaries*, such as Algol (β Persei) where one star moves during its orbit, in front of its companion, thus brightening and dimming the light observed. A third type is the *astrometric binary*, such as Sirius (α Canis Majoris) where the companion star may only be detected by its influence on the motion of the primary star. However, this book is concerned with objects that can be observed visually, therefore I will concentrate on double stars that can be split with either the naked eye, or by using some sort of optical equipment.

As in previous sections, some terminology must now be introduced that is specific to double star observation. The brightest of the two stars is usually called the *primary* star, whilst the fainter is called the *secondary* (in some texts it may be called the companion, and either terms will be used throughout this book). This terminology is employed regardless of how massive either star is, or whether the brighter is in fact the less luminous of the two in reality, but just appears brighter as it may be closer.

Perhaps the most important terms used in double star work are the *separation* and *position angle* (PA). The separation is the angular distance between the two stars usually in seconds of arc, and measured from the brightest star to the fainter. The position angle is a somewhat more difficult concept to understand. It is the relative position of one star, usually the secondary, with respect to the primary, and is measured in degrees, with 0° at due north, 90° at due east, 180° due south, 270° at due west, and back to 0°. It is best described by an example; using Fig. 2.2, the double star γ Virginis, with components of magnitude 3.5 and 3.5, has a separation of 1.8″ (arcseconds), at a PA of 267° (epoch 2000.0). Note that the secondary star is the one always placed somewhere on the orbit, the primary star is at the centre of the perpendicular lines, and that the separation and PA of any double star are constantly changing, and should be quoted for the year observed. Some stars, where the period is very long, will have no appreciable change in PA for several years, others, however, will change from year-to-year. Many books that discuss double stars in detail will have similar diagrams for the stars listed, however, to present a similar facility here, would entail a doubling of the books size.[24]

---

[24]Several of the books listed in the appendices will have the double star orbits drawn, which will significantly aid you in determining which star is which.

It is worth mentioning again, that although your optical equipment, including your eyes, should in theory be able to resolve many of the double listed here,[25] there are several factors which will constrain the resolution, i.e., the seeing, light pollution, dark adaption, etc., your temperament. Thus, if you cannot initially resolve a double star, do not panic, but move onto another, and return to the one in question at another date. Also recall that the colours ascribed to the star, will not necessarily be the colour you see. They are just indicators of the general colour, and in fact, as you will see from the text, many observers disagree on several star's colour.

Finally, Table 2.4 shows a brief list of several stars that will, as an initial indicator, help you to determine the resolution of both yourself and your binoculars/telescope.[26] All the positions quoted are for the primary star.

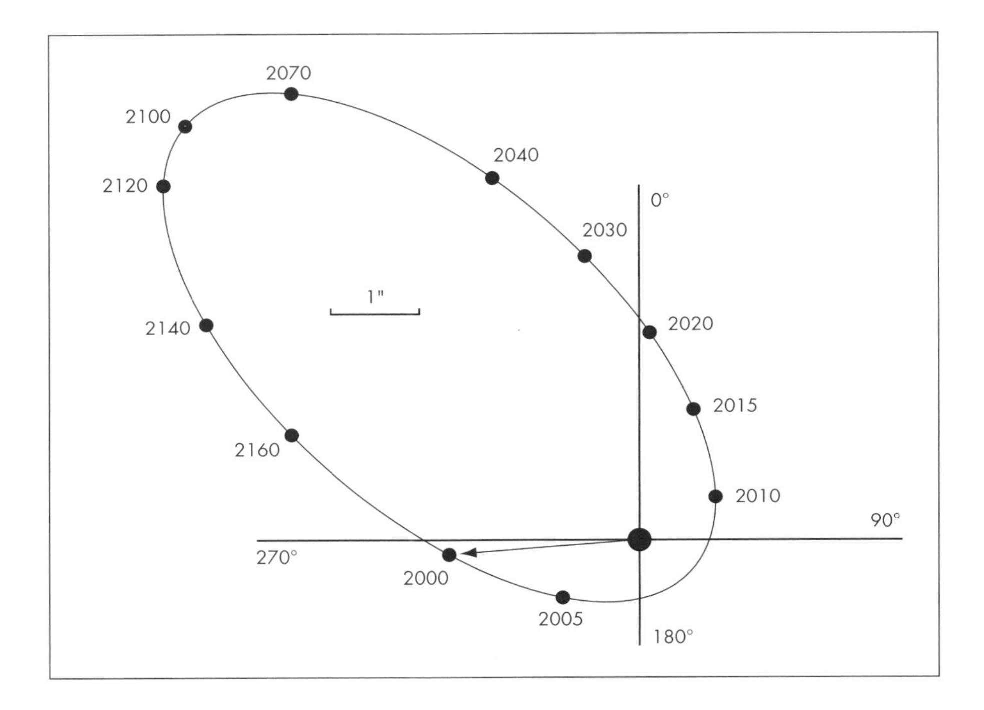

Figure 2.2. γ Virgins.

[25]There are literally thousands of double, triple and multiple star systems in the sky, all within reach of the amateur astronomer and the list that follows is but a taste of what awaits the observer. If your favourite star is omitted, then I apologize, but to include them all would have been impossible.

[26]Note that the position angle and separation are quoted for epoch 2000.0. With double stars which have small periods, these figures will change appreciably.

Table 2.4. Selected double stars

Star		RA	Dec	m1	m2	PA°	Sep″	Aperture (cm)
ζ CMa	Zeta Ursa Majoris	$13^h 23^m 55.5^s$	+54° 55′ 31″	2.5	3.9	152	14.4	naked eye
ε Lyr	Epsilon Lyrae	$18^h 44^m 20.3^s$	+39° 40′ 12″	4.7	4.6	173	207.7	naked eye
o Cep	Omicron Cephei	$23^h 18^m 37.4^s$	+68° 06′ 42″	5.0	7.3	223	2.8	5
ζ Aqr	Zeta Aquarii	$22^h 28^m 49.6^s$	–00° 01′ 13″	4.4	4.6	192	2.1	5
ε Ari	Epsilon Arietis	$02^h 59^m 12.6^s$	+21° 20′ 25″	5.2	5.6	203	1.4	7.5
μ Lib	Mu Librae	$14^h 49^m 19.0^s$	–14° 08′ 56″	5.7	6.7	355	1.8	7.5
θ Aur	Theta Aurigae	$05^h 59^m 43.2^s$	+37° 12′ 45″	2.6	7.1	313	3.6	10
η Ori	Eta Orionis	$05^h 24^m 28.6^s$	–02° 23′ 49″	3.6	4.9	80	1.5	10
ζ Boo	Zeta Boötis	$14^h 41^m 08.8^s$	+13° 43′ 42″	4.5	4.6	300	0.8	15
ι Leo	Iota Leonis	$11^h 23^m 55.4^s$	+10° 31′ 45″	4.1	6.9	116	1.7	20
χ Aql	Chi Aquilae	$23^h 18^m 37.4^s$	+68° 06′ 42″	5.8	6.9	77	0.5	25

January

φ^5 Aurigae	HD 48682	$06^h 46.7^m$	+43° 35′	January 2
5.3, 8.3 m	PA 31°; Sep. 36.2″	Spec. G0[27]		easy[28]
A pair of yellow and blue stars set against a backdrop of faint stars.				

π Canis Majoris	HD 51199	$06^h 55.6^m$	–20° 08′	January 4
4.7, 9.7 m	PA 18°; Sep. 11.6″	Spec. F2		easy
A pale-yellow primary and bluish secondary.				

h 3945[29]	HD 56577	$07^h 16.1^m$	–23° 19′	January 9
4.8, 6.1 m	PA 55°; Sep. 26.6″	Spec. K5 F0		easy
A wonderful dark-orange and blue double star system. Has also been described as gold and blue.				

Struve 1149	Σ1149	$07^h 49.5^m$	+03° 13′	January 18
7.9, 9.6 m	PA 41°; Sep. 21.7″	Spec. G0		easy
A lovely double star system. A yellow primary and blue secondary				

38 Geminorum	ADS 5559[30]	$06^h 54.6^m$	+13° 11′	January 4
4.7, 7.7 m	PA 145°; Sep. 7.1″	Spec. F0		moderate
Easily split using small telescopes. Yellow and blue. Some observers see the secondary as purple.				

μ Canis Majoris	ADS 5605	$06^h 56.1^m$	–14° 03′	January 4
5.3, 8.6 m	PA 340°; Sep. 3.0″	Spec. G5		moderate
Two stars of differing brightness that nevertheless present a glorious double of orange and blue.				

χ Geminorum	ADS 6321	$07^h 44.4^m$	+24° 24′	January 16
3.6, 8.1 m	PA 240°; Sep. 7.1″	Spec. G5		moderate
A bright orange–yellow primary and fainter blue secondary.				

ϕ^2 Cancri	ADS 6815	$08^h 26.8^m$	+26° 56′	January 27
6.3, 6.3 m	PA 218°; Sep. 5.1″	Spec. A2 A2		moderate
A superb pair of twin white stars.				

ε Arietis	ADS 2257	$02^h 59.2^m$	+21° 20′	November 5
5.2, 5.5 m	PA 208°; Sep. 1.5″	Spec. A2		difficult
See November.				

[27] Refers to the spectral classification of the primary star. Two values given for easily resolved double stars.
[28] The terms "easy, moderate and difficult" have a slightly different meaning to that used previously. In the present context it refers to the ability to split the double, and not just its ease of being observed. Although this will also play a significant factor in the use of the definition.
[29] This signifies that it is the 3945th star in the John Herschel catalogue.
[30] The ADS number is the number given in the *New General Catalogue of Double Stars*, which covers as far south as –30°.

θ Aurigae	ADS 4566	$05^h 59.7^m$	+37° 13′	December 21
2.6, 7.1 m	PA 313°; Sep. 3.6″	Spec. A0		v. difficult
See December.				

UV Aurigae		$05^h 21.8^m$	+32° 31′	December 11
7.4–10.6, 11.5 m	PA 4°; Sep. 3.4″	Spec. C6–C8		v. difficult
See December.				

February

ι^1 Cancri	ADS 6988	$08^h 46.7^m$	+28° 46′	February 1
4.0, 6.6 m	PA 307°; Sep. 30.5″	Spec. G5 A5		easy
Spectacular! Gold primary and blue secondary. Can be split with as low a magnification as 12X. But is a challenge for binoculars.				

!6 Leonis	ADS 7416	$09^h 32.0^m$	+09° 43′	February 12
5.2, 8.2 m	PA 75°; Sep. 37.4″	Spec. K0		easy
A stunning gold and blue double star system.				

ο Leonis	ADS 7480	$09^h 41.2^m$	+09° 54′	February 15
3.5, 9.5 m	PA 44°; Sep. 85.4″	Spec. F5		easy
A nice double of yellow and blue stars. Easy to resolve.				

38 Lyncis	ADS 7292	$09^h 18.8^m$	+36° 48′	February 9
3.9, 6.3 m	PA 229°; Sep. 2.7″	Spec. A2		moderate
A nice pair of stars. The primary is white, while some observers describe the secondary as rust-coloured.				

γ Leonis	ADS 7724	$10^h 20.0^m$	+19° 51′	February 25
2.5, 3.6 m	PA 125°; Sep. 4.4″	Spec. K0 G7		easy/moderate
A fine pairing of yellow stars, one deep and the other paler. Easy to find, as it is the brightest star in the curve of the sickle of Leo. [See also Regulus.]				

38 Geminorum	ADS 5559	$06^h 54.6^m$	+13° 11′	January 4
4.7, 7.7 m	PA 145°; Sep. 7.1″	Spec. F0		moderate
See January.				

μ Canis Majoris	ADS 5605	$06^h 56.1^m$	–14° 03′	January 4
5.3, 8.6 m	PA 340°; Sep. 3.0″	Spec. G5		moderate
See January.				

ξ Ursae Majoris	ADS 8119	$11^h 18.2^m$	+31° 32′	March 12
4.3, 4.8 m	PA 273°; Sep. 1.8″	Spec. G0		v. difficult
See March.				

ι Leonis	ADS 8148	$11^h 23.9^m$	+10° 32′	March 13
4.0, 6.7 m	PA 116°; Sep. 1.7″	Spec. F5		v. difficult
See March.				

March

N Hydrae	17 Crt	$11^h 32.3^m$	–29° 16′	March 15
5.8, 5.9 m	PA 210°; Sep. 9.2″	Spec. F6 F7		easy

An easy double to resolve with a small telescope, consisting of unequally bright-yellow stars.

δ Corvi	ADS 8572	$12^h 29.9^m$	–16° 31′	March 30
3.0, 9.2 m	PA 214°; Sep. 24.2″	Spec. A0		easy

Superb double star system consisting of a bright-white primary and fainter pale-blue secondary. First observed in 1823.

ι^1 Cancri	ADS 6988	$08^h 46.7^m$	+28° 46′	February 1
4.0, 6.6 m	PA 307°; Sep. 30.5″	Spec. G5 A5		easy

See February.

!6 Leonis	ADS 7416	$09^h 32.0^m$	+09° 43′	February 12
5.2, 8.2 m	PA 75°; Sep. 37.4″	Spec. K0		easy

See February.

ο Leonis	ADS 7480	$09^h 41.2^m$	+09° 54′	February 15
3.5, 9.5 m	PA 44°; Sep. 85.4″	Spec. F5		easy

See February.

38 Sextantis	ADS 7902	$10^h 43.3^m$	+04° 45′	March 3
36.3, 7.4 m	PA 240°; Sep. 6.8″	Spec. K0		easy/moderate

Easily resolved by small telescopes. A fine system of orange and yellowish stars.

2 CVn	ADS 8489	$12^h 16.1^m$	+40° 40′	March 26
5.8, 8.1 m	PA 260°; Sep. 11.4″	Spec. K5		easy/moderate

An easy double for small telescopes. Yellow primary and pale-blue secondary.

γ Crt	ADS 8153	$11^h 24.9^m$	–17° 41′	March 13
4.1, 9.6 m	PA 96°; Sep. 5.2″	Spec. A5		moderate

An unequally bright double consisting of a white primary and blue secondary.

84 Virginis	ADS 9000	$13^h 43.1^m$	+03° 32′	April 17
5.5, 7.9 m	PA 229°; Sep. 2.9″	Spec. K0		difficult

See April.

ξ Ursae Majoris	ADS 8119	$11^h 18.2^m$	+31° 32′	March 12
4.3, 4.8 m	PA 273°; Sep. 1.8″	Spec. G0		v. difficult

Discovered by William Herschel in 1780, this is a close pair of pale yellow stars. It also has the distinction of being the first binary system to have its orbit calculated by Savary in 1828. Both components are also spectroscopic binaries.

ι Leonis	ADS 8148	$11^h 23.9^m$	+10° 32′	March 13
4.0, 6.7 m	PA 116°; Sep. 1.7″	Spec. F5		v. difficult

A difficult double to resolve owing to its small separation. However, it is widening now (AD 2000.0), and so will get easier to split. Yellow and white.

April

ξ Ursae Majoris	ADS 8891	$13^h\ 23.9^m$	+54° 56′	April 12
2.3, 4.0 m	PA 152°; Sep. 14.4″	Spec. A2 A2		v. easy ©

The famous double *Mizar* and *Alcor* (*80* *U*Ma). Visible to the naked eye. Nice in binoculars. A small telescope will resolve Mizar's 4th-magnitude companion. Alcor and both members of Mizar are themselves spectroscopic binaries. Thus there are six stars in the system. Mizar also has several other distinctions: the first double to be discovered by telescope, by Riccioli in 1650, the first to be photographed, by Bond in 1857, and the first spectroscopic binary detected, by Pickering in 1889.

α CVn	ADS 8706	$12^h\ 56.0^m$	+38° 19′	April 5
2.9, 5.5 m	PA 229°; Sep. 19.4″	Spec. A0		easy

Also known as *Cor Caroli*, the stars of this system are separated by a distance equivalent to 5 solar system widths – 770 astronomical units! The two stars are yellowish in small instruments; however, with large aperture, subtle tints become apparent and have been called flushed white and pale lilac or pale yellow and fawn!

χ Boötis	ADS 9173	$14^h\ 13.5^m$	+51° 47′	April 25
4.6, 6.6 m	PA 236°; Sep. 13.4″	Spec. A5		easy ©

A nice double for small telescopes. The primary is white, although some observers see yellow, and the secondary is blue. The primary is also a variable.

ε Boötis	ADS 9372	$13^h\ 45.0^m$	+27° 04′	April 18
2.5, 4.9 m	PA 339°; Sep. 2.8″	Spec. K0 A0		moderate

A wonderful contrast of gold and green stars. Has also been reported to be yellow and blue. Difficult with apertures of around 7.5 cm, and even a challenge for beginners with apertures of 15.0 cm. With small telescopes a high power is needed to resolve them. Also known as *Mirak*.

2 CVn	ADS 8489	$12^h\ 16.1^m$	+40° 40′	March 26
5.8, 8.1 m	PA 260°; Sep. 11.4″	Spec. K5		easy/moderate

See March.

γ Crt	ADS 8153	$11^h\ 24.9^m$	−17° 41′	March 13
4.1, 9.6 m	PA 96°; Sep. 5.2″	Spec. A5		moderate

See March.

84 Virginis	ADS 9000	$13^h\ 43.1^m$	+03° 32′	April 17
5.5, 7.9 m	PA 229°; Sep. 2.9″	Spec. K0		difficult

A high magnification will split this system into vivid orange and pale yellow stars.

γ Virginis	ADS 8630	$12^h\ 41.7^m$	−01° 27′	April 2
3.5, 3.5 m	PA 259°; Sep. 1.5″	Spec. F0 F0		v. difficult

A very difficult double to resolve owing to the small separation, which will decrease even further until in 2010 they will be 0.9″ apart, and so to all intents and purposes appear single. A nice pair of pale yellow stars.

May

α Librae	h186	$14^h 50.9^m$	−16° 02′	May 4
2.8, 5.2 m	PA 314°; Sep. 231.0″	Spec. A2 F5		v. easy

Can be seen with binoculars as a widely separated pair of white and yellow stars. The primary is itself a spectroscopic binary.

54 Hydrae	ADS 9375	$14^h 46.0^m$	−25° 27′	May 3
5.1, 7.1 m	PA 126°; Sep. 8.6″	Spec. F1 F9		easy

An easy double of yellowish stars, with the primary a pale yellow, while the secondary is a stronger yellow.

ι^1 Librae	ADS 9532	$15^h 12.2^m$	−19° 47′	May 10
4.5, 9.4 m	PA 111°; Sep. 57.8″	Spec. A0		easy

A nice system of white and red stars. There is a much fainter third star, magnitude 11.1, which is a purplish colour. However, this is only 1.9″ away from the primary and exceedingly difficult to detect.

ν^1 Corona Borealis	$15^h 22.4^m$	+33° 48′	May 12
5.4, 5.3 m	PA 165°; Sep. 364.4″	Spec. M K5	easy

A nice binocular pair of orange stars.

β Scorpii	ADS 9913	$16^h 05.4^m$	−19° 48′	May 23
2.6, 4.9 m	PA 21°; Sep. 13.6″	Spec. B1		easy

A good double for small apertures of around 5.0 cm, this has a brilliant blue–white primary with a paler blue secondary. The primary is itself a binary.

$O\Sigma$ 300	ADS 9740	$15^h 40.2^m$	+12° 03′	May 17
6.4, 9.5 m	PA 261°; Sep. 15.3″	Spec. G5		easy/moderate

Located within the faint constellation Serpens Caput. A nice system consisting of yellow and blue stars.

ζ Corona Borealis	ADS 9737	$15^h 39.4^m$	+36° 38′	May 17
5.1, 6.0 m	PA 305°; Sep. 6.3″	Spec. B8		moderate

See May.

ρ Ophiuchi	ADS 10049	$16^h 25.6^m$	−23° 27′	May 28
5.3, 6.0 m	PA 344°; Sep. 3.1″	Spec. B2 B2		moderate

A medium-sized telescope of at least 10.0-cm aperture along with a high magnification is needed to resolve this pair of blue stars.

λ Ophiuchi	ADS 10087	$16^h 30.9^m$	+01° 59′	May 30
4.1, 5.2 m	PA 29°; Sep. 1.4″	Spec. A0		difficult

A nice pairing of white and pale lemon stars. A test for small telescopes.

α Herculis	ADS 10418	$17^h 14.6^m$	+14° 23′	June 10
3.5_v, 5.4_vm	PA 104°; Sep. 4.6″	Spec. M5 G5		difficult

See June.

ν **Scorpii**	ADS 9951	$16^h\ 12.0^m$	–19° 28′	May 25
4.3, 6.8 m 6.4, 7.8 m	PA 03°; Sep. 0.9″ PA 51°; Sep. 2.3″	Spec. B3		v. difficult easy

Another system of double–double stars. The main white pair will appear as two discs in contact, but only under near-perfect conditions; the easier pair are nice yellowish stars. The two components appear as magnitude 4.2 and 6.1, separation 41.1″, and with a PA of 337°. [See Antares.]

June

ο **Ophiuchi**	ADS 10442	$17^h\ 18.0^m$	–24° 17′	June 11
5.4, 6.9 m	PA 355°; Sep. 10.3″	Spec. K2 F6		easy

Located in a field of bright stars, this double makes a nice contrast of orange and yellow.

54 Hydrae	ADS 9375	$14^h\ 46.0^m$	–25° 27′	May 3
5.1, 7.1 m	PA 126°; Sep. 8.6″	Spec. F1 F9		easy

See May.

β **Lyrae**	ADS 11745	$18^h\ 50.1^m$	+33° 22′	July 4
3.4v, 8.6 m	PA 149°; Sep. 45.7″	Spec. B9		easy

See July.

56 Herculis	ADS 10259	$16^h\ 55.0^m$	+25° 44′	June 5
6.1, 10.6 m	PA 93°; Sep. 18.1″	Spec. K0		easy/moderate

The primary is a semi-regular variable star, type M, with a magnitude change of 3.1 to 3.9 and a period of 90 days. It is a wonderful double, with an orange primary and a greenish secondary.

ε **Lyrae**	ADS 11635	$18^h\ 44.3^m$	+39° 40′	July 2
5.4, 6.5 m 5.1, 5.3 m	PA 357°; Sep. 2.6″ PA 94°; Sep. 2.3″	Spec. A2 F4		easy/moderate easy/moderate

See July.

21 Sagittarii	ADS 1325	$18^h\ 25.3^m$	–20° 32′	June 28
4.9, 7.4 m	PA 289°; Sep. 1.8″	Spec. K0 A0		moderate

A rigorous test for small telescopes, and even a challenge for medium apertures, this is a nice contrast of orange and blue stars. Some observers report the secondary as greenish.

23 Aquilae	Σ2492	$19^h\ 18.5^m$	+1° 05′	July 11
5.3, 9.3 m	PA 5°; Sep. 3.1″	Spec. K0		moderate

See July.

α **Herculis**	ADS 10418	$17^h\ 14.6^m$	+14° 23′	June 10
3.5v, 5.4_vm	PA 104°; Sep. 4.6″	Spec. M5 G5		difficult

A lovely colour-contrast double: orange and bluish green. The primary star is itself variable, while the secondary is an unresolvable double.

λ Ophiuchi	ADS 10087	$16^h 30.9^m$	+01° 59′	May 30
4.1, 5.2 m	PA 29°; Sep. 1.4″	Spec. A0		difficult

See May.

July

α Capricorni	ADS 13632	$20^h 18.1^m$	−12° 33′	July 26
3.6, 4.2 m	PA 29°; Sep. 378″	Spec. G9 G3		v. easy

A widely spaced, naked-eye optical double. Both stars are a yellow–orange colour, but when they are viewed through small telescopes it will be seen that both stars are themselves binaries.

ρ Capricorni	ADS 13887	$20^h 28.9^m$	−17° 49′	July 29
5.0, 6.7 m	PA 150°; Sep. 247.6″	Spec. F0 K0		v. easy

Easily seen in binoculars. This attractive pair are coloured yellow and purplish.

ζ Lyrae	ADS 11639	$18^h 44.8^m$	+37° 36′	July 3
4.3, 5.9 m	PA 150°; Sep. 43.7″	Spec. A3 A3		easy

An easy pair of yellowish stars.

β Lyrae	ADS 11745	$18^h 50.1^m$	+33° 22′	July 4
3.4_v, 8.6 m	PA 149°; Sep. 45.7″	Spec. B9		easy

This pair of white stars is a challenging double for binoculars. β^1 is also an eclipsing binary. A fascinating situation occurs owing to the gravitational effects of the components of β^1. The stars are distorted from their spherical shapes into ellipsoids.

11 Aquilae	Σ2424	$18^h 59.1^m$	+13° 37′	July 6
5.2, 8.7 m	PA 286°; Sep. 17.5″	Spec. F5		easy

An optical double with a nice colour contrast – yellow and blue.

β Cygni	ADS 12540	$19^h 30.7^m$	+27° 58′	July 14
3.1, 5.1 m	PA 54°; Sep. 34.4″	Spec. K3 B8		easy

Thought by many to be the finest double in the skies, Albireo is a golden-yellow primary and lovely blue secondary against the backdrop of the myriad fainter stars of the Milky Way. Easy to locate at the foot of the Northern Cross. The colours can be made to appear even more spectacular if you slightly defocus the images. Wonderful!

H N 84	ADS 12750	$19^h 39.4^m$	+16° 34′	July 16
6.5, 8.9 m	PA 302°; Sep. 28.2″	Spec. K5		easy

Located in the constellation of Sagitta, this is a fine double of orange and blue.

α Capricorni	ADS 13675	$20^h 19.4^m$	−19° 07′	July 27
5.5, 9.0 m	PA 179°; Sep. 55.9″	Spec. K4		easy

An easy system to resolve, although the primary is considerably brighter than the secondary. The stars are a yellow and pale blue in colour.

ε Lyrae	ADS 11635	$18^h 44.3^m$	+39° 40′	July 2
5.4, 6.5 m 5.1, 5.3 m	PA 357°; Sep. 2.6″ PA 94°; Sep. 2.3″	Spec. A2 F4		easy/moderate easy/moderate

The famous *Double-Double*, easily split, but to resolve the components of each star, ϵ^1 (magnitude 4.7) and ε^2 (magnitude 4.6), requires a high power. The stars themselves are at a PA 173°, separated by 208″, which is near the naked-eye limit, and some keen-eyed observers report being able to resolve them under perfect seeing. However, there is fierce debate among amateurs – some saying the double is difficult to resolve, others the opposite. All stars are white- or cream-white-coloured. A highlight of the summer sky.

HN 119	ADS 12506	$19^h 29.9^m$	–26° 59′	July 14
65.6, 8.6 m	PA 142°; Sep. 7.8″	Spec. K3		easy/moderate

Known as far back as 1821, this is a nice double of orange and blue stars.

$O\Sigma$ 394	ADS 13240	$20^h 00.2^m$	+36° 25′	July 22
7.1, 9.9 m	PA 294°; Sep. 11.0″	Spec. K0		easy/moderate

A delightful system consisting of an orange primary and a blue secondary.

23 Aquilae	S2492	$19^h 18.5^m$	+1° 05′	July 11
5.3, 9.3 m	PA 5°; Sep. 3.1″	Spec. K0		moderate

A very close pair, but a lovely deep yellow and greenish blue double system

δ Cygni	ADS 12880	$19^h 45.0^m$	+45° 08′	July 18
2.9, 6.3 m	PA 221°; Sep. 2.5″	Spec. B9		difficult

Contrasting reports of this system's colours abound; a blue-white or greenish white primary, and a blue-white or bluish secondary. A test for telescopes of 10.0 cm to 15.0 cm, and exceptional seeing is needed.

β 441 (Vulpecula)	ADS 13648	$20^h 17.5^m$	+29° 09′	July 26
6.2, 10.7 m	PA 66°; Sep. 5.9″	Spec. K0		difficult

Also known as *Burnham 441*, this is a nice contrasting pair of yellow and blue stars.

α Herculis	ADS 10418	$17^h 14.6^m$	+14° 23′	June 10
3.5_v, 5.4_vm	PA 104°; Sep. 4.6″	Spec. M5 G5		difficult

See June.

λ Ophiuchi	ADS 10087	$16^h 30.9^m$	+01° 59′	May 30
4.1, 5.2 m	PA 29°; Sep. 1.4″	Spec. A0		difficult

See May.

August

61 Cygni	ADS 14636	$21^h 06.9^m$	+38° 45′	August 8
5.2, 6.0 m	PA 150°; Sep. 30.3″	Spec. K5 K7		easy

Best seen with binoculars (but sometimes a challenge if conditions are poor), which seem to emphasise the vibrant colours of these stars, both orange-red. Famous for being the first star to have its distance measured by the technique of parallax. The German astronomer Friedrich Bessel determined its distance to be 10.3 l.y.; modern measurements give a figure of 11.36. Also has an unseen third component, which has the mass of 8 Jupiters. Has a very large proper motion.

β Cephei	ADS 15032	$21^h 28.7^m$	+70° 34′	August 13
3.2, 7.9 m	PA 249°; Sep. 13.3″	Spec. B2		easy ©

Also known as *Alfirk*, it is a Cepheid variable. The system is a nice white and blue double. Using a large-aperture telescope, the secondary takes on a definite green tint.

-	Σ2841	$21^h 54.3^m$	+19° 43′	August 20
6.4, 7.9 m	PA 110°; Sep. 22.3″	Spec. K0		easy

Easily seen in binoculars. Yellow primary and greenish secondary. In the constellation Pegasus.

-	Σ2848	$21^h 58.0^m$	+05° 56′	August 21
7.2, 7.5 m	PA 56°; Sep. 10.7″	Spec. A2		easy

Both stars are a pale tinted yellow. In the constellation Pegasus.

Σ2894 Lac	ADS 15828	$22^h 18.9^m$	+37° 46′	August 26
6.1, 8.3 m	PA 194°; Sep. 15.6″	Spec. F0		easy

A nice system of yellow and blue stars. In the constellation Lacerta.

94 Aquarii	HD 219834	$22^h 19.1^m$	–13° 28′	August 26
5.3, 7.3 m	PA 350°; Sep. 12.7″	Spec. G5		easy

A lovely double, yellowish red and pale green.

Aquarius	Σ2838	$21^h 54.7^m$	–03° 18′	August 20
6.3, 9.1 m	PA 184°; Sep. 17.6″	Spec. F8		easy/moderate

A yellow and bluish pair of stars with a background of many faint stars.

γ Delphini	ADS 14279	$20^h 46.7^m$	+16° 07′	August 3
4.3, 5.5 m	PA 268°; Sep. 9.6″	Spec. K2 F8		easy/moderate

Easily resolved with a small telescope, this is a beautiful double with a yellow primary and a rare green secondary.

12 Aquarii	Σ2745	$21^h 04.1^m$	–05° 49′	August 7
5.9, 7.3 m	PA 192°; Sep. 2.8″	Spec. F5 A3		moderate

A close pair of pale blue and yellow stars.

29 Aquarii	S 802[31]	$22^h 02.4^m$	–16° 58′	August 22
7.2, 7.4 m	PA 244°; Sep. 3.7″	Spec. A2		moderate

A high power is needed to split this pair of white stars.

41 Aquarii	ADS 15753	$22^h 14.3^m$	–21° 04′	August 25
7.1, 7.1 m	PA 114°; Sep. 5.0″	Spec. G8		moderate

A lovely double system consisting of contrasting gold and blue stars.

[31]This signifies that it is the 802nd star in the John South catalogue.

ζ Aquarii	ADS 15971	$22^h 28.8^m$	−00° 01′	August 28
4.3, 4.5 m	PA 192°; Sep. 2.1″	Spec. F2		moderate

Good conditions and optics are needed to be able to split this double star, which is white and white. The central star of the *Water Jar* asterism in Aquarius.

δ Cygni	ADS 12880	$19^h 45.0^m$	+45° 08′	July 18
2.9, 6.3 m	PA 221°; Sep. 2.5″	Spec. B9		difficult

See July.

β 441 (Vulpecula)	ADS 13648	$20^h 17.5^m$	+29° 09′	July 26
6.2, 10.7 m	PA 66°; Sep. 5.9″	Spec. K0		difficult

See July.

α Herculis	ADS 10418	$17^h 14.6^m$	+14° 23′	June 10
3.5_v, 5.4_vm	PA 104°; Sep. 4.6″	Spec. M5 G5		difficult

See June.

September

τ Aquarii	ADS 16268	$22^h 47.7^m$	−14° 03′	September 2
5.8, 9.0 m	PA 121°; Sep. 23.7″	Spec. B9		easy

A double that exhibits many colours to different observers; yellowish and orange, blue-white and greenish, white and yellow, white and pale red.

Herschel 975	ADS 16268	$22^h 47.7^m$	−14° 03′	September 2
5.6, 9.5 m	PA 243°; Sep. 51.0″	Spec. B9		easy

A system with a large magnitude difference. A white primary and pale blue secondary. In the constellation Lacerta.

42 Piscium		$00^h 22.4^m$	+13° 29′	September 26
6.2, 10.1 m	PA 324°; Sep. 28.5″	Spec. K0		easy

An easy double star system to split. Orange primary and blue secondary.

51 Piscium		$00^h 22.4^m$	+06° 57′	September 29
5.7, 9.5 m	PA 83°; Sep. 27.5″	Spec. A0		easy

A wonderful double star system. Bluish white primary and greenish secondary.

57 Pegasi	HD 218634	$23^h 09.5^m$	+08° 41′	September 8
5.1, 9.7 m	PA 198°; Sep. 32.6″	Spec. M		easy/moderate

A lovely system, easily resolved in small telescopes. Orange primary and bluish secondary.

107 Aquarii	H II 24	$23^h 46.0^m$	−18° 41′	September 17
5.7, 6.7 m	PA 136°; Sep. 6.6″	Spec. A5		easy/moderate

A close double with colours of pale yellow and a definite bluish white.

σ Cassiopeiae	ADS 17140	$23^h 59.0^m$	+55° 45′	September 20
5.0, 7.1 m	PA 326°; Sep. 3.0″	Spec. B2		moderate ©

Located within a nice star field, a bluish and yellow system. Also described as green and blue.

Struve 3050	Σ3050	$23^h 59.5^m$	+33° 43′	September 20
6.6, 6.6 m	PA 335°; Sep. 1.7″	Spec. F8		moderate

A very close double that shows the two yellow stars almost touching.

Groombridge 34	ADS 246	$00^h 17.9^m$	+44° 00′	September 25
8.2, 10.6 m	PA 62°;Sep. 40.0″	Spec. M2		mod/difficult

A red dwarf binary system, with a large proper motion that can be plotted over several years. Discovered in 1860.

δ Cygni	ADS 12880	$19^h 45.0^m$	+45° 08′	July 18
2.9, 6.3 m	PA 221°; Sep. 2.5″	Spec. B9		difficult

See July.

β 441 (Vulpecula)	ADS 13648	$20^h 17.5^m$	+29° 09′	July 26
6.2, 10.7 m	PA 66°; Sep. 5.9″	Spec. K0		difficult

See July.

α Herculis	ADS 10418	$17^h 14.6^m$	+14° 23′	June 10
3.5_v, 5.4_vm	PA 104°; Sep. 4.6″	Spec. M5 G5		difficult

See June.

October

η Cassiopeiae	AADS 671	$00^h 49.1^m$	+57° 49′	October 3
3.4, 7.5 m	PA 317°; Sep. 12.9″	Spec. G0		easy ©

Discovered by William Herschel in 1779. Another system which has differing colours reported. The primary has been described as gold, yellow and topaz, while the secondary has been called orange, red and purple. Has an apparently near-circular orbit.

26 Ceti	S84	$01^h 03.8^m$	+01° 22′	October 7
6.2, 8.6 m	PA 253°; Sep. 16.0″	Spec. F0		easy

A nice system, reported as yellow and lilac.

γ Arietis	ADS 1507	$01^h 53.5^m$	+19° 18′	October 19
4.8, 4.8 m	PA 0°; Sep. 7.8″	Spec. A0		easy

Discovered by Robert Hooke in 1664. A lovely pair of equally bright bluish white stars.

λ Arietis	ADS 1563	$01^h 57.9^m$	+23° 35′	October 20
4.9, 7.7 m	PA 46°; Sep. 37.4″	Spec. A5		easy

An easy pair to split in binoculars. Pale yellow and pale blue.

66 Ceti	ADS 1703	02^h 12.8^m	–02° 24′	October 24
5.7, 7.5 m	PA 234°; Sep. 16.5″	Spec. G0		easy

With small aperture, the colours are yellow and blue, but with large aperture, they are reported to be topaz and violet!

α Ursae Minoris	ADS 1477	02^h 31.8^m	+89° 16′	October 29
2.0, 8.2 m	PA 218°; Sep. 18.4″	Spec. F8		easy ©

Possibly the most famous star in the entire sky. *Polaris*, or the *Pole Star*, is located less than a degree from the celestial pole, and is a nice double consisting of a yellowish primary and a faint whitish blue secondary. The primary is also a Population II cepheid variable, and a spectroscopic binary. Although claims have been made to the effect that the system can be resolved in an aperture as small as 4.0 cm, at least 6.0 cm will be required to split it clearly.

33 Arietis	HD 16628	02^h 40.7^m	+27° 04′	October 31
5.5, 8.4 m	PA 0°; Sep. 28.6″	Spec. A2		easy

An easy pair to split. Pale yellow and pale blue.

55 Piscium	ADS 558	00^h 39.9^m	+21° 26′	October 1
5.4, 8.7 m	PA 194°; Sep. 6.5″	Spec. K0		moderate

A lovely system. Vivid yellow primary and blue secondary.

1 Arietis	ADS 1457	01^h 50.1^m	+22° 17′	October 18
6.2, 7.2 m	PA 166°; Sep. 2.8″	Spec. F5		moderate

A nice system consisting of yellow and faint blue. Test for 5.0 cm.

ι Trianguli	ADS 1697	02^h 12.4^m	+30° 18′	October 24
5.3, 6.9 m	PA 71°; Sep. 3.9″	Spec. G0		moderate

A nice yellow and blue system. First observed by Herschel in 1781. Both stars are themselves spectroscopic binaries.

84 Ceti	Σ295	02^h 41.2^m	–00° 42′	October 31
5.8, 9.0 m	PA 310°; Sep. 4.0″	Spec. F5		moderate

A lovely double star system, yellow primary and reddish secondary.

36 Andromedae	ADS 755	00^h 55.0^m	+23° 38′	October 4
6.0, 6.4 m	PA 313°; Sep. 0.9″	Spec. K0		difficult

This pair of brilliant yellow stars is a test for large amateur telescopes. Discovered by F. Struve in 1836.

ε Trianguli	HD 12471	02^h 03.0^m	+33° 17′	October 22
5.4, 11.4 m	PA 118°; Sep. 3.9″	Spec. A2		difficult

A difficult double because of the large magnitude difference. Blue white primary and white secondary.

47 Tauri	HD 26722	04^h 13.9^m	+09° 16′	November 24
4.9, 7.4 m	PA 351°; Sep. 1.1″	Spec. G5		v. difficult

See November.

November

η Persei	ADS 2157	$02^h 50.7^m$	+55° 54′	November 3
3.8, 8.5 m	PA 300°; Sep. 28.3″	Spec. K0		easy ©
Magnificent! Gold primary and blue secondary.				

Struve 390	S390	$03^h 30.0^m$	+55° 27′	November 13
5.1, 9.5 m	PA 159°; Sep. 14.8″	Spec. A2		easy ©
Two stars of very different brightness; the primary is white, but the secondary is a lovely purple colour.				

φ Tauri	ADS 3137	$04^h 20.4^m$	+27° 21′	November 26
5.0, 8.4 m	PA 250°; Sep. 52.1″	Spec. K0		easy
Easily seen in binoculars. Bright yellow primary and blue secondary.				

o^2 Eradini	ADS 3093	$04^h 15.2^m$	–07° 39′	November 24
4.4, 9.5 m	PA 104°; Sep. 83″	Spec. WD		easy/moderate
A challenge to split with binoculars. What makes this system so interesting is that the secondary is the brightest *white dwarf* star visible from Earth.				

–	Σ326	$02^h 55.6^m$	+26° 52′	November 4
7.6, 9.8 m	PA 220°; Sep. 5.9″	Spec. K0 K5		moderate
A nice double star system. Orange and red stars.				

30 Tauri	HD 23793	$03^h 48.3^m$	+11° 09′	November 17
5.1, 10.2 m	PA 59°; Sep. 9.0″	Spec. B3		moderate
An interesting system lying in a star field. A bluish white and reddish double. Observers have reported the primary as green and pale yellow, and the secondary as purple.				

32 Eridani	HD24555	$03^h 54.3^m$	–02° 57′	November 19
4.8, 6.1 m	PA 347°; Sep. 6.8″	Spec. G5 A2		moderate
A yellow and white double star system. Has also been described as yellow and blue, topaz and bright green.				

39 Eridani	HD 26846	$04^h 14.4^m$	–10° 15′	November 24
5.0, 8.0 m	PA 146°; Sep. 6.4″	Spec. K0		moderate
Situated in a nice star field, a nice orange and white double star system.				

ε Arietis	ADS 2257	$02^h 59.2^m$	+21° 20′	November 5
5.2, 5.5 m	PA 208°; Sep. 1.5″	Spec. A2		difficult
A test for your optics. Both white stars are nearly equally bright. Test for 7.5 cm.				

36 Andromedae	ADS 755	$00^h 55.0^m$	+23° 38′	October 4
6.0, 6.4 m	PA 313°; Sep. 0.9″	Spec. K0		difficult
See October.				

47 Tauri	HD 26722	$04^h 13.9^m$	+09° 16′	November 24
54.9, 7.4 m	PA 351°; Sep. 1.1″	Spec. G5		v. difficult
A high magnification is needed to split this system. Both stars appear yellow.				

December

11 & 12 Camelopardalis		05h 06.1m	+58° 58′	December 7
5.4, 6.5 m	PA 8°; Sep. 108.5″	Spec. B3 K0		easy ©
Easily spotted in binoculars; a pair of white and deep yellow stars.				

S473	ADS 3883	05h 17.1m	−15° 13′	December 10
6.7, 8.7 m	PA 305°; Sep. 20.6″	Spec. B8.5		easy
Suitable for small telescopes; white primary and blue secondary.				

–	Σ698	05h 25.2m	+34° 51′	December 12
6.6, 8.7 m	PA 345°; Sep. 31.2″	Spec. K0		easy
Located in a nice star field; both stars are yellow–orange.				

La 1	ADS 4260	05h 39.7m	−20° 26′	December 16
6.9, 7.9 m	PA 123°; Sep. 11.0″	Spec. B8		easy
A rare treat – white and purple stars!				

γ Leporis	ADS 4334	05h 44.5m	−22° 27′	December 17
3.7, 6.3 m	PA 350°; Sep. 96.3″	Spec. F8 G5		easy
A glorious system. An easy object for even the smallest telescope. Bright yellow and pale tinted orange. Test for 7.5 cm.				

41 Aurigae	ADS 4773	06h 11.6m	+48° 43′	December 24
6.3, 7.0 m	PA 356°; Sep. 7.7″	Spec. A0		easy
A nice white and bluish white double star system.				

15 Geminorum	h 70	06h 27.8m	+20° 47′	December 28
6.6, 8.0 m	PA 204°; Sep. 25.1″	Spec. K0		easy
A double star system of yellow and blue stars.				

ν¹ Canis Majoris	ADS 5253	06h 36.7m	−18° 40′	December 30
5.8, 8.5 m	PA 262°; Sep. 17.5″	Spec. G5 G0		easy
An easily resolved pair of yellow stars.				

ω Aurigae	ADS 3572	04h 59.3m	+37° 53′	December 5
5.0, 8.0 m	PA 359°; Sep. 5.4″	Spec. A0		moderate
Stars appear white and blue in small telescopes, but have shown subtle tints in larger instruments.				

h3750	ADS 3930	05h 20.4m	−21° 14′	December 11
4.7, 8.4 m	PA 282°; Sep. 4.2″	Spec. A0		moderate
A lovely yellow and blue system. Has been described as "most beautiful".				

–	Σ326	02h 55.6m	+26° 52′	November 4
7.6, 9.8 m	PA 220°; Sep. 5.9″	Spec. K0 K5		moderate
See November.				

30 Tauri	HD 23793	03^h 48.3^m	+11° 09′	November 17
5.1, 10.2 m	PA 59°; Sep. 9.0″	Spec. B3		moderate
See November.				

32 Eridani	HD24555	03^h 54.3^m	−02° 57′	November 19
4.8, 6.1 m	PA 347°; Sep. 6.8″	Spec. G5 A2		moderate
See November.				

39 Eridani	HD 26846	04^h 14.4^m	−10° 15′	November 24
5.0, 8.0 m	PA 146°; Sep. 6.4″	Spec. K0		moderate
See November.				

ε Arietis	ADS 2257	02^h 59.2^m	+21° 20′	November 5
5.2, 5.5 m	PA 208°; Sep. 1.5″	Spec. A2		difficult
See November.				

χ Leporis	ADS 3800	05^h 13.2^m	−12° 56′	December 9
4.5, 7.4 m	PA 358°; Sep. 2.6″	Spec. B8		v. difficult
A pair of white stars. Will require the best conditions and optics.				

UV Aurigae		05^h 21.8^m	+32° 31′	December 11
7.4–10.6, 11.5 m	PA 4°; Sep. 3.4″	Spec. C6–C8		v. difficult
Difficult to locate owing to its variable nature. It is a carbon star, coupled with a B-type giant star. Persevere and you will be rewarded by a lovely combination of orange and blue stars.				

θ Aurigae	ADS 4566	05^h 59.7^m	+37° 13′	December 21
2.6, 7.1 m	PA 313°; Sep. 3.6″	Spec. A0		v. difficult
The spectrum shows strong lines of silicon. With small telescopes excellent conditions and superb optics are required to see these two bluish white stars. Test for 10 cm.				

2.8 Triple, Quadruple and Multiple Stars

There are also several beautiful triple, quadruple and multiple stars systems within reach of modest telescopes. A few of theses are presented. The same nomenclature applies as in the double star section.

January

17 Canis Majoris	ADS 5585	06^h 55.0^m	−20° 24′	January 4
5.8, 9.3, 9.0 m	PA 147°; Sep. 44.4″	Spec. A2 K5		easy
A nice triple system. Wonderful colour contrasts of white and orange–red stars.				

Struve 1369	ADS 7438	$09^h 35.4^m$	±39° 57′	February 13
7.9, 8.0, 8.7 m	PA 148°; Sep. 24.7″	Spec. F2 G0		easy

See February.

Trapezium	*θ* Orionis	$05^h 35.3^m$	–05° 23′	December 14
5.1, 6.7 m	PA 241°; Sep. 13.4″	B0 B0		easy
6.7, 7.9 m		O6 B0		

See December.

Trapezium	*θ* Orionis	$05^h 35.3^m$	–05° 23′	December 14
5.1, 6.7 m	PA 241°; Sep. 13.4″	B0 B0		easy
6.7, 7.9 m		O6 B0		

See December.

***σ* Orionis**	ADS 4241	$05^h 38.7^m$	–02° 36′	December 15
4.0, 10.3 m	PA 238°; Sep. 11.4″	Spec. O9	B3 A2	easy
7.5, 6.5 m				

See December.

***ε* Monocerotis**	ADS 5012	$06^h 23.8^m$	+04° 36′	December 27
4.5, 6.5, 5.6 m	PA 27°; Sep. 13.4″	Spec. A5 A5		easy

See December.

***β* Monocerotis**	ADS 5107	$06^h 28.8^m$	–07° 02′	December 28
4.7, 5.2, 6.1 m	PA 132°; Sep. 7.3″	Spec. B2 B2		easy

See December.

Burnham 324	ADS 5498	$06^h 49.7^m$	–24° 05′	January 2
6.3, 7.6, 8.6 m	PA 206°; Sep. 1.8″	Spec. A0 A2		easy/moderate

A nice multiple star system in the constellation of Canis Major, consisting or many white and blue stars.

Struve 1245	ADS 6886	$08^h 35.8^m$	+06° 37′	January 29
6.0, 7.2, 10.7,	PA 25°; Sep. 10.3″	Spec. F6 G5		easy/moderate

A lovely multiple star system in the constellation Cancer. The triple aspect is seen in small telescopes. Yellow, pale yellow and white stars. Depending on the telescope used, you will see a double, a triple or a multiple star system!

***OΣ* 147**	ADS 5188	$06^h 34.3^m$	+38° 05′	December 29
6.6, 10, 10.6 m	PA 73°; Sep. 43.2″	Spec. K0		moderate

See December.

Struve 939	S 939	$06^h 35.9^m$	+05° 19′	December 30
8.3, 9.2, 9.4 m	PA 106°; Sep. 30.2″	Spec. B5 B8		moderate

See December.

***τ* Canis Majoris**	ADS 5977	$07^h 18.7^m$	–24° 57′	January 10
4.4, 10.5, 11.2 m	PA 90°; Sep. 80.2″	Spec. O9		difficult

Wonderful! This triple is within the open cluster *NGC 2362*, and so its yellow and blue components are set against a glorious backdrop of faint stars.

η **Orionis**	ADS 4002	$05^h 24.5^m$	–02° 24′	December 12
3.6, 5.0, 10.1 m	PA 80°; Sep. 1.5″	Spec. B1		difficult
See December.				

λ **Orionis**	ADS 4179	$05^h 31.1^m$	+09° 56′	December 13
3.5, 5.6, 11.1, 11.1 m	PA 43°; Sep. 4.4″	Spec. O8 B0		difficult
See December.				

February

Struve 1369	ADS 7438	$09^h 35.4^m$	±39° 57′	February 13
7.9, 8.0, 8.7 m	PA 148°; Sep. 24.7″	Spec. F2 G0		easy
Located in the constellation Lynx, this is an easily resolved system of yellow stars, along with faint white stars.				

17 Canis Majoris	ADS 5585	$06^h 55.0^m$	–20° 24′	January 4
5.8, 9.3, 9.0 m	PA 147°; Sep. 44.4″	Spec. A2 K5		easy
See January.				

Trapezium	θ Orionis	$05^h 35.3^m$	–05° 23′	December 14
5.1, 6.7 m 6.7, 7.9 m	PA 241°; Sep. 13.4″	B0 B0 O6 B0		easy
See December.				

σ **Orionis**	ADS 4241	$05^h 38.7^m$	–02° 36′	December 15
4.0, 10.3 m 7.5, 6.5 m	PA 238°; Sep. 11.4″	Spec. O9 B3 A2		easy
See December.				

ε **Monocerotis**	ADS 5012	$06^h 23.8^m$	+04° 36′	December 27
4.5, 6.5, 5.6 m	PA 27°; Sep. 13.4″	Spec. A5 A5		easy
See December.				

β **Monocerotis**	ADS 5107	$06^h 28.8^m$	–07° 02′	December 28
4.7, 5.2, 6.1 m	PA 132°; Sep. 7.3″	Spec. B2 B2		easy
See December.				

Burnham 324	ADS 5498	$06^h 49.7^m$	–24° 05′	January 2
6.3, 7.6, 8.6 m	PA 206°; Sep. 1.8″	Spec. A0 A2		easy/moderate
See January.				

Struve 1604	ADS 8440	$12^h 09.5^m$	–11° 51′	March 24
6.8, 8.5, 9.1 m	PA 89°; Sep. 9.9″	Spec. G0		moderate
See March.				

Struve 1245	ADS 6886	$08^h 35.8^m$	+06° 37′	January 29
6.0, 7.2, 10.7, 12.2, 8.8 m	PA 25°; Sep. 10.3″	Spec. F6 G5		easy/moderate/ difficult
See January.				

$O\Sigma$ **147**	ADS 5188	$06^h\ 34.3^m$	+38° 05′	December 29
6.6, 10, 10.6 m	PA 73°; Sep. 43.2″	Spec. K0		moderate
See December.				

Struve 939	Σ 939	$06^h\ 35.9^m$	+05° 19′	December 30
8.3, 9.2, 9.4 m	PA 106°; Sep. 30.2″	Spec. B5 B8		moderate
See December.				

τ **Canis Majoris**	ADS 5977	$07^h\ 18.7^m$	−24° 57′	January 10
4.4, 10.5, 11.2 m	PA 90°; Sep. 80.2″	Spec. O9		difficult
See January.				

η **Orionis**	ADS 4002	$05^h\ 24.5^m$	−02° 24′	December 12
3.6, 5.0, 10.1 m	PA 80°; Sep. 1.5″	Spec. B1		difficult
See December.				

λ **Orionis**	ADS 4179	$05^h\ 31.1^m$	+09° 56′	December 13
3.5, 5.6, 11.1, 11.1 m	PA 43°; Sep. 4.4″	Spec. O8 B0		difficult
See December.				

March

Struve 1369	ADS 7438	$09^h\ 35.4^m$	±39° 57′	February 13
7.9, 8.0, 8.7 m	PA 148°; Sep. 24.7″	Spec. F2 G0		easy
See February.				

17 Canis Majoris	ADS 5585	$06^h\ 55.0^m$	−20° 24′	January 4
5.8, 9.3, 9.0 m	PA 147°; Sep. 44.4″	Spec. A2 K5		easy
See January.				

Trapezium	θ Orionis	$05^h\ 35.3^m$	−05° 23′	December 14
5.1, 6.7 m 6.7, 7.9 m	PA 241°; Sep. 13.4″	B0 B0 O6 B0		easy
See December.				

σ **Orionis**	ADS 4241	$05^h\ 38.7^m$	−02° 36′	December 15
4.0, 10.3 m 7.5, 6.5 m	PA 238°; Sep. 11.4″	Spec. O9 B3 A2		easy
See December.				

ε **Monocerotis**	ADS 5012	$06^h\ 23.8^m$	+04° 36′	December 27
4.5, 6.5, 5.6 m	PA 27°; Sep. 13.4″	Spec. A5 A5		easy
See December.				

β **Monocerotis**	ADS 5107	$06^h\ 28.8^m$	−07° 02′	December 28
4.7, 5.2, 6.1 m	PA 132°; Sep. 7.3″	Spec. B2 B2		easy
See December.				

Burnham 324	ADS 5498	$06^h 49.7^m$	–24° 05′	January 2
6.3, 7.6, 8.6 m	PA 206°; Sep. 1.8″	Spec. A0 A2		easy/moderate
See January.				

Struve 1604	ADS 8440	$12^h 09.5^m$	–11° 51′	March 24
6.8, 8.5, 9.1 m	PA 89°; Sep. 9.9″	Spec. G0		moderate
Another rare but lovely triple star system which forms an equilateral triangle. In the constellation Corvus.				

Burnham 800	β800[32]	$13^h 16.9^m$	+17° 01′	April 10
6.6, 9.7, 10.4 m	PA 106°; Sep. 6.8″	Spec. K0		moderate
See April.				

Struve 1245	ADS 6886	$08^h 35.8^m$	+06° 37′	January 29
6.0, 7.2, 10.7, 12.2, 8.8 m	PA 25°; Sep. 10.3″	Spec. F6 G5		easy/moderate/ difficult!
See January.				

OΣ 147	ADS 5188	$06^h 34.3^m$	+38° 05′	December 29
6.6, 10, 10.6 m	PA 73°; Sep. 43.2″	Spec. K0		moderate
See December.				

Struve 939	Σ939	$06^h 35.9^m$	+05° 19′	December 30
8.3, 9.2, 9.4 m	PA 106°; Sep. 30.2″	Spec. B5 B8		moderate
See December.				

35 Comae Berenices	ADS 8695	$12^h 53.3^m$	+21° 14′	April 5
5.1, 7.2, 9.1 m	PA 182°; Sep. 1.2″	Spec. K0		difficult
See April.				

ι Boötis	ADS 9198	$14^h 16.2^m$	+51° 22′	April 26
4.9, 7.5, 12.6 m	PA 33°; Sep. 38.5″	Spec. A5		difficult ©
See April.				

τ Canis Majoris	ADS 5977	$07^h 18.7^m$	–24° 57′	January 10
4.4, 10.5, 11.2 m	PA 90°; Sep. 80.2″	Spec. 09		difficult
See January.				

η Orionis	ADS 4002	$05^h 24.5^m$	–02° 24′	December 12
3.6, 5.0, 10.1 m	PA 80°; Sep. 1.5″	Spec. B1		difficult
See December.				

λ Orionis	ADS 4179	$05^h 31.1^m$	+09° 56′	December 13
3.5, 5.6, 11.1, 11.1 m	PA 43°; Sep. 4.4″	Spec. O8 B0		difficult
See December.				

[32]This signifies that it is the 800th object in the S.W. Burnham catalogue.

April

Struve 1369	ADS 7438	$09^h 35.4^m$	±39° 57′	February 13
7.9, 8.0, 8.7 m	PA 148°; Sep. 24.7″	Spec. F2 G0		easy

See February.

17 Canis Majoris	ADS 5585	$06^h 55.0^m$	−20° 24′	January 4
5.8, 9.3, 9.0 m	PA 147°; Sep. 44.4″	Spec. A2 K5		easy

See January.

16 & 17 Draconis	ADS 10129	$16^h 36.2^m$	+52° 55′	May 31
5.4, 5.5, 6.4 m	PA 194°; Sep. 90.3″	Spec. A2 A0		easy ©

See May.

Burnham 324	ADS 5498	$06^h 49.7^m$	−24° 05′	January 2
6.3, 7.6, 8.6 m	PA 206°; Sep. 1.8″	Spec. A0 A2		easy/moderate

See January.

Struve 1604	ADS 8440	$12^h 09.5^m$	−11° 51′	March 24
6.8, 8.5, 9.1 m	PA 89°; Sep. 9.9″	Spec. G0		moderate

See March.

Struve 1245	ADS 6886	$08^h 35.8^m$	+06° 37′	January 29
6.0, 7.2, 10.7, 12.2, 8.8 m	PA 25°; Sep. 10.3″	Spec. F6 G5		easy/moderate/ difficult!

See January.

Burnham 800	β800[33]	$13^h 16.9^m$	+17° 01′	April 10
6.6, 9.7, 10.4 m	PA 106°; Sep. 6.8″	Spec. K0		moderate

In the constellation of Coma Berenices lies this lovely triple star system of orange, pale red and white stars.

55 Serpentis	ADS 9584	$15^h 19.3^m$	+01° 46′	May 12
5.1, 10.1, 9.1 m	PA 36°; Sep. 11.2″	Spec. G0		moderate

See May.

35 Comae Berenices	ADS 8695	$12^h 53.3^m$	+21° 14′	April 5
5.1, 7.2, 9.1 m	PA 182°; Sep. 1.2″	Spec. K0		difficult

A fine triple star system, consisting of one yellow, and two very rare purple stars.

ι Boötis	ADS 9198	$14^h 16.2^m$	+51° 22′	April 26
4.9, 7.5, 12.6 m	PA 33°; Sep. 38.5″	Spec. A5		difficult ©

A nice triple star system, although difficult to see in small instruments. The brighter members are yellow and blue coloured stars.

[33]This signifies that it is the 800th object in the S.W. Burnham catalogue.

σ Coronae Borealis	ADS 9979	$16^h 14.7^m$	+33° 52′	May 26
5.6, 6.6, 13.1, 10.6 m	PA 236°; Sep. 7.1″	Spec. G0		difficult
See May.				

λ Ophiuchi	ADS 10087	$16^h 30.9^m$	+01° 59′	May 30
4.2, 5.2, 11.1, 9.5 m	PA 30°; Sep. 1.5″	Spec. A0		difficult
See May.				

May

16 & 17 Draconis	ADS 10129	$16^h 36.2^m$	+52° 55′	May 31
5.4, 5.5, 6.4 m	PA 194°; Sep. 90.3″	Spec. A2 A0		easy ©
An easily split triple star system. The two main stars can be seen in binoculars. However, the third star will require a small telescope in order to be seen. Three white stars.				

Struve 1369	ADS 7438	$09^h 35.4^m$	±39° 57′	February 13
7.9, 8.0, 8.7 m	PA 148°; Sep. 24.7″	Spec. F2 G0		easy
See February.				

17 Canis Majoris	ADS 5585	$06^h 55.0^m$	–20° 24′	January 4
5.8, 9.3, 9.0 m	PA 147°; Sep. 44.4″	Spec. A2 K5		easy
See January.				

S5 Serpentis	ADS 9584	$15^h 19.3^m$	+01° 46′	May 12
5.1, 10.1, 9.1 m	PA 36°; Sep. 11.2″	Spec. G0		moderate
A triple system of unequally bright stars, which always seems to enhance colour contrast. It consists of a pair of yellow and red stars with a faint white companion.				

β Serpentis	Σ 1970	$15^h 46.2^m$	+15° 25′	May 1
3.7, 9.9, 10.7 m	PA 265°; Sep. 30.6″	Spec. A2		moderate
A nice system for small telescopes. The stars are coloured lemon, blue and white.				

Burnham 800	β800	$13^h 16.9^m$	+17° 01′	April 10
6.6, 9.7, 10.4 m	PA 106°; Sep. 6.8″	Spec. K0		moderate
See April.				

μ Herculis	ADS 10786	$17^h 46.5^m$	+27° 43′	June 18
6.3.4, 10.1 m	PA 247°; Sep. 33.8″	Spec. G5		moderate
See June.				

Struve 2306	Σ 2306	$18^h 22.2^m$	–15° 05′	June 27
67.9, 8.6, 9.0 m	PA 221°; Sep. 10.2″	Spec. F5		moderate
See June.				

Struve 1604	ADS 8440	$12^h 09.5^m$	–11° 51′	March 24
6.8, 8.5, 9.1 m	PA 89°; Sep. 9.9″	Spec. G0		moderate
See March.				

μ Boötis	ADS 9626	$15^h 24.5^m$	+37° 23′	May 13
4.3, 7.0, 7.6 m	PA 8°; Sep. 2.3″	Spec. F0		difficult

A nice triple star system of close stars consisting of a pale yellow primary and pale yellow and orange companions. Discovered by G. Struve in 1835.

ξ Scorpii	Σ 1998	$16^h 04.4^m$	−11° 22′	May 23
44.8, 5.1, 7.3 m	PA 61°; Sep. 0.5″	Spec. A0		difficult

A triple star system that is a test for medium-aperture telescopes. Consists of two yellow stars and a fainter blue companion, although reports describe the primary as golden in colour.

σ Coronae Borealis	ADS 9979	$16^h 14.7^m$	+33° 52′	May 26
5.6, 6.6, 13.1, 10.6 m	PA 236°; Sep. 7.1″	Spec. G0		difficult

A quadruple star system. The two main stars are easily seen, coloured pale and deep yellow, although some observers report that the colour contrast is slight. The remaining two companion stars are, however, very faint.

λ Ophiuchi	ADS 10087	$16^h 30.9^m$	+01° 59′	May 30
4.2, 5.2, 11.1, 9.5 m	PA 30°; Sep. 1.5″	Spec. A0		difficult

A nice though difficult quadruple star system. The white and yellow primaries are a good test for small telescopes.

June

Struve 2445	Σ 2445	$19^h 04.6^m$	+23° 20′	July 8
7.2, 8.9, 8.9 m	PA 263°; Sep. 12.6″	Spec. B3		easy

See July.

β Capricorni	Σ 152	$20^h 21.0^m$	−14° 47′	July 27
43.4, 6.2, 9.0 m	PA 267°; Sep. 205.3″	Spec. K0 B9		easy

See July.

16 & 17 Draconis	ADS 10129	$16^h 36.2^m$	+52° 55′	May 31
5.4, 5.5, 6.4 m	PA 194°; Sep. 90.3″	Spec. A2 A0		easy ©

See May.

μ Herculis	ADS 10786	$17^h 46.5^m$	+27° 43′	June 18
6.3.4, 10.1 m	PA 247°; Sep. 33.8″	Spec. G5		moderate

A triple star system consisting of a yellow primary and two faint red dwarfs. Discovered by William Herschel in 1781.

Struve 2306	Σ 2306	$18^h 22.2^m$	−15° 05′	June 27
67.9, 8.6, 9.0 m	PA 221°; Sep. 10.2″	Spec. F5		moderate

Located in the constellation Scutum, this is a wonderful triple star system of delicately coloured stars. Observers have reported the primary as gold or copper-coloured and the secondary as cobalt blue or blue. The blue secondary will need a high magnification in order to split it.

α **Lyrae**	ADS 11510	18^h 36.9^m	+38° 47′	July 1
0.0, 9.5, 11.0, 9.5 m	PA 173°; Sep. 62.8″	Spec. A0		moderate
See July.				

54 Sagittarii	ADS 12767	19^h 40.7^m	−16° 18′	July 17
35.4, 11.9, 8.9 m	PA 38°; Sep. 274″	Spec. A0 K F		moderate
See July.				

5 Serpentis	ADS 9584	15^h 19.3^m	+01° 46′	May 12
5.1, 10.1, 9.1 m	PA 36°; Sep. 11.2″	Spec. G0		moderate
See May.				

β **Serpentis**	Σ 1970	15^h 46.2^m	+15° 25′	May 18
3.7, 9.9, 10.7 m	PA 265°; Sep. 30.6″	Spec. A2		moderate
See May.				

Burnham 800	β800	13^h 16.9^m	+17° 01′	April 10
6.6, 9.7, 10.4 m	PA 106°; Sep. 6.8″	Spec. K0		moderate
See April.				

ξ **Scorpii**	Σ 1998	16^h 04.4^m	−11° 22′	May 23
44.8, 5.1, 7.3 m	PA 61°; Sep. 0.5″	Spec. A0		difficult
See May.				

σ **Coronae Borealis**	ADS 9979	16^h 14.7^m	+33° 52′	May 26
5.6, 6.6, 13.1, 10.6 m	PA 236°; Sep. 7.1″	Spec. G0		difficult
See May.				

λ **Ophiuchi**	ADS 10087	16^h 30.9^m	+01° 59′	May 30
4.2, 5.2, 11.1, 9.5 m	PA 30°; Sep. 1.5″	Spec. A0		difficult
See May.				

July

Struve 2445	Σ 2445	19^h 04.6^m	+23° 20′	July 8
7.2, 8.9, 8.9 m	PA 263°; Sep. 12.6″	Spec. B3		easy
Located in the constellation Vulpecula, this is a nice triple for small telescopes or binoculars. Blue stars and white ones.				

β **Capricorni**	Σ 152	20^h 21.0^m	−14° 47′	July 27
43.4, 6.2, 9.0 m	PA 267°; Sep. 205.3″	Spec. K0 B9		easy
A fine triple star system which can be easily resolved by binoculars. A nice colour contrast of a yellow primary with blue and pale yellow secondaries.				

16 & 17 Draconis	ADS 10129	16^h 36.2^m	+52° 55′	May 31
5.4, 5.5, 6.4 m	PA 194°; Sep. 90.3″	Spec. A2 A0		easy ©
See May.				

α Lyrae	ADS 11510	$18^h\ 36.9^m$	+38° 47′	July 1
0.0, 9.5, 11.0, 9.5 m	PA 173°; Sep. 62.8″	Spec. A0		moderate

A very famous star, Vega, it is the brightest in the summer sky, and has a wonderful steely blue–white colour. Has many faint companions which are not physically associated. It was the first star to be photographed (1850), and recent measurements of infrared radiation from it indicate possible proto-planetary material surrounding the star. A solar system in formation?

54 Sagittarii	ADS 12767	$19^h\ 40.7^m$	−16° 18′	July 17
5.4, 11.9, 8.9 m	PA 38°; Sep. 274″	Spec. A0 K F		moderate

A wonderfully coloured triple star system, with a yellow-orangeish primary, a pale blue secondary and pale yellow companion.

μ Herculis	ADS 10786	$17^h\ 46.5^m$	+27° 43′	June 18
6.3.4, 10.1 m	PA 247°; Sep. 33.8″	Spec. G5		moderate

See June.

Struve 2306	Σ 2306	$18^h\ 22.2^m$	−15° 05′	June 27
67.9, 8.6, 9.0 m	PA 221°; Sep. 10.2″	Spec. F5		moderate

See June.

5 Serpentis	ADS 9584	$15^h\ 19.3^m$	+01° 46′	May 12
5.1, 10.1, 9.1 m	PA 36°; Sep. 11.2″	Spec. G0		moderate

See May.

β Serpentis	Σ 1970	$15^h\ 46.2^m$	+15° 25′	May 1
3.7, 9.9, 10.7 m	PA 265°; Sep. 30.6″	Spec. A2		moderate

See May.

Burnham 800	β800	$13^h\ 16.9^m$	+17° 01′	April 10
6.6, 9.7, 10.4 m	PA 106°; Sep. 6.8″	Spec. K0		moderate

See April.

ε Equulei	ADS 14499	$20^h\ 59.1^m$	+04° 18′	August 6
6.0, 6.3, 7.1 m	PA 70°; Sep. 10.7″	Spec. F5		difficult

See August.

ξ Scorpii	Σ 1998	$16^h\ 04.4^m$	−11° 22′	May 23
44.8, 5.1, 7.3 m	PA 61°; Sep. 0.5″	Spec. A0		difficult

See May.

σ Coronae Borealis	ADS 9979	$16^h\ 14.7^m$	+33° 52′	May 26
5.6, 6.6, 13.1, 10.6 m	PA 236°; Sep. 7.1″	Spec. G0		difficult

See May.

λ Ophiuchi	ADS 10087	$16^h\ 30.9^m$	+01° 59′	May 30
4.2, 5.2, 11.1, 9.5 m	PA 30°; Sep. 1.5″	Spec. A0		difficult

See May.

August

Struve 2445	Σ 2445	$19^h 04.6^m$	+23° 20′	July 8
7.2, 8.9, 8.9 m	PA 263°; Sep. 12.6″	Spec. B3		easy
See July.				

β Capricorni	Σ 152	$20^h 21.0^m$	−14° 47′	July 27
43.4, 6.2, 9.0 m	PA 267°; Sep. 205.3″	Spec. K0 B9		easy
See July.				

16 & 17 Draconis	ADS 10129	$16^h 36.2^m$	+52° 55′	May 31
5.4, 5.5, 6.4 m	PA 194°; Sep. 90.3″	Spec. A2 A0		easy ©
See May.				

α Lyrae	ADS 11510	$18^h 36.9^m$	+38° 47′	July 1
0.0, 9.5, 11.0, 9.5 m	PA 173°; Sep. 62.8″	Spec. A0		moderate
See July.				

54 Sagittarii	ADS 12767	$19^h 40.7^m$	−16° 18′	July 17
5.4, 11.9, 8.9 m	PA 38°; Sep. 274″	Spec. A0 K F		moderate
See July.				

μ Herculis	ADS 10786	$17^h 46.5^m$	+27° 43′	June 18
6.3.4, 10.1 m	PA 247°; Sep. 33.8″	Spec. G5		moderate
See June.				

Struve 2306	Σ 2306	$18^h 22.2^m$	−15° 05′	June 27
67.9, 8.6, 9.0 m	PA 221°; Sep. 10.2″	Spec. F5		moderate
See June.				

5 Serpentis	ADS 9584	$15^h 19.3^m$	+01° 46′	May 12
5.1, 10.1, 9.1 m	PA 36°; Sep. 11.2″	Spec. G0		moderate
See May.				

β Serpentis	Σ 1970	$15^h 46.2^m$	+15° 25′	May 18
3.7, 9.9, 10.7 m	PA 265°; Sep. 30.6″	Spec. A2		moderate
See May.				

ε Equulei	ADS 14499	$20^h 59.1^m$	+04° 18′	August 6
6.0, 6.3, 7.1 m	PA 70°; Sep. 10.7″	Spec. F5		difficult
A very difficult triple star system. The two brightest members are very close at the moment, and will remain so for quite some time, and so will appear as an elongated blob for quite some time, even under high magnification. The third member of the system is blue, and in contrast to the yellow of the main stars.				

ξ Scorpii	Σ 1998	$16^h 04.4^m$	−11° 22′	May 23
44.8, 5.1, 7.3 m	PA 61°; Sep. 0.5″	Spec. A0		difficult
See May.				

σ Coronae Borealis	ADS 9979	$16^h 14.7^m$	+33° 52′	May 26
5.6, 6.6, 13.1, 10.6 m	PA 236°; Sep. 7.1″	Spec. G0		difficult
See May.				

λ Ophiuchi	ADS 10087	$16^h 30.9^m$	+01° 59′	May 30
4.2, 5.2, 11.1, 9.5 m	PA 30°; Sep. 1.5″	Spec. A0		difficult
See May.				

September

Struve 2445	Σ 2445	$19^h 04.6^m$	+23° 20′	July 8
7.2, 8.9, 8.9 m	PA 263°; Sep. 12.6″	Spec. B3		easy
See July.				

β Capricorni	Σ 152	$20^h 21.0^m$	–14° 47′	July 27
43.4, 6.2, 9.0 m	PA 267°; Sep. 205.3″	Spec. K0 B9		easy
See July.				

α Lyrae	ADS 11510	$18^h 36.9^m$	+38° 47′	July 1
0.0, 9.5, 11.0, 9.5 m	PA 173°; Sep. 62.8″	Spec. A0		moderate
See July.				

54 Sagittarii	ADS 12767	$19^h 40.7^m$	–16° 18′	July 17
5.4, 11.9, 8.9 m	PA 38°; Sep. 274″	Spec. A0 K F		moderate
See July.				

μ Herculis	ADS 10786	$17^h 46.5^m$	+27° 43′	June 18
6.3.4, 10.1 m	PA 247°; Sep. 33.8″	Spec. G5		moderate
See June.				

Struve 2306	Σ 2306	$18^h 22.2^m$	–15° 05′	June 27
67.9, 8.6, 9.0 m	PA 221°; Sep. 10.2″	Spec. F5		moderate
See June.				

γ Ceti	ADS 2080	$02^h 43.3^m$	+03° 14′	November 1
3.5, 7.3, 10.1 m	PA 294°; Sep. 2.8″	Spec. A2		moderate
See November.				

σ^2 Eridani	ADS 3093	$04^h 15.2^m$	–07° 39′	November 24
4.4, 9.4, 11.2 m	PA 347°; Sep. 7.6″	Spec. G5		moderate/difficult
See November.				

ε Equulei	ADS 14499	$20^h 59.1^m$	+04° 18′	August 6
6.0, 6.3, 7.1 m	PA 70°; Sep. 10.7″	Spec. F5		difficult
See August.				

η Cassiopeiae	ADS 1860	02^h 29.1^m	+67° 24′	October 28
4.6, 6.9, 8.4 m	PA 230°; Sep. 2.5″	Spec. A5		difficult ©

See October.

β Camelopardalis	HD 31910	05^h 03.4^m	+60° 27′	December 6
4.0, 8.6, 11.2 m	PA 208°; Sep. 80.8″	Spec. G0 A5		easy ©

See December.

Trapezium	*θ* Orionis	05^h 35.3^m	–05° 23′	December 14
5.1, 6.7 m 6.7, 7.9 m	PA 241°; Sep. 13.4″	B0 B0 O6 B0		easy

See December.

σ Orionis	ADS 4241	05^h 38.7^m	–02° 36′	December 15
4.0, 10.3 m 7.5, 6.5 m	PA 238°; Sep. 11.4″	Spec. O9 B3 A2		easy

See December.

ε Monocerotis	ADS 5012	06^h 23.8^m	+04° 36′	December 27
4.5, 6.5, 5.6 m	PA 27°; Sep. 13.4″	Spec. A5 A5		easy

See December.

β Monocerotis	ADS 5107	06^h 28.8^m	–07° 02′	December 28
4.7, 5.2, 6.1 m	PA 132°; Sep. 7.3″	Spec. B2 B2		easy

See December.

14 Aurigae	ADS 3824	05^h 15.4^m	+32° 31′	December 9
5.1, 11.1, 7.4 m	PA 352°; Sep. 14.6″	Spec. A2		easy/moderate

See December.

OΣ 147	ADS 5188	06^h 34.3^m	+38° 05′	December 29
6.6, 10, 10.6 m	PA 73°; Sep. 43.2″	Spec. K0		moderate

See December.

Struve 939	Σ 939	06^h 35.9^m	+05° 19′	December 30
8.3, 9.2, 9.4 m	PA 106°; Sep. 30.2″	Spec. B5 B8		moderate

See December.

ζ Orionis	ADS 4263	05^h 40.8^m	–01° 57′	December 16
1.9, 4.0, 9.9 m	PA 165°; Sep. 2.3″	Spec. O9 B0		moderate

See December.

γ Ceti	ADS 2080	02^h 43.3^m	+03° 14′	November 1
3.5, 7.3, 10.1 m	PA 294°; Sep. 2.8″	Spec. A2		moderate

See November.

σ^2 **Eridani**	ADS 3093	$04^h\ 15.2^m$	–07° 39′	November 24
4.4, 9.4, 11.2 m	PA 347°; Sep. 7.6″	Spec. G5		mod/difficult
See November.				

ι **Orionis**	ADS 4193	$05^h\ 35.4^m$	–05° 55′	December 15
2.8, 7.3, 11.1 m	PA 141°; Sep. 11.3″	Spec. O9 B7		mod/difficult
See December.				

η **Orionis**	ADS 4002	$05^h\ 24.5^m$	–02° 24′	December 12
3.6, 5.0, 10.1 m	PA 80°; Sep. 1.5″	Spec. B1		difficult
See December.				

λ **Orionis**	ADS 4179	$05^h\ 31.1^m$	+09° 56′	December 13
3.5, 5.6, 11.1, 11.1 m	PA 43°; Sep. 4.4″	Spec. O8 B0		difficult
See December.				

ε **Equulei**	ADS 14499	$20^h\ 59.1^m$	+04° 18′	August 6
6.0, 6.3, 7.1 m	PA 70°; Sep. 10.7″	Spec. F5		difficult
See August.				

η **Cassiopeiae**	ADS 1860	$02^h\ 29.1^m$	+67° 24′	October 28
4.6, 6.9, 8.4 m	PA 230°; Sep. 2.5″	Spec. A5		difficult ©
Thought by many to be one of the loveliest triple star systems in the entire sky. A brilliant yellowish-white primary with bluish companions. The primary is also a variable of the *Alpha Canum Venaticorum* type, with a magnitude range of only 0.03.				

November

β **Camelopardalis**	HD 31910	$05^h\ 03.4^m$	+60° 27′	December 6
4.0, 8.6, 11.2 m	PA 208°; Sep. 80.8″	Spec. G0 A5		easy ©
See December.				

Trapezium	θ Orionis	$05^h\ 35.3^m$	–05° 23′	December 14
5.1, 6.7 m 6.7, 7.9 m	PA 241°; Sep. 13.4″	B0 B0 O6 B0		easy
See December.				

σ **Orionis**	ADS 4241	$05^h\ 38.7^m$	–02° 36′	December 15
4.0, 10.3 m 7.5, 6.5 m	PA 238°; Sep. 11.4″	Spec. O9 B3 A2		easy
See December.				

ε **Monocerotis**	ADS 5012	$06^h\ 23.8^m$	+04° 36′	December 27
4.5, 6.5, 5.6 m	PA 27°; Sep. 13.4″	Spec. A5 A5		easy
See December.				

β Monocerotis	ADS 5107	06^h 28.8^m	–07° 02′	December 28
4.7, 5.2, 6.1 m	PA 132°; Sep. 7.3″	Spec. B2 B2		easy

See December.

14 Aurigae	ADS 3824	05^h 15.4^m	+32° 31′	December 9
5.1, 11.1, 7.4 m	PA 352°; Sep. 14.6″	Spec. A2		easy/moderate

See December.

γ Ceti	ADS 2080	02^h 43.3^m	+03° 14′	November 1
3.5, 7.3, 10.1 m	PA 294°; Sep. 2.8″	Spec. A2		moderate

In medium-aperture telescopes (between 20.0 and about 25 cm) this triple appears as a lovely white, yellow and faint red system, although the latter has been called tawny or dusky!

$O\Sigma$ 147	ADS 5188	06^h 34.3^m	+38° 05′	December 29
6.6, 10, 10.6 m	PA 73°; Sep. 43.2″	Spec. K0		moderate

See December.

Struve 939	Σ 939	06^h 35.9^m	+05° 19′	December 30
8.3, 9.2, 9.4 m	PA 106°; Sep. 30.2″	Spec. B5 B8		moderate

See December.

ζ Orionis	ADS 4263	05^h 40.8^m	–01° 57′	December 16
1.9, 4.0, 9.9 m	PA 165°; Sep. 2.3″	Spec. O9 B0		moderate

See December.

σ^2 Eridani	ADS 3093	04^h 15.2^m	–07° 39′	November 24
4.4, 9.4, 11.2 m	PA 347°; Sep. 7.6″	Spec. G5		moderate/difficult

A triple star system consisting of a creamy yellowish star along with pale blue companions. What is exceptional about the two fainter companions is that they are a white dwarf and a red dwarf. The white dwarf is the brightest visible from the Earth, with a mass equal to that of the Sun, although only 17,000 miles across. The red dwarf has a mass only one-fifth that of the Earth.

η Orionis	ADS 4002	05^h 24.5^m	–02° 24′	December 12
3.6, 5.0, 10.1 m	PA 80°; Sep. 1.5″	Spec. B1		difficult

See December.

λ Orionis	ADS 4179	05^h 31.1^m	+09° 56′	December 13
3.5, 5.6, 11.1, 11.1 m	PA 43°; Sep. 4.4″	Spec. O8 B0		difficult

See December.

η Cassiopeiae	ADS 1860	02^h 29.1^m	+67° 24′	October 28
4.6, 6.9, 8.4 m	PA 230°; Sep. 2.5″	Spec. A5		difficult ©

See October.

December

β Camelopardalis	HD 31910	$05^h 03.4^m$	+60° 27′	December 6
4.0, 8.6, 11.2 m	PA 208°; Sep. 80.8″	Spec. G0 A5		easy ©

A nice triple system of yellow and blue stars. What makes this system so memorable is that it is seen against the dark nebula of the Milky Way.

Trapezium	θ Orionis	$05^h 35.3^m$	−05° 23′	December 14
5.1, 6.7 m 6.7, 7.9 m	PA 241°; Sep. 13.4″	B0 B0 O6 B0		easy

Probably the most famous multiple star system in the sky. Always a test for small telescopes. The four stars which make up the famous quadrilateral are set among the wispy embrace of the *Orion Nebula*, M42, one of the most magnificent sites in any telescope. Very young stars recently formed from the material in the nebula, and so should all appear bright white, but the nebula itself probably affects the light that is observed, so the stars appear as off-white, delicately tinted yellowish and bluish. Other observers have reported the colours as pale white, faint lilac, garnet and reddish! It is believed that these four stars contribute nearly all the radiation that makes the Orion Nebula glow. Well worth spending an entire evening just observing the region. A glorious sight! [See the relevant entry on emission nebulae in Chapter 4.]

σ Orionis	ADS 4241	$05^h 38.7^m$	−02° 36′	December 15
4.0, 10.3 m 7.5, 6.5 m	PA 238°; Sep. 11.4″	Spec. O9 B3 A2		easy

A multiple star system of white and bluish stars.

ε Monocerotis	ADS 5012	$06^h 23.8^m$	+04° 36′	December 27
4.5, 6.5, 5.6 m	PA 27°; Sep. 13.4″	Spec. A5 A5		easy

A lovely triple system of pale yellow stars along with a very faintly tinted blue companion. Set against the star fields of the Milky Way.

β Monocerotis	ADS 5107	$06^h 28.8^m$	−07° 02′	December 28
4.7, 5.2, 6.1 m	PA 132°; Sep. 7.3″	Spec. B2 B2		easy

A magnificent triple, first discovered in 1781 by Herschel. All the stars are a lovely steely blue–white in colour. What makes this system so unique is that all the stars are very nearly equal in brightness.

17 Canis Majoris	ADS 5585	$06^h 55.0^m$	−20° 24′	January 4
5.8, 9.3, 9.0 m	PA 147°; Sep. 44.4″	Spec. A2 K5		easy

See January.

14 Aurigae	ADS 3824	$05^h 15.4^m$	+32° 31′	December 9
5.1, 11.1, 7.4 m	PA 352°; Sep. 14.6″	Spec. A2		easy/moderate

A nice system in all telescopes. Yellow, blue and white stars.

Burnham 324	ADS 5498	$06^h 49.7^m$	−24° 05′	January 2
6.3, 7.6, 8.6 m	PA 206°; Sep. 1.8″	Spec. A0 A2		easy/moderate

See January.

*O*Σ **147**	ADS 5188	$06^h\ 34.3^m$	+38° 05′	December 29
6.6, 10, 10.6 m	PA 73°; Sep. 43.2″	Spec. K0		moderate
Located in the constellation Auriga, this is a wonderful triple star system forming a triangle of yellow and blue stars.				

Struve 939	Σ 939	$06^h\ 35.9^m$	+05° 19′	December 30
8.3, 9.2, 9.4 m	PA 106°; Sep. 30.2″	Spec. B5 B8		moderate
A fascinating but rare triple system, with all its members forming a nearly perfect equilateral triangle.				

ζ **Orionis**	ADS 4263	$05^h\ 40.8^m$	−01° 57′	December 16
1.9, 4.0, 9.9 m	PA 165°; Sep. 2.3″	Spec. O9 B0		moderate
A nice triple system of blue and white and very pale red stars. Located among and near several bright and dark nebulae.				

γ **Ceti**	ADS 2080	$02^h\ 43.3^m$	+03° 14′	November 1
3.5, 7.3, 10.1 m	PA 294°; Sep. 2.8″	Spec. A2		moderate
See November.				

ι **Orionis**	ADS 4193	$05^h\ 35.4^m$	−05° 55′	December 15
2.8, 7.3, 11.1 m	PA 141°; Sep. 11.3″	Spec. O9 B7		mod/difficult
A nice colour-contrasted triple system and also a test for small telescopes. The stars are coloured white with delicately tinted blue and red companions.				

σ^2 **Eridani**	ADS 3093	$04^h\ 15.2^m$	−07° 39′	November 24
4.4, 9.4, 11.2 m	PA 347°; Sep. 7.6″	Spec. G5		moderate/difficult
See November.				

η **Orionis**	ADS 4002	$05^h\ 24.5^m$	−02° 24′	December 12
3.6, 5.0, 10.1 m	PA 80°; Sep. 1.5″	Spec. B1		difficult
Also known as *Dawes 5*, this is a wonderful system. Even under high magnification, the two brighter members will appear as two white discs in contact. η *Orionis* is also a spectroscopic binary.				

λ **Orionis**	ADS 4179	$05^h\ 31.1^m$	+09° 56′	December 13
3.5, 5.6, 11.1, 11.1 m	PA 43°; Sep. 4.4″	Spec. O8 B0		difficult
A nice quadruple star system of white and blue stars. Various observers have reported them as yellowish and purple and pale white and violet.				

τ **Canis Majoris**	ADS 5977	$07^h\ 18.7^m$	−24° 57′	January 10
4.4, 10.5, 11.2 m	PA 90°; Sep. 80.2″	Spec. O9		difficult
See January.				

This stellar swarm is M80 (NGC 6093), one of the densest of the 147 known globular star clusters in the Milky Way galaxy. Located about 28,000 light years from Earth, M80 contains hundreds of thousands of stars, held together by their mutual gravitational attraction.

Globular clusters are particularly useful for studying stellar evolution, since all the stars in the cluster are the same age (about 15 billion years), but cover a range of stellar masses. Every star visible in this image is either more highly evolved than our own Sun, or in a few rare cases more massive than it. By analysing the Hubble Space Telescope's Wide Field and Planetary Camera 2 (WFPC2) images, including images taken through a filter, astronomers have found a large population of "blue stragglers" in the core of the cluster. These appear to be unusually young and more massive than the other stars in a globular cluster.

However, stellar collisions can occur in dense regions like the core of M80 and, in some cases, the collisions can result in the merging of two stars to produce an unusually massive single star, which mimics a normal, young star.

M80 was not previously known to contain blue stragglers but is now believed to contain more than twice as many as any other globular cluster surveyed with the HST. Based on the number of blue stragglers, the stellar collision rate in the core of M80 appears to be exceptionally high!

This high-resolution image was created from two separate pointings of HST. The two datasets that were used to create the Heritage image of NGC 6093 had similar but not identical images.

Hubble Heritage Team (AURA/STScI/NASA)

Chapter 3 Star Clusters

3.1 Introduction

Having looked at stars as individual objects in the previous chapter, let's now turn our attention to *groups* of stars, or star *clusters* as they are called.

There are so many star clusters that they provide a rich and diverse selection of observing targets with the naked eye, binoculars or telescopes. Some clusters may have only be a few members (some open clusters), while others may contain several thousand stars – or in some cases even a million – as is the case for many globular clusters. Many display a jewel-box array of colours, while others may contain only steely blue stars or brilliant white ones. Some types of clusters contain only very young stars, perhaps only a few million years old, while other kinds will contain stars that are very old, perhaps several billion years.

The clusters I've selected here are only a representative few of the hundreds that are available to the amateur astronomer.

Also included in the chapter are a few *stellar associations*, which at first glance may seem a strange inclusion as they are not, in the correct sense, star clusters at all. Nevertheless, they are included for the reason that they are the relatives of star clusters, and an important aspect of stellar cluster evolution. As usual, a few words on origin, evolution and structure are appropriate before I describe the objects.

3.2 Open Star Clusters

A casual glance at the night sky may lead you into believing that stars are solitary, isolated objects, but in fact no star is born in isolation. The process of star birth takes place in large interstellar clouds of dust and gas that, depending upon the cloud's size, can give rise to anything from a few dozen to many thousands of young stars. Over time, however, the stellar nursery of young stars will gradually disperse. Calculations predict that massive stars have much shorter life spans than smaller, less massive ones, so you can easily see that some stars, the more massive ones, do not live long enough to escape their birthplace, whereas, a smaller star, say, of solar mass size, will in most cases easily escape from its stellar birthplace.

It's worth noting, in relation to stars of mass about equal to that of the Sun, that where there may be several thousand of the objects, the combined gravitational attraction of so many stars may slow down the dispersion of the group. It really depends on the star-density and mass of the particular cluster. Thus the conclusion is that the most dense or closely packed clusters, which contain solar-mass-sized stars, will be the ones that contain the oldest population of stars, while the most open clusters will have the youngest star population.

Open clusters, or *galactic clusters*, as they are sometimes called, are collections of young stars, containing anything from maybe a dozen members to hundreds. A few of them, for example, Messier 11 in Scutum, contains an impressive number of stars, equalling that of globular clusters, while others seem little more than a faint grouping set against the background star field. Such is the variety of open clusters that they come in all shapes and sizes. Several are over a degree in size and their full impact can only be appreciated by using binoculars, as a telescope has too narrow a field of view. An example of such a large cluster is Messier 44 in Cancer. Then there are tiny clusters, seemingly nothing more than compact multiple stars, as is the case with IC 4996 in Cygnus. In some cases all the members of the cluster are equally bright, such as Caldwell 71 in Puppis, but there are others which consist of only a few bright members accompanied by several fainter companions, as is the case of Messier 29 in Cygnus. The stars which make up an open cluster are called Population I stars, which are metal-rich and usually to be found in or near the spiral arms of the Galaxy.

The size of a cluster can vary from a few dozen light years across, as in the case of NGC 255 in Cassiopeia, to about 70 l.y. across, as in either component of Caldwell 14, the Perseus Double Cluster.

The reason for the varied and disparate appearances of open clusters is the circumstances of their births. It is the interstellar cloud which determines both the number and type of stars that are born within it. Factors such as the size, density, turbulence, temperature and magnetic field all play a role as the deciding parameters in star birth. In the case of *giant molecular clouds*, or GMCs, the conditions can give rise to both O- and B-type giant stars along with solar-type dwarf stars – whereas *in small molecular clouds* (SMCs) only solar-type stars will be formed, with none of the luminous B-type stars. An example of an SMC is the Taurus Dark Cloud, which lies just beyond the Pleiades.

An interesting aspect of open clusters is their distribution in the night sky. You may be forgiven in thinking that they are randomly distributed across the sky, but surveys show that although well over a thousand clusters have been discovered, only a few are observed to be at distances greater than 25° above or below the galactic equator. Some parts of the sky are very rich in clusters – Cassiopeia and Puppis – and this is due to the absence of dust lying along these lines of sight, allowing us to see across the spiral plane of our Galaxy. Many of the clusters mentioned here actually lie in different spiral arms, and so as you observe them you are actually looking at different parts of the spiral structure of our Galaxy.

I mentioned earlier that stars are not born in isolation. However, this does not mean that they are all born simultaneously. The more massive a star the faster it contracts and becomes stable, thus joining the main sequence. This results in some clusters having bright young O and B main sequence stars, while at the same time containing low-mass members which may still be in the process of gravitational contraction, for example the star cluster at the centre of the Lagoon Nebula. In a few cases, the star production in a cluster is at a very early stage, with only a few stars visible, the majority still in the process of contraction and hidden within the interstellar cloud.

A perfect example of such a process is the open cluster within Messier 42, the Orion Nebula. The stars within the cluster, the Trapezium, are the brightest, youngest and most massive stars in what will eventually become a large cluster containing many A-, F- and G-type stars. However, the majority are blanketed by the dust and gas clouds, and are only detectable by their infrared radiation.

As time passes, the dust and gas surrounding a new cluster will be blown away by the radiation from the O-type stars, resulting in the cluster becoming visible in its entirety, such as in the case of the young cluster Caldwell 76 in Scorpius.

Once a cluster has formed it will remain more or less unchanged for at least a few million years, but then changes within the cluster may occur. Two processes are responsible for changes with any given cluster. The evolution of open clusters depends on both the initial stellar content of the group and the ever pervasive pull of gravity. If a cluster contains O-, B- and A-type stars, then these stars will eventually become supernovae, leaving the cluster with slower-evolving, less massive and less luminous members of type A and M stars. A famous example of such a cluster is Caldwell 94, the Jewel Box in Crux, which is a highlight of the southern sky, and, alas, unobservable to northern hemisphere observers. However, these too will become supernovae, with the result that the most luminous members of a cluster will, one by one, disappear over time. This doesn't necessarily mean the demise of a cluster, especially those that have many tens or hundreds of members. But some, which consist of only a few bright stars, will seem to meld into the background star field. However, even those clusters that have survived the demise of their brighter members will eventually begin to feel the effect of a force which pervades everywhere – the Galaxy's gravitational field. As time passes, the cluster will be affected by the influence of other globular clusters and the interstellar matter itself, as well as the tidal force of the Galaxy. The cumulative affect of all these encounters will result in some of the less massive members of the cluster acquiring enough velocity to escape from the cluster. Thus, given enough time, a cluster will fade and disperse. (Take heart, as this isn't likely to happen in the near future so that you would notice: the Hyades star cluster, even after having lost most of its K- and M-type dwarf stars, is still with us, after 600 million years!)

For the amateur, observing open clusters is a very rewarding experience, as they are readily observable, from naked-eye clusters to those visible only in larger telescopes. Happily, many of them are best viewed by binoculars, especially the larger clusters that are of an appreciable angular size. Furthermore, nearly all have double or triple stars within the cluster, and so regardless of magnification there is always something of interest to be seen.

From the preceding chapter you will know that colour in observed stars is best seen when contrasted with a companion or companions. Thus an open cluster presents a perfect opportunity for observing star colours. Many clusters, such as the ever and rightly popular Pleiades, are all a lovely steely blue colour. On the other hand, Caldwell 10 in Cassiopeia has contrasting bluish stars along with a nice orange star. Other clusters have a solitary yellowish or ruddy orange star along with fainter white ones, such as Messier 6 in Scorpius. An often striking characteristic of open clusters is the apparent chains of stars that are seen. Many clusters have stars that arc across apparently empty voids, as in Messier 41 in Canis Major.

Another word for a very small, loose group of stars is an *asterism*. In some cases there may only be 5 to 6 stars within the group. I've given a few examples below.

Because open clusters display such a wealth of characteristics, different parameters are assigned to a cluster which describe its shape and content. For instance, a designation called the *Trumpler* type is often used. It is a three-part designation that describes the cluster's degree of concentration, that is, from a packed cluster to one that is evenly distributed, the range in brightness of the stars within the cluster, and finally the richness of the cluster, from poor (less than 50 stars) to rich (more than 100). The full classification is:

Trumpler classification for star clusters

Concentration

I Detached – strong concentration of stars towards the centre.
II Detached – weak concentration of stars towards the centre.
III Detached – no concentration of stars towards the centre.
IV Poor detachment from background star field.

Range of brightness

1 Small range.
2 Moderate range.
3 Large range.

Richness of cluster

p Poor (with less than 50 stars).
m Moderate (with 50–100 stars).
r Rich (with more than 100 stars).
n Cluster within nebulosity.

Actually, in my experience this designation is of limited use as some clusters that appear very rich when seen in good binoculars are disappointing when viewed using a telescope. Additionally, the use of high magnification and large aperture will often make a poor cluster appear very rich. However, for completeness, I am using the designation as a rough guide as to what you can expect to see under perfect conditions with perfect optics, and so on.

Two further and final points need to be mentioned which can often cause problems: the *magnitude* and *size* of the cluster. The quoted magnitude of a cluster may be the result of only a few bright stars, or on the other hand may be the result of a large number of faint stars. Also, the diameter of a cluster is often misleading, as in most cases it has been calculated from photographic plates, which, as experienced amateurs will know, bear little resemblance to what is seen at the eyepiece.

Although magnitudes and diameters may be quoted in the text, do treat them with a certain amount of caution.

In the descriptions given below, the first line lists the name, the position and the approximate midnight transit time, the second line the visual magnitude (this is the combined magnitude of all stars in cluster), object size in arc minutes (⊕), the approximate number of stars in the cluster (bear in mind that the number of stars seen will depend on magnification and aperture, and will increase when large apertures are used, thus the number quoted is an estimate using modest aperture), the Trumpler designation and the level of difficulty (based on the magnitude, size and ease of finding the cluster).

January

Messier 41	NGC 2287	$06^h\ 47.0^m$	–20° 44′	January 2
4.5 m	⊕38′	70	II 3 m	easy

Easily visible to the naked eye on very clear nights as a cloudy spot slightly larger than the full moon. Nicely resolved in binoculars, it becomes very impressive with medium aperture, with many double and multiple star combinations. Contains blue B-type giant stars as well as several K-type giants. Current research indicates that the cluster is about 100 million years old and occupies a volume of space 80 l.y. in diameter.

–	NGC 2301	06h 51.8m	+00° 28′	January 3
6.0 m	⊕ 12′	70	I 3 m	easy

Superb! A very striking cluster. In binoculars, a north–south chain of 8th- and 9th-magnitude stars is revealed, marked at its midway point by a faint haze of unresolvable stars. With large aperture, there is a colourful trio of red, gold and blue stars at the cluster's centre.

Collinder 121	–	06h 54.2m	–24° 38′	January 3
2.6 m	⊕ 50′	20	III 3 p	easy

A very large cluster, but one that is difficult to locate because of the plethora of stars in the background. At the northern border of the cluster is o Canis Majoris. Best seen with large binoculars or low-power telescopes.

Messier 50	NGC 2323	07h 03.2m	–08° 20′	January 6
5.9 m	⊕ 16′	80	II 3 m	easy

The only Messier object in Monoceros and one that is often overlooked by amateurs. Discovered by Cassini, this is a fine, heart-shaped cluster easily seen in binoculars, and visible to the naked eye on clear nights. Within the large, bright and irregular cluster of blue stars is a striking red star. What makes the cluster particularly challenging is that the area of the sky where it resides is full of small stellar groupings and asterisms. The question often arises, where does the cluster end and the background star field begin?

Herschel 34	NGC 2353	07h 14.6m	–10° 18′	January 9
7.1 m	⊕ 20′	30	II 2 p	easy

A cluster of stars best seen in a telescope. It includes many orange 6th-magnitude stars.

Caldwell 58	NGC 2360	07h 17.8m	–15° 37′	January 9
7.2 m	⊕ 12′	80	II 2 m	easy

A beautiful open cluster, irregularly shaped and very rich. There are many faint stars, however, so the cluster needs moderate-aperture telescopes for these to be resolved, although it will appear as a faint blur in binoculars. This is believed to be an old cluster with an estimated age of around 1.3 billion years.

Caldwell 64	NGC 2362	07h 18.8m	–24° 57′	January 10
4.1 m	⊕ 8′	60	I 3 p n	easy

A very nice cluster, tightly packed and easily seen. With small binoculars the glare from τ CMa trends to overwhelm the majority of stars, although it itself is a nice star, with two bluish companion stars (recent research indicates that the star is a quadruple system). But the cluster becomes truly impressive with telescopic apertures; the bigger the aperture, the more stunning the vista. It is believed to be very young – only a couple of million years old – and thus has the distinction of being the youngest cluster in our Galaxy. Contains O- and B-type giant stars.

Messier 48	NGC 2548	08h 13.8m	−05° 48′	January 24
5.8 m	⊕ 55′	80	I 3 r	easy

Located in a rather empty part of the constellation Hydra, this is believed to be the missing Messier object. It is a nice cluster in both binoculars and small telescopes. In the former, about a dozen stars are seen, with a pleasing triangular asterism at its centre, while the latter will show a rather nice but large group of about 50 stars. Many amateurs often find the cluster difficult to locate for the reason mentioned above, but also for the fact that within a few degrees of M48 is another nameless, but brighter, cluster of stars which is often mistakenly identified as M48. Some observers claim that this nameless group of stars is in fact the correct missing Messier object, and not the one which now bears the name.

Messier 44	NGC 2632	08h 40.1m	+19° 59′	January 30
3.1 m	⊕ 95′	60	II 2 m	easy

A famous cluster, called Praesepe (the Manger) or the Beehive. One of the largest and brightest open clusters from the viewpoint of an observer. An old cluster, about 700 million years, distance 500 l.y., with the same space motion and velocity as the Hyades, which suggests a common origin for the two clusters. A nice triple star, Burnham 584, is located within M44, located just south of the cluster's centre. A unique Messier object in that it is brighter than the stars of the constellation within which it resides. Owing to its large angular size in the sky, it is best seen through binoculars or a low-power eyepiece.

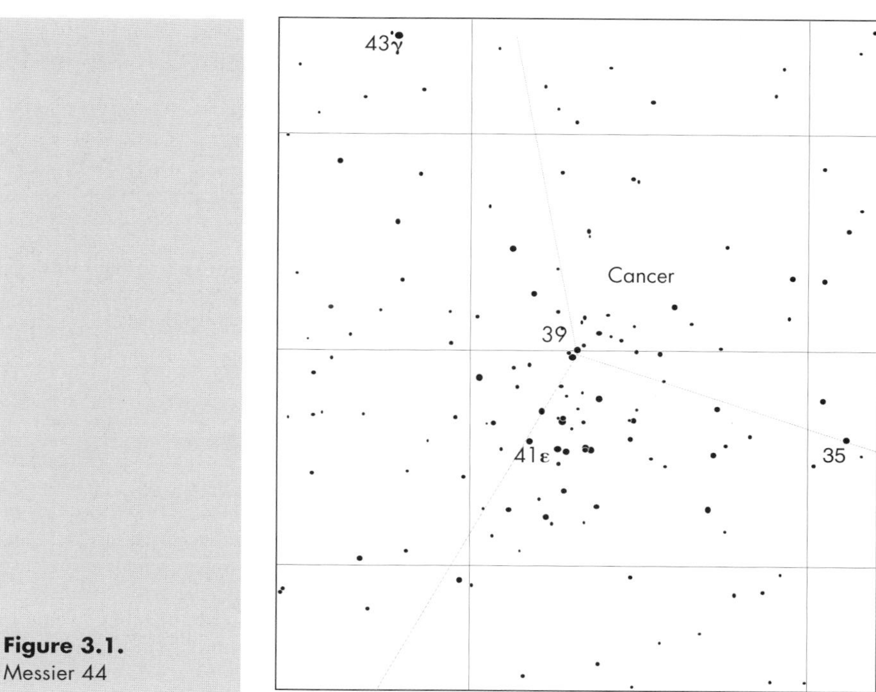

Figure 3.1.
Messier 44

Melotte 28	NGC 1746	$05^h\,03.6^m$	+23° 49′	December 6
6.1 m	⊕ 42′	20	III 2 p	easy
See December.				

Herschel 33	NGC 1857	$05^h\,20.2^m$	+39° 21′	December 11
7.0 m	⊕ 6′	35	II 2 m	easy
See December.				

Collinder 464		$05^h\,22.0^m$	+73° 00′	December 11
4.2 m	⊕ 120′	50	II 2 m	easy ©
See December.				

Messier 38	NGC 1912	$05^h\,28.7^m$	+35° 50′	December 13
6.4 m	⊕ 21′	75	III 2 m	easy
See December.				

Collinder 69	–	$05^h\,35.1^m$	+09° 56′	December 14
2.8 m	⊕ 65′	20	II 3 p n	easy
See December.				

Collinder 73	NGC 1981	$05^h\,35.2^m$	–04° 26′	December 14
4.6 m	⊕ 25′	20	III 2 p n	easy
See December.				

Messier 36	NGC 1960	$05^h\,36.1^m$	+34° 08′	December 15
6.0 m	⊕ 20′	70	II 3 m	easy
See December.				

Messier 37	NGC 2099	$05^h\,52.4^m$	+32° 33′	December 19
5.6 m	⊕ 20′	150	II 1 r	easy
See December.				

Collinder 38	NGC 2169	$06^h\,08.4^m$	+13° 57′	December 23
5.9 m	⊕ 6′	30	I 3 p n	easy
See December.				

Messier 35	NGC 2168	$06^h\,08.9^m$	+24° 20′	December 23
5.1 m	⊕ 28′	200	III 2 m	easy
See December.				

Caldwell 50	NGC 2244	$06^h\,32.4^m$	+04° 52′	December 29
4.8 m	⊕ 24′	35	–	easy
See December.				

Collinder 112	NGC 2264	$06^h\,41.1^m$	+09° 53′	December 31
3.9 m	⊕ 30′	70	IV 3 p	easy
See December.				

Caldwell 54	NGC 2506	08h 00.2m	−10° 47′	January 20
7.6 m	⊕ 7′	100	I 2 r	medium

A nice rich and concentrated cluster best seen with a telescope, but one that is often overlooked owing to its faintness even though it is just visible in binoculars. Includes many 11th- and 12th-magnitude stars. It is a very old cluster, around 2 billion years, and contains several *blue stragglers*. These are old stars that nevertheless have the spectrum signatures of young stars. This paradox was solved when research indicated that the young-looking stars are the result of a merger of two old stars.

Herschel 59	NGC 1664	04h 51.1m	+43° 42′	December 3
7.6 m	⊕ 13′	25	III 1 p	medium

See December.

Collinder 81	NGC 2158	06h 07.5m	+24° 06′	December 23
8.6 m	⊕ 5′	70	II 3 r	medium

See December.

Herschel 21	NGC 2266	06h 43.2m	+26° 58′	January 1
9.5 m	⊕ 6′	50	II 2 m	difficult

A pleasant cluster, though difficult to locate, consisting of over 50 stars, tightly compressed.

Herschel 61	NGC 1778	05h 08.1m	+37° 03′	December 8
7.7 m	⊕ 6′	30	III 2 p	difficult

See December.

Herschel 68	NGC 2126	06h 03.0m	+49° 54′	December 22
10.2 m	⊕ 6′	40	II 1 p	difficult ©

See December.

Herschel 13	NGC 2204	06h 15.7m	−18° 39′	December 25
8.6 m	⊕ 13′	80	III 3 m	difficult

See December.

February

Messier 41	NGC 2287	06h 47.0m	−20° 44′	January 2
4.5 m	⊕ 38′	70	II 3 m	easy

See January.

–	NGC 2301	06h 51.8m	+00° 28′	January 3
6.0 m	⊕ 12′	70	I 3 m	easy

See January.

Collinder 121	–	$06^h 54.2^m$	–24° 38′	January 3
2.6 m	⊕ 50′	20	III 3 p	easy
See January.				

Messier 50	NGC 2323	$07^h 03.2^m$	–08° 20′	January 6
5.9 m	⊕ 16′	80	II 3 m	easy
See January.				

Herschel 34	NGC 2353	$07^h 14.6^m$	–10° 18′	January 9
7.1 m	⊕ 20′	30	II 2 p	easy
See January.				

Caldwell 58	NGC 2360	$07^h 17.8^m$	–15° 37′	January 9
7.2 m	⊕ 12′	80	II 2 m	easy
See January.				

Caldwell 64	NGC 2362	$07^h 18.8^m$	–24° 57′	January 10
4.1 m	⊕ 8′	60	I 3 p n	easy
See January.				

Messier 48	NGC 2548	$08^h 13.8^m$	–05° 48′	January 24
5.8 m	⊕ 55′	80	I 3 r	easy
See January.				

Messier 44	NGC 2632	$08^h 40.1^m$	+19° 59′	January 30
3.1 m	⊕ 95′	60	II 2 m	easy
See January.				

Melotte 111	–	$12^h 25.0^m$	+26°	March 28
1.8 m	⊕ 275′	75	II 3 p	easy
See March.				

Upgren 1	–	$12^h 35.0^m$	+36° 18′	March 31
6.6 m	⊕ 10′	10	IV 2 p	easy
See March.				

Melotte 28	NGC 1746	$05^h 03.6^m$	+23° 49′	December 6
6.1 m	⊕ 42′	20	III 2 p	easy
See December.				

Herschel 33	NGC 1857	$05^h 20.2^m$	+39° 21′	December 11
7.0 m	⊕ 6′	35	II 2 m	easy
See December.				

Collinder 464	–	$05^h 22.0^m$	+73° 00′	December 11
4.2 m	⊕ 120′	50	II 2 m	easy ©
See December.				

Messier 38	NGC 1912	05h 28.7m	+35° 50′	December 13
6.4 m	⊕ 21′	75	III 2 m	easy

See December.

Collinder 69	–	05h 35.1m	+09° 56′	December 14
2.8 m	⊕ 65′	20	II 3 p n	easy

See December.

Collinder 73	NGC 1981	05h 35.2m	–04° 26′	December 14
4.6 m	⊕ 25′	20	III 2 p n	easy

See December.

Messier 36	NGC 1960	05h 36.1m	+34° 08′	December 15
6.0 m	⊕ 12′	70	II 3 m	easy

See December.

Messier 37	NGC 2099	05h 52.4m	+32° 33′	December 19
5.6 m	⊕ 20′	150	II 1 r	easy

See December.

Collinder 38	NGC 2169	06h 08.4m	+13° 57′	December 23
5.9 m	⊕ 6′	30	I 3 p n	easy

See December.

Messier 35	NGC 2168	06h 08.9m	+24° 20′	December 23
5.1 m	⊕ 28′	200	III 2 m	easy

See December.

Caldwell 50	NGC 2244	06h 32.4m	+04° 52′	December 29
4.8 m	⊕ 24′	35	–	easy

See December.

Collinder 112	NGC 2264	06h 41.1m	+09° 53′	December 31
3.9 m	⊕ 30′	70	IV 3 p	easy

See December.

Caldwell 54	NGC 2506	08h 00.2m	–10° 47′	January 20
7.6 m	⊕ 7′	100	I 2 r	medium

See January.

Herschel 59	NGC 1664	04h 51.1m	+43° 42′	December 3
7.6 m	⊕ 13′	25	III 1 p	medium

See December.

Collinder 81	NGC 2158	06h 07.5m	+24° 06′	December 23
8.6 m	⊕ 5′	70	II 3 r	medium

See December.

Messier 67	NGC 2682	08h 50.4m	+11° 49′	February 2
6.9 m	⊕ 30′	200	II 2 m	difficult

Often overlooked owing to its proximity to M44, it is nevertheless very pleasing. However, the stars it is composed of are faint ones, so in binoculars it will be unresolved, and seen as a faint misty glow. At a distance of 2500 l.y., it is believed to be very old, possibly 9 billion years, and thus has had time to move from the Galactic Plane, the usual abode of open clusters, to a distance of about 1600 l.y. off the plane.

Herschel 21	NGC 2266	06h 43.2m	+26° 58′	January 1
9.5 m	⊕ 6′	50	II 2 m	difficult

See January.

Herschel 61	NGC 1778	05h 08.1m	+37° 03′	December 8
7.7 m	⊕ 6′	30	III 2 p	difficult

See December.

Herschel 68	NGC 2126	06h 03.0m	+49° 54′	December 22
10.2 m	⊕ 6′	40	II 1 p	difficult ©

See December.

Herschel 13	NGC 2204	06h 15.7m	–18° 39′	December 25
8.6 m	⊕ 13′	80	III 3 m	difficult

See December.

March

Melotte 111	–	12h 25.0m	+26°	March 28
1.8 m	⊕ 275′	75	II 3 p	easy

Also known as the Coma Star Cluster, this is a large and impressive cluster of 5th- and 6th-magnitude stars, spanning about 5°. Owing to its large size, it is only worth observing with binoculars because telescope observation will loose the clustering effect. Believed to be 400 million years old and 260 l.y. distant, it is the third nearest cluster. Because of its extremely weak gravitational field the cluster may be on the verge of complete disruption. Paradoxically, although this cluster is visible to the naked eye, it has neither a Messier nor an NGC designation.

Upgren 1	–	12h 35.0m	+36° 18′	March 31
6.6 m	⊕ 10′	10	IV 2 p	easy

Almost unknown to amateurs, this is a fairly inconspicuous cluster of about 10 stars. Binoculars show about 7 of these. Although it has been shown to be a cluster, it appears to many observers as nothing more than an *asterism*.

Messier 41	NGC 2287	06h 47.0m	–20° 44′	January 2
4.5 m	⊕ 38′	70	II 3 m	easy
See January.				

–	NGC 2301	06h 51.8m	+00° 28′	January 3
6.0 m	⊕ 12′	70	I 3 m	easy
See January.				

Collinder 121	–	06h 54.2m	–24° 38′	January 3
2.6 m	⊕ 50′	20	III 3 p	easy
See January.				

Messier 50	NGC 2323	07h 03.2m	–08° 20′	January 6
5.9 m	⊕ 16′	80	II 3 m	easy
See January.				

Herschel 34	NGC 2353	07h 14.6m	–10° 18′	January 9
7.1 m	⊕ 20′	30	II 2 p	easy
See January.				

Caldwell 58	NGC 2360	07h 17.8m	–15° 37′	January 9
7.2 m	⊕ 12′	80	II 2 m	easy
See January.				

Caldwell 64	NGC 2362	07h 18.8m	–24° 57′	January 10
4.1 m	⊕ 8′	60	I 3 p n	easy
See January.				

Messier 48	NGC 2548	08h 13.8m	–05° 48′	January 24
5.8 m	⊕ 55′	80	I 3 r	easy
See January.				

Messier 44	NGC 2632	08h 40.1m	+19° 59′	January 30
3.1 m	⊕ 95′	60	II 2 m	easy
See January.				

Melotte 111	–	12h 25.0m	+26°	March 28
1.8 m	⊕ 275′	75	II 3 p	easy
See March.				

Upgren 1	–	12h 35.0m	+36° 18′	March 31
6.6 m	⊕ 10′	10	IV 2 p	easy
See March.				

Caldwell 54	NGC 2506	$08^h 00.2^m$	–10° 47′	January 20
7.6 m	⊕ 7′	100	I 2 r	medium
See January.				

Herschel 21	NGC 2266	$06^h 43.2^m$	+26° 58′	January 1
9.5 m	⊕ 6′	50	II 2 m	difficult
See January.				

Messier 67	NGC 2682	$08^h 50.4^m$	+11° 49′	February 2
6.9 m	⊕ 30′	200	II 2 m	difficult
See February.				

April

Messier 41	NGC 2287	$06^h 47.0^m$	–20° 44′	January 2
4.5 m	⊕ 38′	70	II 3 m	easy
See January.				

–	NGC 2301	$06^h 51.8^m$	+00° 28′	January 3
6.0 m	⊕ 12′	70	I 3 m	easy
See January.				

Collinder 121	–	$06^h 54.2^m$	–24° 38′	January 3
2.6 m	⊕ 50′	20	III 3 p	easy
See January.				

Messier 50	NGC 2323	$07^h 03.2^m$	–08° 20′	January 6
5.9 m	⊕ 16′	80	II 3 m	easy
See January.				

Herschel 34	NGC 2353	$07^h 14.6^m$	–10° 18′	January 9
7.1 m	⊕ 20′	30	II 2 p	easy
See January.				

Caldwell 58	NGC 2360	$07^h 17.8^m$	–15° 37′	January 9
7.2 m	⊕ 12′	80	II 2 m	easy
See January.				

Caldwell 64	NGC 2362	$07^h 18.8^m$	–24° 57′	January 10
4.1 m	⊕ 8′	60	I 3 p n	easy
See January.				

Messier 48	NGC 2548	$08^h 13.8^m$	–05° 48′	January 24
5.8 m	⊕ 55′	80	I 3 r	easy
See January.				

Messier 44	NGC 2632	$08^h 40.1^m$	+19° 59′	January 30
3.1 m	⊕ 95′	60	II 2 m	easy
See January.				

Melotte 111	–	$12^h 25.0^m$	+26°	March 28
1.8 m	⊕ 275′	75	II 3 p	easy
See March.				

Upgren 1	–	$12^h 35.0^m$	+36° 18′	March 31
6.6 m	⊕ 10′	10	IV 2 p	easy
See March.				

Caldwell 75	NGC 6124	$16^h 25.6^m$	–40° 40′	May 28
5.8 m	⊕ 29′	75	I 3 r	easy
See May.				

Caldwell 76	NGC 6231	$16^h 54.0^m$	–41° 48′	June 5
2.6 m	⊕ 14′	100	I 3 p	easy
See June.				

Trumpler 24	Harvard 12	$16^h 57.0^m$	–40° 40′	June 5
8.6 m	⊕ 60′	100	IV 2 p n	easy
See June.				

Messier 6	NGC 6405	$17^h 40.1^m$	–32° 13′	June 16
4.2 m	⊕ 33′	100	II 3 r	easy
See June.				

–	IC 4665	$17^h 46.3^m$	+05° 43′	June 18
4.2 m	⊕ 40′	30	III 2 m	easy
See June.				

Messier 7	NGC 6475	$17^h 53.9^m$	–34° 49′	June 20
3.3 m	⊕ 80′	80	I 3 r	easy
See June.				

Messier 23	NGC 6494	$17^h 56.8^m$	–19° 01′	June 20
5.5 m	⊕ 27′	100	II 2 r	easy
See June.				

Messier 21	NGC 6531	$18^h 04.6^m$	–22° 30′	June 22
5.9 m	⊕ 14′	60	I 3 r	easy
See June.				

Messier 24	–	18^h 16.5^m	–18° 50′	June 25
2.5 m	⊕ 95′ × 35′	–		easy
See June.				

Messier 16	NGC 6611	18^h 18.8^m	–13° 47′	June 26
6.0 m	⊕ 22′	50	II 3 m n	easy
See June.				

Messier 18	NGC 6613	18^h 19.9^m	–17° 08′	June 26
6.9 m	⊕ 10′	30	II 3 p n	easy
See June.				

Herschel 72	NGC 6633	18^h 27.7^m	+06° 34′	June 28
4.6 m	⊕ 27′	25	III 2 m	easy
See June.				

Messier 25	IC 4725	18^h 31.6^m	–19° 15′	June 29
4.6 m	⊕ 32′	40	I 3 m	easy
See June.				

Collinder 309	NGC 6192	16^h 40.3^m	–43° 22′	June 1
8.5 m	⊕ 7′	40	I 2 r	medium
See June.				

Trumpler 26	Harvard 15	17^h 28.5^m	–29° 29′	June 13
9.5 m	⊕ 17′	45	II 1 m	medium
See June.				

Caldwell 54	NGC 2506	08^h 00.2^m	–10° 47′	January 20
7.6 m	⊕ 7′	100	I 2 r	medium
See January.				

Messier 67	NGC 2682	08^h 50.4^m	+11° 49′	February 2
6.9 m	⊕ 30′	200	II 2 m	difficult
See February.				

Herschel 7	NGC 6520	18^h 03.4^m	–27° 54′	June 22
7.5 m	⊕ 7′	30	I 2 r n	difficult
See June.				

Herschel 21	NGC 2266	06^h 43.2^m	+26° 58′	January 1
9.5 m	⊕ 6′	50	II 2 m	difficult
See January.				

May

Caldwell 75	NGC 6124	$16^h 25.6^m$	–40° 40′	May 28
5.8 m	⊕ 29′	75	I 3 r	easy

A very nice rich cluster, suitable for large binoculars and small telescopes. There is a chain of stars at its southern edge, and a tightly grouped collection of 5 bright stars at its centre. It also contains several nice star chains and a few red-tinted stars. It is relatively close, at a distance of around 1500 l.y.

Melotte 111	–	$12^h 25.0^m$	+26°	March 28
1.8 m	⊕ 275′	75	II 3 p	easy

See March.

Upgren 1	–	$12^h 35.0^m$	+36° 18′	March 31
6.6 m	⊕ 10′	10	IV 2 p	easy

See March.

Caldwell 76	NGC 6231	$16^h 54.0^m$	–41° 48′	June 5
2.6 m	⊕ 14′	100	I 3 p	easy

See June.

Trumpler 24	Harvard 12	$16^h 57.0^m$	–40° 40′	June 5
8.6 m	⊕ 60′	100	IV 2 p n	easy

See June.

Messier 6	NGC 6405	$17^h 40.1^m$	–32° 13′	June 16
4.2 m	⊕ 33′	100	II 3 r	easy

See June.

–	IC 4665	$17^h 46.3^m$	+05° 43′	June 18
4.2 m	⊕ 40′	30	III 2 m	easy

See June.

Messier 7	NGC 6475	$17^h 53.9^m$	–34° 49′	June 20
3.3 m	⊕ 80′	80	I 3 r	easy

See June.

Messier 23	NGC 6494	$17^h 56.8^m$	–39° 01′	June 20
5.5 m	⊕ 27′	100	II 2 r	easy

See June.

Messier 21	NGC 6531	$18^h 04.6^m$	–22° 30′	June 22
5.9 m	⊕ 14′	60	I 3 r	easy

See June.

Messier 24	–	18h 16.5m	–18° 50′	June 25
2.5 m	⊕ 95′ × 35′	–	easy	
See June.				

Messier 16	NGC 6611	18h 18.8m	–13° 47′	June 26
6.0 m	⊕ 22′	50	II 3 m n	easy
See June.				

Messier 18	NGC 6613	18h 19.9m	–17° 08′	June 26
6.9 m	⊕ 10′	30	II 3 p n	easy
See June.				

Herschel 72	NGC 6633	18h 27.7m	+06° 34′	June 28
4.6 m	⊕ 27′	25	III 2 m	easy
See June.				

Messier 25	IC 4725	18h 31.6m	–19° 15′	June 29
4.6 m	⊕ 32′	40	I 3 m	easy
See June.				

Collinder 399	–	19h 25.4m	+20° 11′	July 13
3.6 m	⊕ 60′	35	III 3 m	easy
See July.				

Stock 1		19h 35.8m	+25° 13′	July 16
5.3 m	⊕ 60′	40	III 2 m	easy
See July.				

Collinder 413	NGC 6871	20h 05.9m	+35° 47′	July 23
5.2 m	⊕ 20′	30	II 2 p n	easy
See July.				

Caldwell 37	NGC 6885	20h 12.0m	+26° 29′	July 25
5.9 m	⊕ 7′	25	III 2 p	easy
See July.				

Messier 29	NGC 6913	20h 23.9m	+38° 32′	July 28
6.6 m	⊕ 7′	80	I 2 m n	easy
See July.				

Herschel 8	NGC 6940	20h 34.6m	+28° 18′	July 30
6.3 m	⊕ 32′	75	III 2 r	easy
See July.				

Collinder 309	NGC 6192	$16^h 40.3^m$	–43° 22′	June 1
8.5 m	⊕ 7′	40	I 2 r	medium
See June.				

Trumpler 26	Harvard 15	$17^h 28.5^m$	–29° 29′	June 13
9.5 m	⊕ 17′	45	II 1 m	medium
See June.				

Messier 26	NGC 6694	$18^h 45.2^m$	–09° 24′	July 3
8.0 m	⊕ 14′	30	I 2 m	medium
See July.				

Messier 11	NGC 6705	$18^h 51.1^m$	–06° 16′	July 4
5.8 m	⊕ 13′	200	I 2 r	medium
See July.				

Collinder 392	NGC 6709	$18^h 51.5^m$	+10° 21′	July 4
6.7 m	⊕ 13′	45	III 2 m	medium.
See July.				

–	NGC 6755	$19^h 07.8^m$	+04° 14′	July 8
7.5 m	⊕ 15′	100	II 2 r	medium
See July.				

Collinder 402	NGC 6811	$19^h 36.9^m$	+46° 23′	July 16
6.8 m	⊕ 21′	65	IV 3 p	medium
See July.				

Collinder 403	NGC 6819	$19^h 41.3^m$	+40° 11′	July 17
7.3 m	⊕ 10′	100	I 1 r	medium
See July.				

Harvard 20	–	$19^h 53.1^m$	+18° 20′	July 20
7.7 m	⊕ 10′	20	IV 2 p	medium
See July.				

Herschel 7	NGC 6520	$18^h 03.4^m$	–27° 54′	June 22
7.5 m	⊕ 7′	30	I 2 r n	difficult
See June.				

–	NGC 6791	$19^h 20.7^m$	+37° 51′	July 12
9.5 m	⊕ 15′	250	I 2 r	difficult
See July.				

-	NGC 6939	20h 31.4m	+60° 38′	July 30
7.8 m	⊕ 7.0′	50	I 1 m	difficult ©

See July

June

Caldwell 76	NGC 6231	16h 54.0m	−41° 48′	June 5
2.6 m	⊕ 14′	100	I 3 p	easy

A superb cluster located in an awe-inspiring region of the sky. Brighter by 2.5 magnitudes than its northern cousins, the double cluster in Perseus. The cluster is full of spectacular stars: very hot and luminous O-type and B0-type giants and supergiants, a couple of Wolf–Rayet stars, and ξ^{1} Scorpii, which is a B1.5 Ia extreme supergiant star with a luminosity nearly 280,000 times that of the Sun ! The cluster is thought to be a member of the stellar association Sco OB1, with an estimated age of 3 million years. A wonderful object in binoculars and telescopes, the cluster contains many blue, orange and yellow stars. It lies between μ^{1+2} Scorpii and ξ^{1} Scorpii, an area rich in spectacular views. A good cluster to test the technique of averted vision, where many more stars will jump into view. Observe and enjoy.

Trumpler 24	Harvard 12	16h 57.0m	−40° 40′	June 5
8.6 m	⊕ 60′	100	IV 2 p n	easy

A loose and scattered cluster, set against the backdrop of the Milky Way. It is, along with nearby Collinder 316, the core of the Scorpius OB1 stellar association.

Messier 6	NGC 6405	17h 40.1m	−32° 13′	June 16
4.2 m	⊕ 33′	100	II 3 r	easy

Also known as the Butterfly Cluster. Easily seen with the naked eye as a dim patch of light. It is in my opinion one of the few stellar objects that actually looks like the entity after which it is named. A fine sight in binoculars, it contains the lovely orange-tinted star BM Scorpii east of its centre. This star is a semi-regular variable, period 850 days, which changes from magnitude 5.5 to 7. Surrounding it are many nice steely blue–white stars. Believed to be at a distance of 1590 l.y.

-	IC 4665	17h 46.3m	+05° 43′	June 18
4.2 m	⊕ 40′	30	III 2 m	easy

A naked-eye object under perfect seeing conditions, this large cluster appears as a hazy spot measuring over two full Moon diameters. With binoculars, nearly 30 blue–white 6th-magnitude stars can be seen. Its position in a sparse area of the sky emphasizes the cluster, even though it is not a particularly dense collection of stars.

Messier 7	NGC 6475	17h 53.9m	−34° 49′	June 20
3.3 m	⊕ 80′	80	I 3 r	easy

An enormous and spectacular cluster. It presents a fine spectacle in binoculars and telescopes, containing over 80 blue–white and pale yellow stars. It is only just over 800 l.y. away, but is over 200 million years old. Many of the stars are around 6th and 7th magnitude, and thus should be resolvable with the naked eye. Try it and see.

Messier 23	NGC 6494	$17^h 56.8^m$	−19° 01′	June 20
5.5 m	⊕ 27′	100	II 2 r	easy

Often overlooked because it lies in an area studded with celestial showpieces, this is a wonderful cluster which is equally impressive seen in telescopes or binoculars, but the latter will only show a few of the brighter stars shining against a misty glow of fainter stars. Full of double stars and star chains.

Messier 21	NGC 6531	$18^h 04.6^m$	−22° 30′	June 22
5.9 m	⊕ 14′	60	I 3 r	easy

An outstanding cluster for small telescopes and binoculars. A compact, symmetrical cluster of bright stars with a nice double system of 9th and 10th magnitude located at its centre. Very close to the Trifid Nebula. In the cluster is the grouping called Webb's Cross, which consists of several stars of 6th and 7th magnitude arranged in a cross. Several amateurs report that some stars within the cluster show definite tints of blue, red and yellow. Can you see them?

Messier 24[1]	–	$18^h 16.5^m$	−18° 50′	June 25
2.5 m	⊕ 95′ × 35′	–		easy

Another superb object for binoculars. This is the Small Sagittarius Star Cloud, visible to the naked eye on clear nights, and nearly four times the angular size of the Moon. The cluster is in fact part of the Norma Spiral Arm of our Galaxy, located about 15,000 l.y. from us. The faint background glow from innumerable unresolved stars is a backdrop to a breathtaking display of 6th- to 10th-magnitude stars. It also includes several dark nebulae, which adds to the three-dimensional impression. Many regard the cluster as truly a showpiece of the sky. Spend a long time observing this jewel!

Messier 16	NGC 6611	$18^h 18.8^m$	−13° 47′	June 26
6.0 m	⊕ 22′	50	II 3 m n	easy

A fine large cluster easily seen with binoculars. It is about 7000 l.y. away, located in the Sagittarius–Carina Spiral Arm of the Galaxy. Its hot O-type stars provide the energy for the Eagle Nebula, within which the cluster is embedded. A very young cluster of only 800,000 years, with a few at 50,000 years old.

Messier 18	NGC 6613	$18^h 19.9^m$	−17° 08′	June 26
6.9 m	⊕ 10′	30	II 3 p n	easy

A small and unremarkable Messier object, and perhaps the most often ignored, this little cluster, containing many 9th-magnitude stars, is still worth observing. Best seen with binoculars or low-power telescopes. A double star is located within the cluster.

Herschel 72	NGC 6633	$18^h 27.7^m$	+06° 34′	June 28
4.6 m	⊕ 27′	25	III 2 m	easy

Bordering on naked-eye visibility, this bright, large but loose cluster is perfect for binoculars and small telescopes. The stars are a lovely bluish-white set against the faint glow of the unresolved members. At the northern periphery of the cluster is a small but nice triple star system.

[1]Located within Sagittarius are numerous open clusters. Only the brightest are listed here.

Messier 25	IC 4725	$18^h 31.6^m$	−19° 15′	June 29
4.6 m	⊕ 32′	40	I 3 m	easy

Visible to the naked eye, this is a pleasing cluster suitable for binocular observation. It contains several star chains and is also noteworthy for small areas of dark nebulosity that seem to blanket out areas within the cluster, but you will need perfect conditions to appreciate these. Unique for two reasons: it is the only Messier object referenced in the Index Catalogue (IC), and is one of the few clusters to contain a cepheid-type variable star – U Sagittarii. The star displays a magnitude change from 6.3 to 7.1 over a period of 6 days and 18 hours.

Caldwell 75	NGC 6124	$16^h 25.6^m$	−40° 40′	May 28
5.8 m	⊕ 29′	75	I 3 r	easy

See May.

Collinder 399	–	$19^h 25.4^m$	+20° 11′	July 13
3.6 m	⊕ 60′	35	III 3 m	easy

See July.

Stock 1	–	$19^h 35.8^m$	+25° 13′	July 16
5.3 m	⊕ 60′	40	III 2 m	easy

See July.

Collinder 413	NGC 6871	$20^h 05.9^m$	+35° 47′	July 23
5.2 m	⊕ 20′	30	II 2 p n	easy

See July.

Caldwell 37	NGC 6885	$20^h 12.0^m$	+26° 29′	July 25
5.9 m	⊕ 7′	25	III 2 p	easy

See July.

Messier 29	NGC 6913	$20^h 23.9^m$	+38° 32′	July 28
6.6 m	⊕ 7′	80	I 2 m n	easy

See July.

Herschel 8	NGC 6940	$20^h 34.6^m$	+28° 18′	July 30
6.3 m	⊕ 32′	75	III 2 r	easy

See July.

Messier 73	NGC 6994	$20^h 59.0^m$	−12° 38′	August 6
8.9 m	⊕ 2.8′	4	IV 1 p	easy

See August.

Messier 39	NGC 7092	$21^h 32.2^m$	+48° 26′	August 14
4.6 m	⊕ 31′	30	III 2 m	easy ©

See August.

Caldwell 16	NGC 7243	$22^h\ 15.3^m$	+49° 53′	August 25
6.4 m	⊕ 21′	40	IV 2 p	easy ©

See August.

Collinder 309	NGC 6192	$16^h\ 40.3^m$	–43° 22′	June 1
8.5 m	⊕ 7′	40	I 2 r	medium

A fine cluster of around three dozen stars. Suitable for telescopes only, as it contains many 11th-magnitude and fainter stars.

Trumpler 26	Harvard 15	$17^h\ 28.5^m$	–29° 29′	June 13
9.5 m	⊕ 17′	45	II 1 m	medium

A nice cluster of bright stars located with many more fainter ones. Best seen at medium magnification in a telescope. Difficult for binoculars.

Messier 26	NGC 6694	$18^h\ 45.2^m$	–09° 24′	July 3
8.0 m	⊕ 14′	30	I 2 m	medium

See July.

Messier 11	NGC 6705	$18^h\ 51.1^m$	–06° 16′	July 4
5.8 m	⊕ 13′	200	I 2 r	medium

See July.

Collinder 392	NGC 6709	$18^h\ 51.5^m$	+10° 21′	July 4
6.7 m	⊕ 13′	45	III 2 m	medium.

See July.

–	NGC 6755	$19^h\ 07.8^m$	+04° 14′	July 8
7.5 m	⊕ 15′	100	II 2 r	medium

See July.

Collinder 402	NGC 6811	$19^h\ 36.9^m$	+46° 23′	July 16
6.8 m	⊕ 21′	65	IV 3 p	medium

See July.

Collinder 403	NGC 6819	$19^h\ 41.3^m$	+40° 11′	July 17
7.3 m	⊕ 10′	100	I 1 r	medium

See July.

Harvard 20	–	$19^h\ 53.1^m$	+18° 20′	July 20
7.7 m	⊕ 10′	20	IV 2 p	medium

See July.

–	IC 1396	$21^h\ 39.1^m$	+57° 30′	August 16
3.7 m	⊕ 50′	40	II m n	medium ©

See August.

Herschel 7	NGC 6520	$18^h\ 03.4^m$	–27° 54′	June 22
7.5 m	⊕ 7′	30	I 2 r n	difficult

This cluster, although fairly bright, is situated within the Great Sagittarius Star Cloud, and thus makes positive identification difficult. It contains about three dozen faint stars and locating it is a test of an observer's skill.

–	NGC 6791	$19^h\ 20.7^m$	+37° 51′	July 12
9.5 m	⊕ 15′	250	I 2 r	difficult

See July.

–	NGC 6939	$20^h\ 31.4^m$	+60° 38′	July 30
7.8 m	⊕ 7.0′	50	I 1 m	difficult ©

See July.

Collinder 445	IC 1434	$22^h\ 10.5^m$	+52° 50′	August 24
9.0 m	⊕ 7′	40	II 1 p	difficult ©

See August.

July

Collinder 399	–	$19^h\ 25.4^m$	+20° 11′	July 13
3.6 m	⊕ 60′	35	III 3 m	easy

Also know as The Coathanger or Brocchi's Cluster. Often overlooked by observers, this is a large, dissipated cluster easily seen with binoculars; indeed, several of the brightest members should be visible with the naked eye. It contains a nice orange-tinted star and several blue tinted stars. Its 3 dozen members are set against a background filled with fainter stars. Well worth observing during warm summer evenings.

Stock 1	–	$19^h\ 35.8^m$	+25° 13′	July 16
5.3 m	⊕ 60′	40	III 2 m	easy

An enormous cluster best seen in binoculars, although it is difficult to estimate where the cluster ends and the background stars begin.

Collinder 413	NGC 6871	$20^h\ 05.9^m$	+35° 47′	July 23
5.2 m	⊕ 20′	30	II 2 p n	easy

A nice cluster that is easily seen in small telescopes. It appears as a enhancement of the background Milky Way. Binoculars will show several stars of 7th to 9th magnitude surrounded by fainter members.

Caldwell 37	NGC 6885	$20^h\ 12.0^m$	+26° 29′	July 25
5.9 m	⊕ 7′	25	III 2 p	easy

An irregular cluster containing many 9th- to 13th-magnitude stars. One of a number of objects which apparently have been unknown to the amateur astronomer. Visible in binoculars as a hazy blur. It is located next to (within?) the cluster NGC 6882 and thus is not easily delineated. An old cluster with an estimated age of around 1 billion years, with recent measurements by Hipparchos placing it at a distance of 1140 l.y.

Messier 29	NGC 6913	20h 23.9m	+38° 32′	July 28
6.6 m	⊕ 7′	80	I 2 m n	easy

A very small cluster and one of only two Messier objects in Cygnus. It contains only about a dozen stars visible with small instruments, and even then benefits from a low magnification. However, studies show that it contains many more bright B0-type giant stars, which are obscured by dust. Without this, the cluster would be a very spectacular object.

Herschel 8	NGC 6940	20h 34.6m	+28° 18′	July 30
6.3 m	⊕ 32′	75	III 2 r	easy

A beautiful cluster, which although visible in binoculars, is best appreciated with a telescope. It contains the semi-variable star FG Vulpeculae near its centre that has a nice reddish-orange tint.

Caldwell 76	NGC 6231	16h 54.0m	–41° 48′	June 5
2.6 m	⊕ 14′	100	I 3 p	easy

See June.

Trumpler 24	Harvard 12	16h 57.0m	–40° 40′	June 5
8.6 m	⊕ 60′	100	IV 2 p n	easy

See June.

Messier 6	NGC 6405	17h 40.1m	–32° 13′	June 16
4.2 m	⊕ 33′	100	II 3 r	easy

See June.

–	IC 4665	17h 46.3m	+05° 43′	June 18
4.2 m	⊕ 40′	30	III 2 m	easy

See June.

Messier 7	NGC 6475	17h 53.9m	–34° 49′	June 20
3.3 m	⊕ 80′	80	I 3 r	easy

See June.

Messier 23	NGC 6494	17h 56.8m	–19° 01′	June 20
5.5 m	⊕ 27′	100	II 2 r	easy

See June.

Messier 21	NGC 6531	18h 04.6m	–22° 30′	June 22
5.9 m	⊕ 14′	60	I 3 r	easy

See June.

Messier 24	–	18h 16.5m	–18° 50′	June 25
2.5 m	⊕ 95′× 35′	–		easy

See June.

Messier 16	NGC 6611	18^h 18.8^m	–13° 47′	June 26
6.0 m	⊕ 22′	50	II 3 m n	easy
See June.				

Messier 18	NGC 6613	18^h 19.9^m	–17° 08′	June 26
6.9 m	⊕ 10′	30	II 3 p n	easy
See June.				

Herschel 72	NGC 6633	18^h 27.7^m	+06° 34′	June 28
4.6 m	⊕ 27′	25	III 2 m	easy
See June.				

Messier 25	IC 4725	18^h 31.6^m	–19° 15′	June 29
4.6 m	⊕ 32′	40	I 3 m	easy
See June.				

Caldwell 75	NGC 6124	16^h 25.6^m	–40° 40′	May 28
5.8 m	⊕ 29′	75	I 3 r	easy
See May.				

Messier 73	NGC 6994	20^h 59.0^m	–12° 38′	August 6
8.9 m	⊕ 2.8′	4	IV 1 p	easy
See August.				

Messier 39	NGC 7092	21^h 32.2^m	+48° 26′	August 14
4.6 m	⊕ 31′	30	III 2 m	easy ©
See August.				

Caldwell 16	NGC 7243	22^h 15.3^m	+49° 53′	August 25
6.4 m	⊕ 21′	40	IV 2 p	easy ©
See August.				

Messier 52	NGC 7654	23^h 24.2^m	+61° 35′	September 11
6.9 m	⊕ 12′	100	I 2 r	easy ©
See September.				

Blanco 1	–	00^h 04.3^m	–29° 56′	September 22
4.5 m	⊕ 90′	30	III 2 m	easy
See September.				

Herschel 78	NGC 129	00^h 29.9^m	+60° 14′	September 28
6.5 m	⊕ 21′	30	IV 2 p	easy ©
See September.				

Messier 26	NGC 6694	$18^h 45.2^m$	−09° 24′	July 3
8.0 m	⊕ 14′	30	I 2 m	medium

This is a small but rich cluster containing 11th- and 12th-magnitude stars, set against a haze of unresolved stars. This makes it unsuitable for binoculars, as it will be only a hazy small patch of light, and so apertures of 10 cm and more will be needed to appreciate it in any detail.

Messier 11	NGC 6705	$18^h 51.1^m$	−06° 16′	July 4
5.8 m	⊕ 13′	200	I 2 r	medium

Also known as the Wild Duck Cluster, this is a gem of an object. Although it is visible with binoculars as a small, tightly compact group, reminiscent of a globular cluster, they do not do it justice. With telescopes, however, its full majesty becomes apparent. Containing many hundreds of stars, it is a very impressive cluster. It takes high magnification well, where many more of its 700 members become visible. At the top of the cluster is a glorious pale yellow tinted star. The British amateur astronomer Michael Hurrell called this "one of the most impressive and beautiful celestial objects in the entire sky".

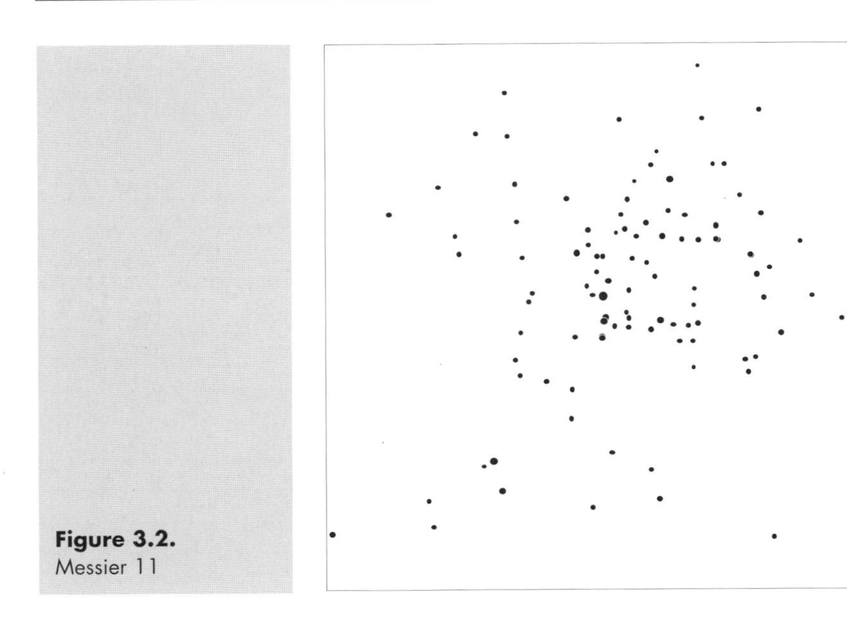

Figure 3.2. Messier 11

Collinder 392	NGC 6709	$18^h 51.5^m$	+10° 21′	July 4
6.7 m	⊕ 13′	45	III 2 m	medium.

A difficult object to locate for binoculars as it will be unresolvable. Nevertheless this presents a good challenge for you to hone your observing skill. For small telescopes it presents a nice rich cluster.

–	NGC 6755	$19^h 07.8^m$	+04° 14′	July 8
7.5 m	⊕ 15′	100	II 2 r	medium

An easily found cluster in small telescopes, standing out from the background star field.

Collinder 402	NGC 6811	$19^h 36.9^m$	+46° 23′	July 16
6.8 m	⊕ 21′	65	IV 3 p	medium

One of many clusters in Cygnus, this is a coarse open cluster of many 10th-magnitude and fainter stars. The cluster has caught the attention of many amateurs, as it has been described as looking like a "smoke ring". The stars resemble the ring, while the interior is very dark. Apparently, the feature is easier to see with a small telescope than with a large one.

Collinder 403	NGC 6819	$19^h 41.3^m$	+40° 11′	July 17
7.3 m	⊕ 10′	100	I 1 r	medium

A rich cluster located within and contrasting with, the Milky Way. Contains many 11th-magnitude stars, and thus an observing challenge. The cluster is very old at over 3 billion years.

Harvard 20	–	$19^h 53.1^m$	+18° 20′	July 20
7.7 m	⊕ 10′	20	IV 2 p	medium

A somewhat difficult binocular object, as the stars are of 12th and 13th magnitude, and spread out without any noticeable concentration. The two 9th-magnitude members are easily spotted, however.

Collinder 309	NGC 6192	$16^h 40.3^m$	–43° 22′	June 1
8.5 m	⊕ 7′	40	I 2 r	medium

See June.

Trumpler 26	Harvard 15	$17^h 28.5^m$	–29° 29′	June 13
9.5 m	⊕ 17′	45	II 1 m	medium

See June.

–	IC 1396	$21^h 39.1^m$	+57° 30′	August 16
3.7 m	⊕ 50′	40	II m n	medium ©

See August.

Herschel 69	NGC 7686	$23^h 30.2^m$	+49° 08′	September 13
5.6 m	⊕ 14′	20	IV 1 p	medium ©

See September.

Herschel 30	NGC 7789	$23^h 57.0^m$	+56° 44′	September 20
6.9 m	⊕ 15′	300	II 1 r	medium ©

See September.

King 14	–	$00^h 31.9^m$	+63° 10′	September 28
8.5 m	⊕ 7′	20	III 2 p	medium ©

See September.

–	NGC 6791	$19^h 20.7^m$	+37° 51′	July 12
9.5 m	⊕ 15′	250	I 2 r	difficult

A rich cluster of faint stars. It contains many faint 11th-magnitude stars and so poses an observing challenge.

–	NGC 6939	$20^h 31.4^m$	+60° 38′	July 30
7.8 m	⊕ 7.0′	50	I 1 m	difficult ©

A moderately bright and small cluster, unresolvable in binoculars. A challenge as the brightest member is only of 11.9 magnitude. In telescopes of aperture 10 cm, it will appear as a small hazy spot with just a few very faint stars resolved.

Herschel 7	NGC 6520	$18^h 03.4^m$	–27° 54′	June 22
7.5 m	⊕ 7′	30	I 2 r n	difficult

See June.

Collinder 445	IC 1434	$22^h 10.5^m$	+52° 50′	August 24
9.0 m	⊕ 7′	40	II 1 p	difficult ©

See August.

King 12	–	$23^h 53.0^m$	+61° 58′	September 19
9.0 m	⊕ 2′	15	I 2 p	difficult ©

See September.

Herschel 35	NGC 136	$00^h 31.5^m$	+61° 32′	September 28
– m	⊕ 1.2′	20	II 2 p	difficult ©

See September.

August

Messier 73	NGC 6994	$20^h 59.0^m$	–12° 38′	August 6
8.9 m	⊕ 2.8′	4	IV 1 p	easy

Something of an enigma. It shouldn't really be classified as an open cluster, as it consists of only four stars! Perhaps originally catalogued when Messier was having a bad day. Still nice, though. A small grouping of stars like this is often called an *asterism*. At the time of writing no research has been done on the stars, and it is still not known whether they are related to each other, or are the result of a chance alignment.

Messier 39	NGC 7092	$21^h 32.2^m$	+48° 26′	August 14
4.6 m	⊕ 31′	30	III 2 m	easy ©

A nice cluster in binoculars, it lies at a distance of 840 l.y. About two dozen stars are visible, ranging from 7th to 9th magnitude. What makes this cluster so distinctive is the lovely colour of the stars – steely blue – and the fact that it is nearly perfectly symmetrical, having a triangular shape. There is also a nice double star at the centre of the cluster.

Caldwell 16	NGC 7243	22h 15.3m	+49° 53′	August 25
6.4 m	⊕ 21′	40	IV 2 p	easy ©

Set against the backdrop of the Milky Way, this large, irregular cluster nevertheless stands out quite well. Several of the stars are visible in binoculars, but the remainder blur in the background star field. A nice object in an otherwise empty part of the sky – if you overlook the fact that it is located within the Milky Way!

Collinder 399	–	19h 25.4m	+20° 11′	July 13
3.6 m	⊕ 60′	35	III 3 m	easy

See July.

Stock 1	–	19h 35.8m	+25° 13′	July 16
5.3 m	⊕ 60′	40	III 2 m	easy

See July.

Collinder 413	NGC 6871	20h 05.9m	+35° 47′	July 23
5.2 m	⊕ 20′	30	II 2 p n	easy

See July.

Caldwell 37	NGC 6885	20h 12.0m	+26° 29′	July 25
5.9 m	⊕ 7′	25	III 2 p	easy

See July.

Messier 29	NGC 6913	20h 23.9m	+38° 32′	July 28
6.6 m	⊕ 7′	80	I 2 m n	easy

See July.

Herschel 8	NGC 6940	20h 34.6m	+28° 18′	July 30
6.3 m	⊕ 32′	75	III 2 r	easy

See July.

Caldwell 76	NGC 6231	16h 54.0m	–41° 48′	June 5
2.6 m	⊕ 14′	100	I 3 p	easy

See June.

Trumpler 24	Harvard 12	16h 57.0m	–40° 40′	June 5
8.6 m	⊕ 60′	100	IV 2 p n	easy

See June.

Messier 6	NGC 6405	17h 40.1m	–32° 13′	June 16
4.2 m	⊕ 33′	100	II 3 r	easy

See June.

–	IC 4665	17h 46.3m	+05° 43′	June 18
4.2 m	⊕ 40′	30	III 2 m	easy

See June.

Messier 7	NGC 6475	17^h 53.9^m	–34° 49′	June 20
3.3 m	⊕ 80′	80	I 3 r	easy
See June.				

Messier 23	NGC 6494	17^h 56.8^m	–19° 01′	June 20
5.5 m	⊕ 27′	100	II 2 r	easy
See June.				

Messier 21	NGC 6531	18^h 04.6^m	–22° 30′	June 22
5.9 m	⊕ 14′	60	I 3 r	easy
See June.				

Messier 24	–	18^h 16.5^m	–18° 50′	June 25
2.5 m	⊕ 95′ × 35′	–		easy
See June.				

Messier 16	NGC 6611	18^h 18.8^m	–13° 47′	June 26
6.0 m	⊕ 22′	50	II 3 m n	easy
See June.				

Messier 18	NGC 6613	18^h 19.9^m	–17° 08′	June 26
6.9 m	⊕ 10′	30	II 3 p n	easy
See June.				

Herschel 72	NGC 6633	18^h 27.7^m	+06° 34′	June 28
4.6 m	⊕ 27′	25	III 2 m	easy
See June.				

Messier 25	IC 4725	18^h 31.6^m	–19° 15′	June 29
4.6 m	⊕ 32′	40	I 3 m	easy
See June.				

Messier 52	NGC 7654	23^h 24.2^m	+61° 35′	September 11
6.9 m	⊕ 12′	100	I 2 r	easy ©
See September.				

Blanco 1	–	00^h 04.3^m	–29° 56′	September 22
4.5 m	⊕ 90′	30	III 2 m	easy
See September.				

Herschel 78	NGC 129	00^h 29.9^m	+60° 14′	September 28
6.5 m	⊕ 21′	30	IV 2 p	easy ©
See September.				

Caldwell 13	NGC 457	01h 19.1m	+58° 20	October 11
6.4 m	⊕ 13′	80	I 3 r	easy ©
See October.				

Collinder 33	NGC 752	01h 57.8m	+37° 41′	October 20
5.7 m	⊕ 45′	77	III 1 m	easy
See October.				

Stock 2	02h 15.0m	+59°	16′	October 25
4.4 m	⊕ 60′	50	III 1 m	easy ©
See October.				

Caldwell 14	NGC 869	02h 19.0m	+57° 09′	October 26
5.3 m	⊕ 29′	200	I 3 r	easy ©
See October.				

–	IC 1396	21h 39.1m	+57° 30′	August 16
3.7 m	⊕ 50′	40	II m n	medium ©
Although a telescope of at least 20 cm is needed to really appreciate this cluster, it is nevertheless worth searching out. It lies south of Herschel's Garnet Star and is rich but compressed. What makes this so special, however, is that it is cocooned within a very large and bright nebula.				

Messier 26	NGC 6694	18h 45.2m	–09° 24′	July 3
8.0 m	⊕ 14′	30	I 2 m	medium
See July.				

Messier 11	Ngc 6705	18h 51.1m	–06° 16′	July 4
5.8 M	⊕ 13′	200	I 2 r	medium
See July.				

Collinder 392	NGC 6709	18h 51.5m	+10° 21′	July 4
6.7 m	⊕ 13′	45	III 2 m	medium
See July.				

–	NGC 6755	19h 07.8m	+04° 14′	July 8
7.5 m	⊕ 15′	100	II 2 r	medium
See July.				

Collinder 402	NGC 6811	19h 36.9m	+46° 23′	July 16
6.8 m	⊕ 21′	65	IV 3 p	medium
See July.				

Collinder 403	NGC 6819	$19^h 41.3^m$	+40° 11′	July 17
7.3 m	⊕ 10′	100	I 1 r	medium

Harvard 20	–	$19^h 53.1^m$	+18° 20′	July 20
7.7 m	⊕ 10′	20	IV 2 p	medium

See July.

Collinder 309	NGC 6192	$16^h 40.3^m$	–43° 22′	June 1
8.5 m	⊕ 7′	40	I 2 r	medium

See June.

Trumpler 26	Harvard 15	$17^h 28.5^m$	–29° 29′	June 13
9.5 m	⊕ 17′	45	II 1 m	medium

See June.

Herschel 69	NGC 7686	$23^h 30.2^m$	+49° 08′	September 13
5.6 m	⊕ 14′	20	IV 1 p	medium ©

See September.

Herschel 30	NGC 7789	$23^h 57.0^m$	+56° 44′	September 20
6.9 m	⊕ 15′	300	II 1 r	medium ©

See September.

King 14	–	$00^h 31.9^m$	+63° 10′	September 28
8.5 m	⊕ 7′	20	III 2 p	medium ©

See September.

Herschel 64	NGC 381	$01^h 08.3^m$	+61° 35′	October 8
9.3 m	⊕ 6′	50	III 2 p	medium ©

See October.

Collinder 445	IC 1434	$22^h 10.5^m$	+52° 50′	August 24
9.0 m	⊕ 7′	40	II 1 p	difficult ©

Located within the Milky Way, this is a large but irregular cluster of over 70 stars of 10th magnitude and fainter. Try using a high magnification of, say, 150 to 200×, and also use averted vision. These two factors will almost certainly improve this cluster.

–	NGC 6791	$19^h 20.7^m$	+37° 51′	July 12
9.5 m	⊕ 15′	250	I12 r	difficult

See July.

–	NGC 6939	$20^h 31.4^m$	+60° 38′	July 30
7.8 m	⊕ 7.0′	50	I 1 m	difficult ©

See July.

Herschel 7	NGC 6520	18h 03.4m	−27° 54′	June 22
7.5 m	⊕ 7′	30	I 2 r n	difficult
See June.				

King 12	–	23h 53.0m	+61° 58′	September 19
9.0 m	⊕ 2′	15	I 2 p	difficult ©
See September.				

Herschel 35	NGC 136	00h 31.5m	+61° 32′	September 28
–m	⊕ 1.2′	20	II 2 p	difficult ©
See September.				

Messier 103	NGC 581	01h 33.2m	+60° 42′	October 14
7.4 m	⊕ 6′	25	III 2 p	difficult ©
See October.				

Trumpler 1	Collinder 15	01h 35.7m	+61° 17′	October 15
8.1 m	⊕ 4.5′	20	I 3 p	difficult ©
See October.				

Herschel 49	NGC 637	01h 42.9m	+64° 00′	October 17
8.2 m	⊕ 3.5′	20	I 3 p	difficult ©
See October.				

September

Messier 52	NGC 7654	23h 24.2m	+61° 35′	September 11
6.9 m	⊕ 12′	100	I 2 r	easy ©
A small, rich, and fairly bright cluster. One of the densest north of the celestial equator. Several stars are visible in binoculars, but telescopic apertures are needed to fully appreciate this cluster. It is one of the few clusters that show a distinct colour. Many observers report a faint blue tint to the group, and this along with a fine topaz-coloured (blue) star and several nice yellow and blue stars make it a very nice object to observe. Apparently, it has a star density of the order of 50 stars per cubic parsec!				

Blanco 1	–	00h 04.3m	−29° 56′	September 22
4.5 m	⊕ 90′	30	III 2 m	easy
Located close to the south *galactic pole*, this is an ill-defined and very large cluster. Easily visible in binoculars.				

Herschel 78	NGC 129	00h 29.9m	+60° 14′	September 28
6.5 m	⊕ 21′	30	IV 2 p	easy ©
A bright, open cluster. Irregularly scattered and uncompressed, making it difficult to distinguish from the background. Up to a dozen stars can be seen with binoculars, but many more are visible under telescopic aperture. Under good observing conditions and using averted vision, the unresolved background stars of the cluster can be seen as a faint glow.				

Messier 73	NGC 6994	20ʰ 59.0ᵐ	–12° 38′	August 6
8.9 m	⊕ 2.8′	4	IV 1 p	easy
See August.				

Messier 39	NGC 7092	21ʰ 32.2ᵐ	+48° 26′	August 14
4.6 m	⊕ 31′	30	III 2 m	easy ©
See August.				

Caldwell 16	NGC 7243	22ʰ 15.3ᵐ	+49° 53′	August 25
6.4 m	⊕ 21′	40	IV 2 p	easy ©
See August.				

Collinder 399	–	19ʰ 25.4ᵐ	+20° 11′	July 13
3.6 m	⊕ 60′	35	III 3 m	easy
See July.				

Stock 1	–	19ʰ 35.8ᵐ	+25° 13′	July 16
5.3 m	⊕ 60′	40	III 2 m	easy
See July.				

Collinder 413	NGC 6871	20ʰ 05.9ᵐ	+35° 47′	July 23
5.2 m	⊕ 20′	30	II 2 p n	easy
See July.				

Caldwell 37	NGC 6885	20ʰ 12.0ᵐ	+26° 29′	July 25
5.9 m	⊕ 7′	25	III 2 p	easy
See July.				

Messier 29	NGC 6913	20ʰ 23.9ᵐ	+38° 32′	July 28
6.6 m	⊕ 7′	80	I 2 m n	easy
See July.				

Herschel 8	NGC 6940	20ʰ 34.6ᵐ	+28° 18′	July 30
6.3 m	⊕ 32′	75	III 2 r	easy
See July.				

Caldwell 13	NGC 457	01ʰ 19.1ᵐ	+58° 20	October 11
6.4 m	⊕ 13′	80	I 3 r	easy ©
See October.				

Collinder 33	NGC 752	01ʰ 57.8ᵐ	+37° 41′	October 20
5.7 m	⊕ 45′	77	III 1 m	easy
See October.				

Stock 2		02h 15.0m	+59° 16′	October 25
4.4 m	⊕ 60′	50	III 1 m	easy ©

See October.

Caldwell 14	NGC 869	02h 19.0m	+57° 09′	October 26
5.3 m	⊕ 29′	200	I 3 r	easy ©

See October.

Messier 34	NGC 1039	02h 42.0m	+42° 47′	November 1
5.2 m	⊕ 35′	60	II 3 m	easy

See November.

Stock 23	–	03h 16.3m	+60° 02′	November 9
5.6 m	⊕ 18′	25	III 3 p n	easy ©

See November.

Messier 45	Melotte 22	03h 47.0m	+24° 07′	November 17
1.2 m	⊕ 110′	100	I 3 r	easy

See November.

Caldwell 41	Melotte 25	04h 27.0m	+16° 00′	November 27
0.5 m	⊕ 330′	40	II 3 m	easy

See November.

Herschel 69	NGC 7686	23h 30.2m	+49° 08′	September 13
5.6 m	⊕ 14′	20	IV 1 p	medium ©

A sparse and widely dispersed cluster containing many 10th- and 11th-magnitude stars. Best seen with large-aperture telescopes.

Herschel 30	NGC 7789	23h 57.0m	+56° 44′	September 20
6.9 m	⊕ 15′	300	II 1 r	medium ©

Visible as a hazy spot to the naked eye, and even with small binoculars is never fully resolvable; it is believed to be one of the major omissions from the Messier catalogue. Through a telescope it is seen as a very rich and compressed cluster. With large aperture, the cluster is superb and has been likened to a field of scattered diamond dust. Contains hundreds stars of 10th magnitude and fainter.

King 14	–	00h 31.9m	+63° 10′	September 28
8.5 m	⊕ 7′	20	III 2 p	medium ©

Often overlooked, this cluster is a faint but rich object. With a 10 cm aperture telescope, several stars can be resolved set against a faint glow.

Messier 26	NGC 6694	18h 45.2m	–09° 24′	July 3
8.0 m	⊕ 14′	30	I 2 m	medium

See July.

Messier 11	NGC 6705	$18^h 51.1^m$	–06° 16′	July 4
5.8 m	⊕ 13′	200	I 2 r	medium

See July.

Collinder 392	NGC 6709	$18^h 51.5^m$	+10° 21′	July 4
6.7 m	⊕ 13′	45	III 2 m	medium.

See July.

–	NGC 6755	$19^h 07.8^m$	+04° 14′	July 8
7.5 m	⊕ 15′	100	II 2 r	medium

See July.

Collinder 402	NGC 6811	$19^h 36.9^m$	+46° 23′	July 16
6.8 m	⊕ 21′	65	IV 3 p	medium

See July.

Collinder 403	NGC 6819	$19^h 41.3^m$	+40° 11′	July 17
7.3 m	⊕ 10′	100	I 1 r	medium

See July.

Harvard 20	–	$19^h 53.1^m$	+18° 20′	July 20
7.7 m	⊕ 10′	20	IV 2 p	medium

See July.

–	IC 1396	$21^h 39.1^m$	+57° 30′	August 16
3.7 m	⊕ 50′	40	II m n	medium ©

See August.

Herschel 64	NGC 381	$01^h 08.3^m$	+61° 35′	October 8
9.3 m	⊕ 6′	50	III 2 p	medium ©

See October.

Herschel 47	NGC 1502	$04^h 07.7^m$	+62° 20′	November 22
5.7 m	⊕ 8′	45	II 3 p	medium ©

See November.

King 12	–	$23^h 53.0^m$	+61° 58′	September 19
9.0 m	⊕ 2′	15	I 2 p	difficult ©

A very faint cluster containing many 10th-, 11th- and 12th-magnitude stars. A challenge to deep-sky observers.

Herschel 35	NGC 136	$00^h 31.5^m$	+61° 32′	September 28
–m	å⊕ 1.2′	20	II 2 p	difficult ©

A very small cluster, looking like a tiny sprinkling of diamond dust. Although it can be observed with a 15 cm telescope, it needs a very large aperture of at least 20 cm to be fully resolvable.

Collinder 445	IC 1434	22^h 10.5^m	+52° 50′	August 24
9.0 m	⊕ 7′	40	II 1 p	difficult ©
See August.				

–	NGC 6791	19^h 20.7^m	+37° 51′	July 12
9.5 m	⊕ 15′	250	I 2 r	difficult
See July.				

–	NGC 6939	20^h 31.4^m	+60° 38′	July 30
7.8 m	⊕ 7.0′	50	I 1 m	difficult ©
See July.				

Messier 103	NGC 581	01^h 33.2^m	+60° 42′	October 14
7.4 m	⊕ 6′	25	III 2 p	difficult ©
See October.				

Trumpler 1	Collinder 15	01^h 35.7^m	+61° 17′	October 15
8.1 m	⊕ 4.5′	20	I 3 p	difficult ©
See October.				

Herschel 49	NGC 637	01^h 42.9^m	+64° 00′	October 17
8.2 m	⊕ 3.5′	20	I 3 p	difficult ©
See October.				

October

Caldwell 13	NGC 457	01^h 19.1^m	+58° 20	October 11
6.4 m	⊕ 13′	80	I 3 r	easy ©

This is a wonderful cluster, and can be considered one of the finest in Cassiopeia. Easily seen in binoculars as two southward-arcing chains of stars, surrounded by many fainter components. The gorgeous blue and yellow double φ Cass and a lovely red star, HD 7902, lie within the cluster. Located at a distance of about 8000 l.y., this young cluster is located within the Perseus Spiral Arm of our Galaxy.

Collinder 33	NGC 752	01^h 57.8^m	+37° 41′	October 20
5.7 m	⊕ 45′	77	III 1 m	easy

Best seen in binoculars, or even at low powers in a telescope, this is a large, loosely structured group of stars containing many chains and double stars. Lies about 5° south-south-west of γ Andromedae. Often underrated by observing guides, it is worth seeking out. It is a cluster of intermediate age.

Stock 2		02^h 15.0^m	+59° 16′	October 25
4.4 m	⊕ 60′	50	III 1 m	easy ©

Another undiscovered and passed-over cluster! Wonderful in binoculars and small telescopes. It lies 2° north of its more famous cousin the Double Cluster. At nearly a degree across it contains over fifty 8th-magnitude and fainter stars. Well worth seeking out.

Caldwell 14	NGC 869	02h 19.0m	+57° 09′	October 26
5.3 m	⊕ 29′	200	I 3 r	easy ©
–	NGC 884	02h 22.4m	+57° 07′	
6.1 m	⊕ 29′	115	II 2 p	

Glorious! The famous Double Cluster in Perseus should be on every amateur's observing schedule and is a highlight of the northern hemisphere winter sky. Strangely, never catalogued by Messier. Visible to the naked eye and best seen using a low-power, wide-field optical system. But whatever system is used, the views are marvellous. NGC 869 has around 200 members, while NGC 884 has about 150. Both are composed of A-type and B-type supergiant stars with many nice red giant stars. However, the systems are dissimilar; NGC 869 is 5.6 million years old (at a distance of 7200 l.y.), whereas NGC 884 is younger at 3.2 million (at a distance of 7500 l.y.). But be advised that in astrophysics, especially distance and age determination, there are very large errors!

Also, it was found that nearly half the stars are variables of the type Be, indicating that they are young stars with possible circumstellar discs of dust. Both are part of the Perseus OB1 Association[2] from which the Perseus Spiral Arm of the Galaxy has been named. Don't rush these clusters, but spend a long time observing both of them and the background star fields.

Messier 52	NGC 7654	23h 24.2m	+61° 35′	September 11
6.9 m	⊕ 12′	100	I 2 r	easy ©

See September.

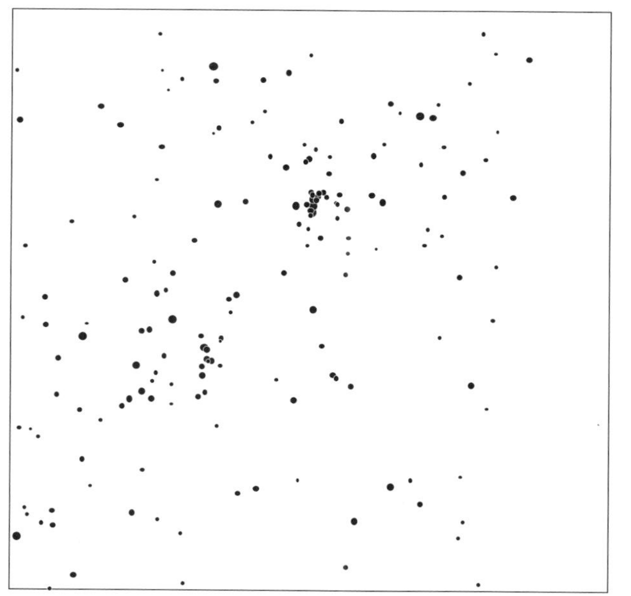

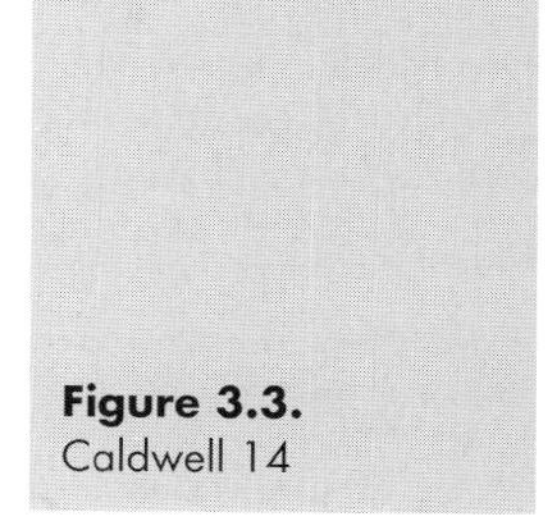

Figure 3.3. Caldwell 14

[2]See Section three in this chapter for a discussion on stellar associations.

Blanco 1	–	$00^h\ 04.3^m$	–29° 56′	September 22
4.5 m	⊕ 90′	30	III 2 m	easy
See September.				

Herschel 78	NGC 129	$00^h\ 29.9^m$	+60° 14′	September 28
6.5 m	⊕ 21′	30	IV 2 p	easy ©
See September.				

Messier 73	NGC 6994	$20^h\ 59.0^m$	–12° 38′	August 6
8.9 m	⊕ 2.8′	4	IV 1 p	easy
See August.				

Messier 39	NGC 7092	$21^h\ 32.2^m$	+48° 26′	August 14
4.6 m	⊕ 31′	30	III 2 m	easy ©
See August.				

Caldwell 16	NGC 7243	$22^h\ 15.3^m$	+49° 53′	August 25
6.4 m	⊕ 21′	40	IV 2 p	easy ©
See August.				

Messier 34	NGC 1039	$02^h\ 42.0^m$	+42° 47′	November 1
5.2 m	⊕ 35′	60	II 3 m	easy
See November.				

Stock 23	–	$03^h\ 16.3^m$	+60° 02′	November 9
5.6 m	⊕ 18′	25	III 3 p n	easy ©
See November.				

Messier 45	Melotte 22	$03^h\ 47.0^m$	+24° 07′	November 17
1.2 m	⊕ 110′	100	I 3 r	easy
See November.				

Caldwell 41	Melotte 25	$04^h\ 27.0^m$	+16° 00′	November 27
0.5 m	⊕ 330′	40	II 3 m	easy
See November.				

Herschel 64	NGC 381	$01^h\ 08.3^m$	+61° 35′	October 8
9.3 m	⊕ 6′	50	III 2 p	medium ©
A faint cluster, but rich and compressed. Can be resolved with an aperture of 10 cm, but with medium aperture, of say, 20 to 25 cm, over 60 stars of 12th and 13th magnitude become visible.				

–	IC 1396	$21^h\ 39.1^m$	+57° 30′	August 16
3.7 m	⊕ 50′	40	II m n	medium ©
See August.				

Herschel 69	NGC 7686	23^h 30.2^m	+49° 08′	September 13
5.6 m	⊕ 14′	20	IV 1 p	medium ©

See September.

Herschel 30	NGC 7789	23^h 57.0^m	+56° 44′	September 20
6.9 m	⊕ 15′	300	II 1 r	medium ©

See September.

King 14	–	00^h 31.9^m	+63° 10′	September 28
8.5 m	⊕ 7′	20	III 2 p	medium ©

See September.

Herschel 47	NGC 1502	04^h 07.7^m	+62° 20′	November 22
5.7 m	⊕ 8′	45	II 3 p	medium ©

See November.

Messier 103	NGC 581	01^h 33.2^m	+60° 42′	October 14
7.4 m	⊕ 6′	25	III 2 p	difficult ©

A nice rich cluster of stars, which is resolvable in small binoculars. Using progressively larger apertures, more and more of the cluster will be revealed (as with most clusters). It has a distinct fan shape, and the star at the top of the fan is Struve 131, a double star with colours reported as pale yellow and blue. Close by is also a rather nice, pale, red-tinted star. The cluster is the last object in Messier's original catalogue.

Trumpler 1	Collinder 15	01^h 35.7^m	+61° 17′	October 15
8.1 m	⊕ 4.5′	20	I 3 p	difficult ©

Even with a telescope of 12 cm aperture, this small and tightly compressed cluster will be a challenge.

Herschel 49	NGC 637	01^h 42.9^m	+64° 00′	October 17
8.2 m	⊕ 3.5′	20	I 3 p	difficult ©

A faint and very condensed cluster. About 10 stars can be seen with a telescope of at least 10 cm aperture, but many more will remain unresolved.

King 12	–	23^h 53.0^m	+61° 58′	September 19
9.0 m	⊕ 2′	15	I 2 p	difficult ©

See September.

Herschel 35	NGC 136	00^h 31.5^m	+61° 32′	September 28
–m	⊕ 1.2′	20	II 2 p	difficult ©

See September.

Collinder 445	IC 1434	22^h 10.5^m	+52° 50′	August 24
9.0 m	⊕ 7′	40	II 1 p	difficult ©

See August.

November

Messier 34	NGC 1039	$02^h 42.0^m$	+42° 47′	November 1
5.2 m	⊕ 35′	60	II 3 m	easy

A nice cluster easily found, about the same size as the full Moon. It can be glimpsed with the naked eye and is best seen with medium-sized binoculars, as a telescope will spread out the cluster and so lessen its impact. At the centre of the cluster is the double star H1123, both members being 8th-magnitude, and of type A0. The pure-white stars are very concentrated toward the cluster's centre, while the fainter members disperse toward its periphery. Thought to be about 200 million years old, lying at a distance of 1500 l.y.

Stock 23	–	$03^h 16.3^m$	+60° 02′	November 9
5.6 m	⊕ 18′	25	III 3 p n	easy ©

A little known cluster on the border of Camelopardalis–Cassiopeia. Binoculars reveal several stars, best viewed in medium-aperture telescopes. It is bright and large but spread out.

Messier 45	Melotte 22	$03^h 47.0^m$	+24° 07′	November 17
1.2 m	⊕ 110′	100	I 3 r	easy

Outstanding! Without a doubt the sky's premier star cluster. The Seven Sisters or Pleiades, is beautiful however you observe it – naked-eye, through binoculars or with a telescope. To see all the members at one go will require binoculars or a rich-field telescope. Consisting of over 100 stars, spanning an area four times that of the full Moon, it will never cease to amaze. It is often stated that from an urban location 6 to 7 stars may be glimpsed with the naked eye. However, it may come as a surprise to many of you that it has 10 stars brighter than 6th magnitude, and that seasoned amateurs with perfect conditions have reported 18 being visible with the naked eye. It lies at a distance of 410 l.y., is about 20 million years old (although some report it as 70 million) and is the 4th-nearest cluster. It contains many stunning blue and white B-type giants.

The cluster contains many double and multiple stars. Under perfect conditions with exceptionally clean optics, the faint nebula NGC 1435, the Merope Nebula surrounding the star of the same name (Merope – 23 Tauri), can be glimpsed, and was described by W. Tempel in 1859 as "a breath on a mirror". However, this and the nebulosity associated with the other Pleiades are not, as they were once thought, to be the remnants of the original progenitor dust and gas cloud. The cluster is just passing through an edge of the Taurus Dark Cloud Complex. It is moving through space at a velocity of about 40 kilometres a second, so by 32,000 AD it will have moved an angular distance equal to that of the full Moon. The cluster contains the stars Pleione, Atlas, Alcyone, Merope, Maia, Electra, Celaeno, Taygeta and Asterope. A true celestial showpiece.

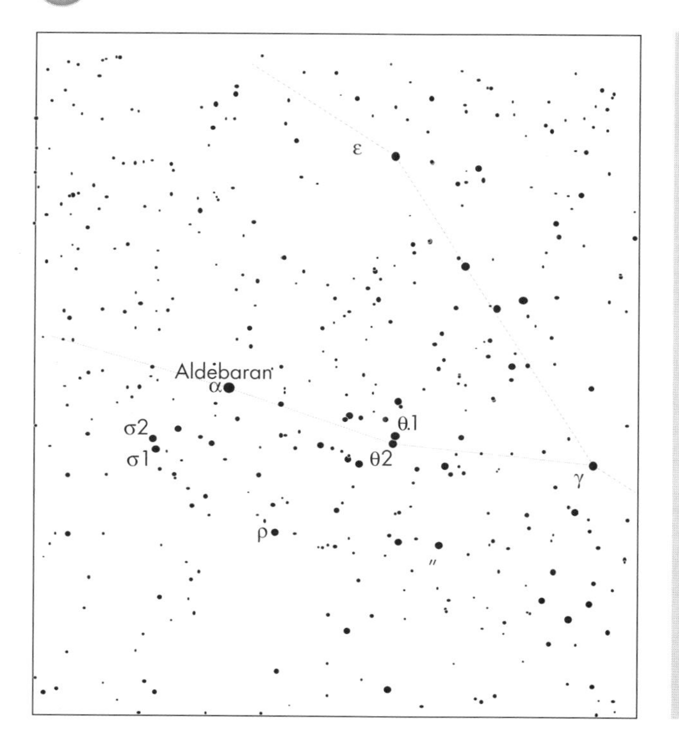

Figure 3.4.
Caldwell 41

Caldwell 41	Melotte 25	04^h 27.0^m	+16° 00′	November 27
0.5 m	⊕ 330′	40	II 3 m	easy

Also known as the Hyades. The nearest cluster after the Ursa Major Moving Stream, lying at a distance of 151 l.y., with an age of about 625 million years. Even though the cluster is widely dispersed both in space and over the sky, it nevertheless is gravitationally bound, with the more massive stars lying at the centre of the cluster.

Best seen with binoculars owing to the large extent of the cluster – over $5^1/_2$°. Hundreds of stars are visible, including the fine orange giant stars γ, δ, ϵ and θ^{-1} Tauri. Aldebaran, the lovely orange K-type giant star, is not a true member of the cluster, but is a foreground star only 70 l.y. away. Visible even from light polluted urban areas – a rarity!

Caldwell 13	NGC 457	01^h 19.1^m	+58° 20	October 11
6.4 m	⊕ 13′	80	I 3 r	easy ©

See October.

Collinder 33	NGC 752	01^h 57.8^m	+37° 41′	October 20
5.7 m	⊕ 45′	77	III 1 m	easy

See October.

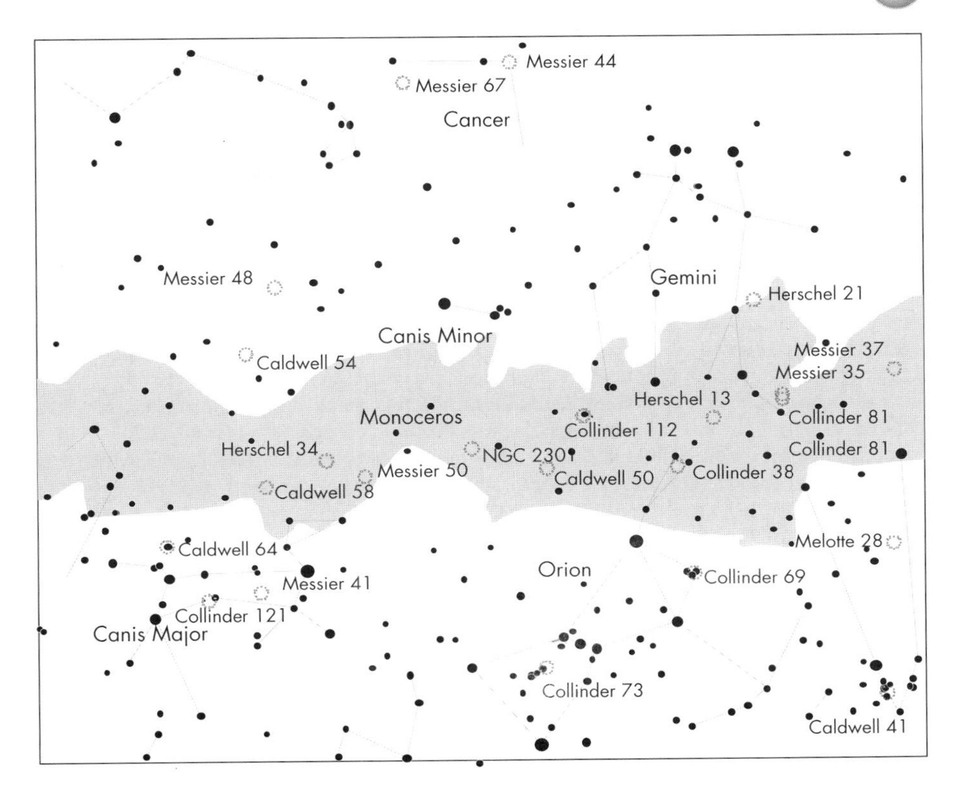

Figure 3.5. Open cluster 1

Stock 2	02^h 15.0^m	+59° 16′	October 25	
4.4 m	⊕ 60′	50	III 1 m	easy ©
See October.				

Caldwell 14	NGC 869	02^h 19.0^m	+57° 09′	October 26
5.3 m	⊕ 29′	200	I 3 r	easy ©
See October.				

Messier 52	NGC 7654	23^h 24.2^m	+61° 35′	September 11
6.9 m	⊕ 12′	100	I 2 r	easy ©
See September.				

Blanco 1	–	00^h 04.3^m	–29° 56′	September 22
4.5 m	⊕ 90′	30	III 2 m	easy
See September.				

Herschel 78	NGC 129	00^h 29.9^m	+60° 14′	September 28
6.5 m	⊕ 21′	30	IV 2 p	easy ©

See September.

Herschel 47	NGC 1502	04^h 07.7^m	+62° 20′	November 22
5.7 m	⊕ 8′	45	II 3 p	medium ©

This cluster is easy to see, but can prove difficult to locate, even though it is in a relatively sparse area of the sky. Visible to the naked eye on clear nights. It is a rich and bright cluster, but small, and may resemble a fan shape, although this does depend on what the observer sees. What do you see? Also contained in the cluster are two multiple stars: Struve 484 and 485. The former is a nice triple system, but the latter is a true spectacle with nine components! Seven of these are visible in a telescope of 10 cm aperture, ranging between 7th and 13th magnitude. The remaining two components, 13.6 and 14.1 magnitude, should be visible in a 20 cm telescope. In addition, the system's brightest component, SZ Camelopardalis, is an eclipsing variable star, which changes magnitude by 0.3 over 2.7 days. What makes this cluster so special is its proximity to the asterism called Kemble's Cascade. This is a long string (2.5°) of 8th-magnitude stars to the north-west of H47. The cascade is best seen in low-power binoculars.

Herschel 64	NGC 381	01^h 08.3^m	+61° 35′	October 8
9.3 m	⊕ 6′	50	III 2 p	medium ©

A faint cluster, but rich and compressed. Can be resolved with an aperture of 10 cm, but at medium aperture, of say, 20 to 25 cm, over 60 stars of 12th and 13th magnitude become visible.

Herschel 69	NGC 7686	23^h 30.2^m	+49° 08′	September 13
5.6 m	⊕ 14′	20	IV 1 p	medium ©

See September.

Herschel 30	NGC 7789	23^h 57.0^m	+56° 44′	September 20
6.9 m	⊕ 15′	300	II 1 r	medium ©

See September.

King 14	–	00^h 31.9^m	+63° 10′	September 28
8.5 m	⊕ 7′	20	III 2 p	medium ©

See September.

Messier 103	NGC 581	01^h 33.2^m	+60° 42′	October 14
7.4 m	⊕ 6′	25	III 2 p	difficult ©

See October.

Trumpler 1	Collinder 15	01^h 35.7^m	+61° 17′	October 15
8.1 m	⊕ 4.5′	20	I 3 p	difficult ©

See October.

Herschel 49	NGC 637	01h 42.9m	+64° 00′	October 17
8.2 m	⊕ 3.5′	20	I 3 p	difficult ©

See October.

King 12	–	23h 53.0m	+61° 58′	September 19
9.0 m	⊕ 2′	15	I 2 p	difficult ©

See September.

Herschel 35	NGC 136	00h 31.5m	+61° 32′	September 28
–m	⊕ 1.2′	20	II 2 p	difficult ©

See September.

December

Melotte 28	NGC 1746	05h 03.6m	+23° 49′	December 6
6.1 m	⊕ 42′	20	III 2 p	easy

Another large and scattered cluster, visible on clear nights with the naked eye. Within the cluster are two other smaller ones, each with its own classification – NGC 1750 and 1758. This phenomenon makes it difficult to determine accurately the true diameter of the cluster.

Herschel 33	NGC 1857	05h 20.2m	+39° 21′	December 11
7.0 m	⊕ 6′	35	II 2 m	easy

A very rich cluster containing several small chains of stars with starless voids located within and around it. The brightest member of the cluster is a nice orange-tinted star, but its glare can overpower the many fainter stars. Some observers try to occult the bright star so that it is obscured, thus allowing the fainter stars to be observed.

Collinder 464		05h 22.0m	+73° 00′	December 11
4.2 m	⊕ 120′	50	II 2 m	easy ©

A large, very rich, irregular open cluster, with the distinction that it is best seen in binoculars, as viewing it in a telescope will dissipate the cluster. Contains many 5th-, 6th- and 7th-magnitude stars.

Messier 38	NGC 1912	05h 28.7m	+35° 50′	December 13
6.4 m	⊕ 21′	75	III 2 m	easy

One of the three Messier clusters in Auriga, and visible to the naked eye. It contains many A-type main sequence and G-type giant stars, with a G0 giant being the brightest, magnitude 7.9. Is elongated in shape with several double stars and voids within it. Seen as a small glow in binoculars, it is truly lovely in small telescopes. It is an old galactic cluster with a star density calculated to be about 8 stars per cubic parsec.

Collinder 69	–	05ʰ 35.1ᵐ	+09° 56′	December 14
2.8 m	⊕ 65′	20	II 3 p n	easy

Perfect for binoculars. This cluster surrounds the 3rd-magnitude stars λ Orionis, and includes φ^{-1} and φ^{-2} Orionis, both 4th-magnitude. Encircling the cluster is the very faint emission nebula Sharpless 2–264, only visible using averted vision and an OIII filter with extremely dark skies.

Collinder 73	NGC 1981	05ʰ 35.2ᵐ	–04° 26′	December 14
4.6 m	⊕ 25′	20	III 2 p n	easy

A nice, bright, coarse cluster, lying about 1° north of M42. Around eight or nine stars can be seen in binoculars, while the remaining stars are a hazy background glow. In moderate telescopes, the most striking feature is two parallel rows of stars.

Messier 36	NGC 1960	05ʰ 36.1ᵐ	+34° 08′	December 15
6.0 m	⊕ 12′	70	II 3 m	easy

About half the size of M38, seen as a glow in binoculars. It is a large, bright cluster. Measurements indicate that it is 10 times farther away then the Pleiades. It contains a nice double star at its centre. Owing to the faintness of its outlying members it is difficult to ascertain where the cluster ends. Visible to naked eye.

Messier 37	NGC 2099	05ʰ 52.4ᵐ	+32° 33′	December 19
5.6 m	⊕ 20′	150	II 1 r	easy

In a word – superb! The finest cluster in Auriga. It really can be likened to a sprinkling of stardust, and some observers liken it to a scattering of gold dust. Contains many A-type stars and several red giants. Visible at all apertures, from a soft glow with a few stars in binoculars to a fine, star-studded field in medium-aperture telescopes. In small telescopes using a low magnification it can appear as a globular cluster. The central star is coloured a lovely deep red, although several observers report it as a much paler red, which may indicate that it is a variable star. Visible to the naked eye.

Collinder 38	NGC 2169	06ʰ 08.4ᵐ	+13° 57′	December 23
5.9 m	⊕ 6′	30	I 3 p n	easy

This is a small but bright cluster. Some observers find it hard to believe that this scattering of stars has been classified as a cluster. Easily visible in binoculars, the stars appear to range in magnitude from about 8 to 10. Also, binoculars will show the four brightest members to be surrounded by faint nebulosity – sometimes!

Messier 35	NGC 2168	06ʰ 08.9ᵐ	+24° 20′	December 23
5.1 m	⊕ 28′	200	III 2 m	easy

One of the most magnificent clusters in the sky. Visible to the naked eye on clear winter nights with a diameter as big as that of the full Moon, and it seems as if the cluster is just beyond being resolved. Many more stars are visible in binoculars set against the hazy glow of unresolved members of the cluster. With telescopes, the magnificence of the cluster becomes apparent, with many curving chains of stars.

Messier 34	NGC 1039	02ʰ 42.0ᵐ	+42° 47′	November 1
5.2 m	⊕ 35′	60	II 3 m	easy

See November.

Stock 23	–	03h 16.3m	+60° 02′	November 9
35.6 m	⊕ 18′	25	III 3 p n	easy ©
See November.				

Messier 45	Melotte 22	03h 47.0m	+24° 07′	November 17
1.2 m	⊕ 110′	100	I 3 r	easy
See November.				

Caldwell 41	Melotte 25	04h 27.0m	+16° 00′	November 27
0.5 m	⊕ 330′	40	II 3 m	easy
See November.				

Messier 41	NGC 2287	06h 47.0m	–20° 44′	January 2
4.5 m	⊕ 38′	70	II 3 m	easy
See January.				

–	NGC 2301	06h 51.8m	+00° 28′	January 3
6.0 m	⊕ 12′	70	I 3 m	easy
See January.				

Collinder 121	–	06h 54.2m	–24° 38′	January 3
2.6 m	⊕ 50′	20	III 3 p	easy
See January.				

Messier 50	NGC 2323	07h 03.2m	–08° 20′	January 6
5.9 m	⊕ 16′	80	II 3 m	easy
See January.				

Herschel 34	NGC 2353	07h 14.6m	–10° 18′	January 9
7.1 m	⊕ 20′	30	II 2 p	easy
See January.				

Caldwell 58	NGC 2360	07h 17.8m	–15° 37′	January 9
7.2 m	⊕ 12′	80	II 2 m	easy
See January.				

Caldwell 64	NGC 2362	07h 18.8m	–24° 57′	January 10
4.1 m	⊕ 8′	60	I 3 p n	easy
See January.				

Messier 48	NGC 2548	08h 13.8m	–05° 48′	January 24
5.8 m	⊕ 55′	80	I 3 r	easy
See January.				

Messier 44	NGC 2632	08h 40.1m	+19° 59′	January 30
3.1 m	⊕ 95′	60	II 2 m	easy

See January.

Herschel 59	NGC 1664	04h 51.1m	+43° 42′	December 3
7.6 m	⊕ 13′	25	III 1 p	medium

A nice bright cluster, but loosely structured and best seen with an aperture of 20 cm. Appears as an enrichment of the background Milky Way star field. There is a 7th-magnitude star within the cluster but is not a true member, and the glare from the star can sometimes make observation of the cluster difficult. Increasing the aperture will show progressively more stars.

Collinder 81	NGC 2158	06h 07.5m	+24° 06′	December 23
8.6 m	⊕ 5′	70	II 3 r	medium

Lying at a distance of 160,00 l.y., this is one of the most distant clusters visible using small telescopes, and lies at the edge of the Galaxy. It needs a 20 cm telescope to be resolved, and even then only a few stars will be visible against a background glow. It is a very tight, compact grouping of stars, and something of an astronomical problem. Some astronomers class it as intermediate between an open cluster and a globular cluster, and it is believed to be about 800 million years old, making it very old as open clusters go.

Herschel 47	NGC 1502	04h 07.7m	+62° 20′	November 22
5.7 m	⊕ 8′	45	II 3 p	medium ©

See November.

Herschel 68	NGC 2126	06h 03.0m	+49° 54′	December 22
10.2 m	⊕ 6′	40	II 1 p	difficult ©

Has been described as diamond dust on black velvet. A very faint but nice cluster, although it can prove a challenge to find.

Herschel 61	NGC 1778	05h 08.1m	+37° 03′	December 8
7.7 m	⊕ 6′	30	III 2 p	difficult

Although a fairly bright cluster, it is so sparse and spread out that it will require some careful observation to be located.

Herschel 13	NGC 2204	06h 15.7m	−18° 39′	December 25
8.6 m	⊕ 13′	80	III 3 m	difficult

A difficult cluster to locate and observe, composed of many faint stars but with a nice orange star at its northern limit.

Messier 103	NGC 581	01h 33.2m	+60° 42′	October 14
7.4 m	⊕ 6′	25	III 2 p	difficult ©

See October.

Trumpler 1	Collinder 15	$01^h\ 35.7^m$	+61° 17′	October 15
8.1 m	⊕ 4.5′	20	13 p	difficult ©
See October.				

Herschel 49	NGC 637	$01^h\ 42.9^m$	+64° 00′	October 17
8.2 m	⊕ 3.5′	20	13 p	difficult ©
See October.				

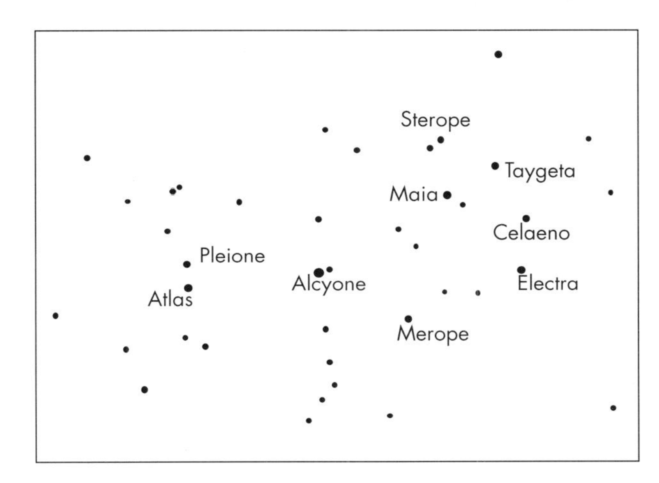

Figure 3.6. Messier 45.

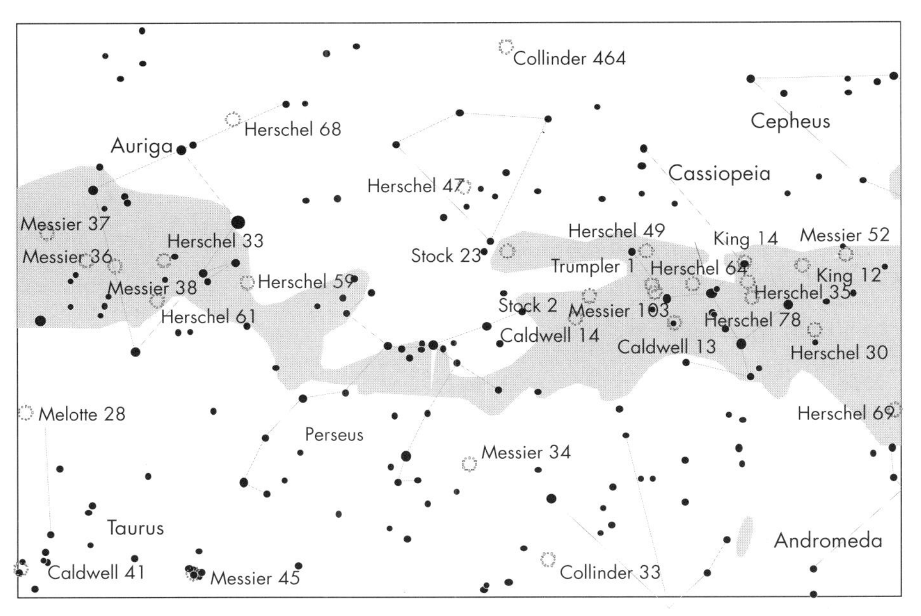

Figure 3.7. Open clusters 2.

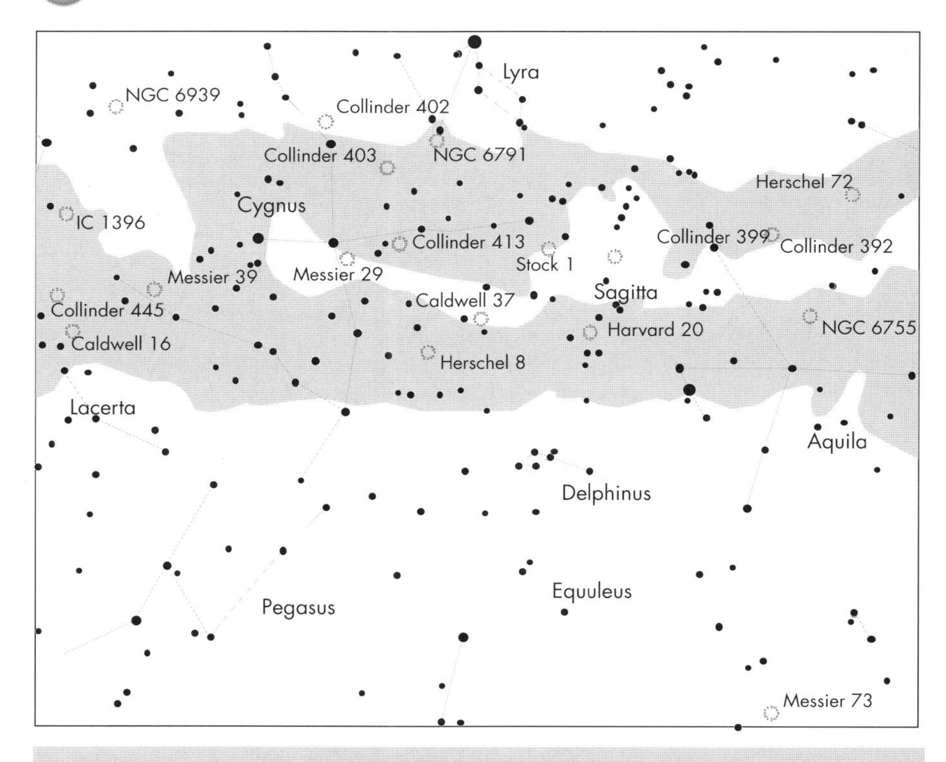

Figure 3.8. Open clusters 3.

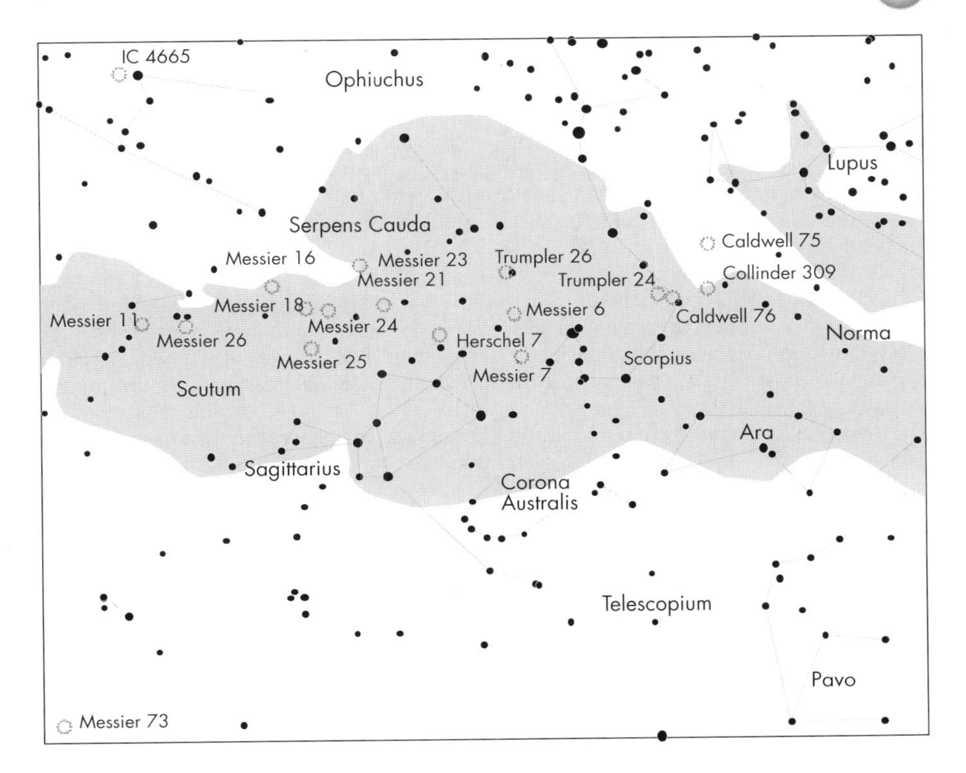

Figure 3.9. Open clusters 4.

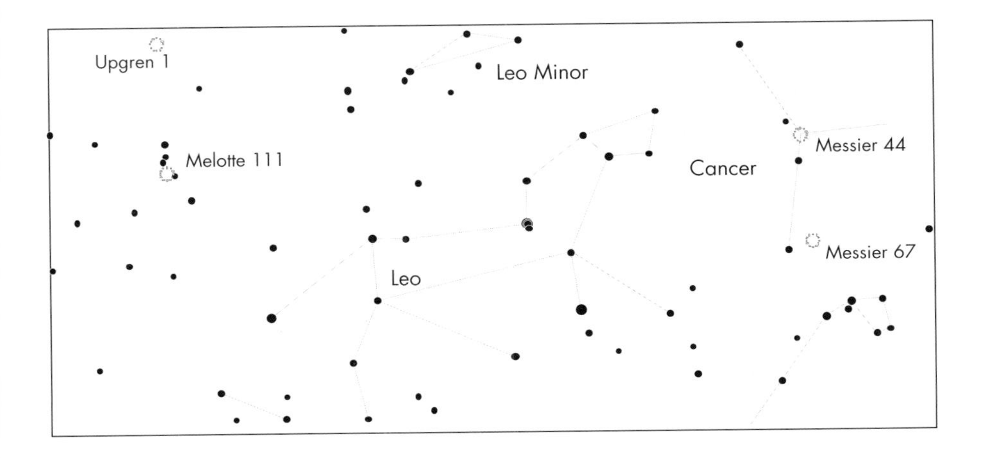

Figure 3.10. Open clusters 5.

3.3 Globular Clusters

The previous section on open clusters dealt with groups of stars that were usually young, have an appreciable angular size and may have a few hundred components. This section deals with clusters that are very old, are compact and may contain up to a million stars, and in some cases even more. These clusters are called *globular clusters*. The stars which make up a globular cluster are called Population II stars. These are metal poor stars and are usually to be found in a spherical distribution around the galactic centre at a radius of about 200 l.y. Furthermore, the number of globular clusters increases significantly the closer one gets to the galactic centre. This means that particular constellations which are located in a direction towards the Galactic bulge have a high concentration of globular clusters within them, such as Sagittarius and Scorpius. Thus, the spring and summer are the most favourable times for viewing these elusive objects.

The origin and evolution of a globular cluster is very different to that of an open or galactic cluster. All the stars in a globular cluster are very old, with the result that any star earlier than a G or F type star will have already left the main sequence and be moving toward the red giant stage of its life. In fact, new star formation no longer takes place within any globular clusters in our Galaxy, and they are believed to be the oldest structures in our Galaxy. In fact, the youngest of the globular clusters is still far older than the oldest open cluster. The origin of the globular clusters is a scene of fierce debate and research with the current models predicting that the globular clusters may have been formed within the proto-galaxy clouds which went to make up our Galaxy.

There are about 150 globular clusters ranging in size from 60 to 150 l.y. in diameter. They all lie at vast distances from the Sun, and are about 60,000 l.y. from the Galactic plane. The nearest globular clusters, for example Caldwell 86 in Ara lies at a distance of over 6000 l.y., and thus the clusters are difficult objects for small telescopes. That is not to say they can't be seen, rather, it means that any structure within the cluster will be difficult to observe. Even the brightest and biggest globular will need apertures of at least 15 cm for individual stars to be resolved. However, if large aperture telescopes are used, then these objects are magnificent. Some globular clusters have dense concentrations towards their centre whilst others may appear as rather compact open clusters. In some cases, it is difficult to say where the globular cluster peters out and the background stars begin.

As in the case of open clusters, there exists a classification system, the Shapley-Sawyer Concentration Class, where Class I globular clusters are the most star-dense, while Class XII is the least star-dense. The ability of an amateur to resolve the stars in a globular actually depends on how condensed the cluster actually is, and so the scheme will be used in the descriptions, but it is only really useful for those amateurs who have large aperture instruments. Nevertheless, the observation of these clusters which are amongst the oldest objects visible to amateurs can provide you with breathtaking, almost three-dimensional aspects, and at the same time, increase and develop your observing skills.

The details listed are of the same order as in open clusters, with the omission of star numbers, and the replacement of Trumpler type with the Shapley-Sawyer concentration class.

January

Messier 68	NGC 4590	$12^h 39.5^m$	−26° 45′	April 1
7.7 m	⊕ 12′		X	easy
See April.				

Messier 79	NGC 1904	$05^h 24.5^m$	−24° 33′	December 12
7.8 m	⊕ 8.5′		V	medium

See December.

–	NGC 4147	$12^h 10.1^m$	+18° 33′	March 25
10.2 m	⊕ 4′		VI	difficult

See March.

Caldwell 25	NGC 2419	$07^h 38.1^m$	+38° 53′	January 15
10.4 m	⊕ 4.1′		II	difficult

Also known as the Intergalactic Wanderer. A difficult object to resolve with any detail; even large telescopes of aperture 40 cm will be unable to resolve any stellar detail. It has been reported as observed with an 8 cm telescope under perfect conditions, but will appear as just a faint, hazy dot. It lies at the vast distance of 250,000 l.y., further even than the Magellanic Clouds, because it has a space velocity in excess of the velocity needed to escape from the gravitational pull of our Galaxy. One of the five most luminous clusters in the Milky Way, and is believed to be very old, at around 12 billion years.

February

Messier 68	NGC 4590	$12^h 39.5^m$	−26° 45′	April 1
7.7 m	⊕ 12′		X	easy

See April.

Caldwell 80	NGC 5139	$13^h 26.8^m$	−47° 29′	April 13
3.5 m	⊕ 36′		VIII	easy

See April.

Messier 3	NGC 5272	$13^h 42.2^m$	+28° 23′	April 17
5.9 m	⊕ 16′		VI	easy

See April.

Messier 53	NGC 5024	$13^h 12.9^m$	+18° 10′	April 9
7.5 m	⊕ 12′		V	medium

See April.

Herschel 9	NGC 5466	$14^h 05.5^m$	+28° 32′	April 23
9.0 m	⊕ 11′		XII	medium

See April.

Messier 79	NGC 1904	$05^h 24.5^m$	−24° 33′	December 12
7.8 m	⊕ 8.5′		V	medium

See December.

Herschel 7	NGC 5053	13h 16.4m	+17° 42′	April 10
10.9 m	⊕ 10′		XI	difficult
See April.				

Caldwell 25	NGC 2419	07h 38.1m	+38° 53′	January 15
10.4 m	⊕ 4.1′		II	difficult
See January.				

–	NGC 4147	12h 10.1m	+18° 33′	March 25
10.2 m	⊕ 4′		VI	difficult
March				

March

Messier 68	NGC 4590	12h 39.5m	–26° 45′	April 1
7.7 m	⊕ 12′		X	easy
See April.				

Caldwell 80	NGC 5139	13h 26.8m	–47° 29′	April 13
3.5 m	⊕ 36′		VIII	easy
See April.				

Messier 3	NGC 5272	13h 42.2m	+28° 23′	April 17
5.9 m	⊕ 16′		VI	easy
See April.				

Messier 53	NGC 5024	13h 12.9m	+18° 10′	April 9
7.5 m	⊕ 12′		V	medium
See April.				

Herschel 9	NGC 5466	14h 05.5m	+28° 32′	April 23
9.0 m	⊕ 11′		XII	medium
See April.				

Messier 79	NGC 1904	05h 24.5m	–24° 33′	December 12
7.8 m	⊕ 8.5′		V	medium
See December.				

Caldwell 25	NGC 2419	07h 38.1m	+38° 53′	January 15
310.4 m	⊕ 4.1′		II	difficult
See January.				

-	NGC 4147	$12^h 10.1^m$	+18° 33′	March 25
10.2 m	⊕ 4′		VI	difficult

This is a faint cluster, hazy in appearance with a star-like core. A challenge for telescopic observers. Several variable stars are also located within the cluster.

Herschel 7	NGC 5053	$13^h 16.4^m$	+17° 42′	April 10
10.9 m	⊕ 10′		XI	difficult

See April.

April

Messier 68	NGC 4590	$12^h 39.5^m$	–26° 45′	April 1
7.7 m	⊕ 12′		X	easy

Appearing only as a small, hazy patch in binoculars, this is a nice cluster in telescopes, with an uneven core and faint halo. Under low magnification, some faint structure or mottling can be glimpsed which under medium to high magnification resolves itself as a myriad assembly of stars. A definite challenge to naked-eye observers, where perfect seeing will be needed. Use averted vision and make sure that your eyes are well and truly dark-adapted.

Caldwell 80	NGC 5139	$13^h 26.8^m$	–47° 29′	April 13
3.5 m	⊕ 36′		VIII	easy

Also known as Omega Centauri. A fabulous cluster and one of the showpieces of the night sky. Visible to the naked eye as a clearly seen, hazy patch of light. A stunning sight in binoculars, and a jaw-dropping spectacle in telescopes.[3] Words do not do it justice, so I won't even attempt to describe it, but will leave it up to you to search out this wonderful object. It contains over a million stars, and some sources put it at having nearly 10 million. It is about 17,000 l.y. away.

Messier 3	NGC 5272	$13^h 42.2^m$	+28° 23′	April 17
5.9 m	⊕ 16′		VI	easy

A splendid object, easily seen in binoculars. A good test for the naked eye. If using giant binoculars with perfect seeing, some stars may be resolved. A beautiful and stunning cluster in telescopes, it easily rivals M13 in Hercules. It definitely shows pale coloured tints, and reported colours include, yellow, blue and even green; in fact, it is often quoted as the most colourful globular in the northern sky. Full of structure and detail including several dark and mysterious tiny dark patches. Many of the stars in the cluster are also variable. One of the three brightest clusters in the northern hemisphere. Located at a distance of about 35,000 l.y.

---

[3]Alas, this wonderful globular cluster is not visible from the UK, or Northern parts of the USA, but I have still included it as it is truly an amazing object to observe, and should you have the chance to see it, then do so.

Messier 53	NGC 5024	13^h 12.9^m	+18° 10′	April 9
7.5 m	⊕ 12′		V	medium

An often ignored globular cluster, which is a shame as it is a nice object. Contains about 100,000 stars, none of which are resolved in binoculars, through which it will appear as a faint hazy patch with a brighter centre, located in a star field. Telescopes show a nice symmetrical glow with a concentrated core. Some observers report a coloured hue to the cluster – what do you see? It stands up nicely to magnification, and indeed is lovely sight in telescopes of aperture 10 cm and greater. Lies at a distance of around 330,000 l.y.

Herschel 9	NGC 5466	14^h 05.5^m	+28° 32′	April 23
9.0 m	⊕ 11′		XII	medium

A challenge for binoculars, and even when located will appear as a faint hazy and small glow. In telescopes the cluster has a resolvable core.

Messier 79	NGC 1904	05^h 24.5^m	–24° 33′	December 12
7.8 m	⊕ 8.5′		V	medium

See December.

Herschel 7	NGC 5053	13^h 16.4^m	+17° 42′	April 10
10.9 m	⊕ 10′		XI	difficult

A very faint and loose cluster containing only 3500 stars, it lies very close to Messier 53, and both can be glimpsed using low power. This is one of those clusters that is very impressive when seen through a large-aperture telescope. Its position in space is also unique, in that it lies about 50,000 l.y. above the galactic plane.

Caldwell 25	NGC 2419	07^h 38.1^m	+38° 53′	January 15
10.4 m	⊕ 4.1′		II	difficult

See January.

–	NGC 4147	12^h 10.1^m	+18° 33′	March 25
10.2 m⊕ 4′		VI	difficult	

See March.

May

Messier 5	NGC 5904	15^h 18.6^m	+02° 05′	May 11
5.7 m	⊕ 17.4′		V	easy

A wonderful cluster, visible to the naked eye on clear nights. Easily seen as a disc with binoculars, and with large telescopes the view is breathtaking – presenting an almost three-dimensional vista. One of the few coloured globulars, with a faint, pale yellow outer region surrounding a blue-tinted interior. It gets even better with higher magnification, as more detail and stars become apparent. Possibly containing over half a million stars, this is one of the finest clusters in the northern hemisphere; many say it is *the* finest.

Messier 80	NGC 6093	16^h 17.0^m	–22° 59′	May 26
7.3 m	⊕ 8.9′		II	easy

Readily detectable in binoculars as a tiny, glowing hazy patch set in a stunning star field, it has a distinctly noticeable brighter core. Telescopes will be needed to resolve its 14th-magnitude stellar core. One of the few globulars to have been the origin of a nova, T Scorpii, when it flared to prominence in 1860, then disappeared back into obscurity within 3 months.

Messier 4	NGC 6121	16^h 23.6^m	–26° 32′	May 28
5.8 m	⊕ 26.3′		IX	easy

A superb object, presenting a spectacle in all optical instruments and even visible to the naked eye. But it does lie very close to the star Antares, so that the glare of the latter may prove a problem in the detection. High-power binoculars will even resolve several stars. Telescopes of all apertures show detail and structure within the cluster, and the use of high magnification will prove beneficial; but what is more noticeable is the bright lane of stars that runs through the cluster's centre. Thought to be the closest globular to the Earth at 6,500 l.y. (although NGC 6397 in Ara may be closer), and about 10 billion years old.

Messier 68	NGC 4590	12^h 39.5^m	–26° 45′	April 1
7.7 m	⊕ 12′		X	easy

See April.

Caldwell 80	NGC 5139	13^h 26.8^m	–47° 29′	April 13
3.5 m	⊕ 36′		VIII	easy

See April.

Messier 3	NGC 5272	13^h 42.2^m	+28° 23′	April 17
5.9 m	⊕ 16′		VI	easy

See April.

Messier 13	NGC 6205	16^h 41.7^m	+36° 28′	June 1
5.7 m	⊕ 16.5′		V	easy

See June.

Messier 12	NGC 6218	16^h 47.2^m	–01° 57′	June 3
6.8 m	⊕ 14.5′		IX	easy

See June.

Messier 10	NGC 6254	16^h 57.1^m	–04° 06′	June 5
6.6 m	⊕ 15′		VII	easy

See June.

Messier 62	NGC 6266	17^h 01.2^m	–30° 07′	June 6
6.7 m	⊕ 14′		IV	easy

See June.

Messier 19	NGC 6273	17^h 02.6^m	−26° 16′	June 7
6.7 m	⊕ 13.5′		VIII	easy

See June.

Messier 9	NGC 6333	17^h 19.2^m	−18° 31′	June 11
7.6 m	⊕ 9.3′		VII	easy

See June.

Messier 22	NGC 6656	18^h 36.4^m	−23° 54′	June 30
5.1 m	⊕ 24′		VII	easy

See June.

Messier 107	NGC 6171	16^h 32.5^m	−13° 03′	May 30
8.1 m	⊕ 10′		X	medium

Often missed off amateurs' observing schedules owing to its faintness. It is nevertheless a pleasant cluster with a mottled disc and brighter core. Not visible with the naked eye, it nevertheless presents a pleasing aspect when medium to high magnification is used. What makes this inconspicuous globular important, however, is that it is one of the very few that seem to be affected by the presence of interstellar dust. Deep imaging has revealed several obscured areas within the cluster, possibly due to dust grains lying between us. This isn't such a surprise, as the globular is located over the hub of the Galaxy in Scorpius.

Messier 53	NGC 5024	13^h 12.9^m	+18° 10′	April 9
7.5 m	⊕ 12′		V	medium

See April.

Herschel 9	NGC 5466	14^h 05.5^m	+28° 32′	April 23
9.0 m	⊕ 11′		XII	medium

See April.

Messier 92	NGC 6341	17^h 17.1^m	+43° 08′	June 10
6.4 m	⊕ 11′		IV	medium

See June.

Messier 14	NGC 6402	17^h 37.6^m	−03° 15′	June 16
7.6 m	⊕ 11.7′		VIII	medium

See June.

Herschel 147	NGC 6304	17^h 14.5^m	−29° 28′	June 18
8.4 m	⊕ 6.9′		VI	medium

See June.

Herschel 49	NGC 6522	18h 03.6m	–30° 02′	June 22
8.6 m	⊕ 5.6′		VI	medium
See June.				

Herschel 12	NGC 6553	18h 09.3m	–25° 54′	June 24
8.1 m	⊕ 8.1′		XI	medium
See June.				

Messier 28	NGC 6626	18h 24.5m	–24° 52′	June 27
6.8 m	⊕ 11.2′		IV	medium
See June.				

Messier 69	NGC 6637	18h 31.4m	–32° 21′	June 29
7.6 m	⊕ 7.1′		V	medium
See June.				

Caldwell 66	NGC 5694	14h 39.6m	–26° 32′	May 1
10.2 m	⊕ 3.6′		VII	difficult

A faint cluster which has a bright core, but an unresolved halo in telescopes of less than 30 cm. An unremarkable object, which you will probably not visit more than once! It is a difficult cluster to locate, especially from the UK. Precise setting circles on your telescope (or of course even a computerised system) will help significantly in finding it. It is actually located on the far side of our Galaxy, at around 110,000 l.y. from the solar system. Research suggests that the cluster may have attained a velocity which will allow it to escape from the gravitational pull of the Galaxy.

Herschel 19	NGC 5897	15h 17.4m	–21° 01′	May 11
8.6 m	⊕ 12.5′		XI	difficult

A very difficult cluster to locate in binoculars owing to its low surface brightness. Even with a telescope, it is not an easy object to observe.

Herschel 7	NGC 5053	13h 16.4m	+17° 42′	April 10
10.9 m	⊕ 10′		XI	difficult
See April.				

Herschel 50	NGC 6229	16h 47.0m	+47° 32′	June 3
9.4 m	⊕ 4.5′		IV	difficult
See June.				

Herschel 200	NGC 6528	18h 04.8m	–30° 03′	June 22
9.5 m	⊕ 3.7′		V	difficult
See June.				

June

Messier 13	NGC 6205	16h 41.7m	+36° 28′	June 1
5.7 m	⊕ 16.5′		V	easy

Also known as the Great Hercules Cluster. A splendid object and the premier cluster of the northern hemisphere. Visible to the naked eye, it has a hazy appearance in binoculars; with telescopes, however, it is magnificent, with a dense core surrounded by a sphere of a diamond-dust-like array of stars. In larger telescopes, several dark bands can be seen bisecting the cluster. It appears bright because is close to us at only 23,000 l.y., and also because it is inherently bright, shining at a luminosity equivalent to over 250,000 Suns. At only 140 l.y. in diameter, the stars must be very crowded, with several stars per cubic light year, a density some 500 times that of our vicinity. All in all a magnificent cluster.

Messier 12	NGC 6218	16h 47.2m	–01° 57′	June 3
6.8 m	⊕ 14.5′		IX	easy

A small cluster that will be a challenge to naked-eye observers. In telescopes of aperture 20 cm and more, this cluster is very impressive, with many stars being resolved against the fainter background of unresolved members. It also contains many faint coloured stars which show up well with telescopes of aperture 10 cm and greater. It is nearly the twin of Messier 10, which is within 3° south-east.

Messier 10	NGC 6254	16h 57.1m	–04° 06′	June 5
6.6 m	⊕ 15′		VII	easy

Similar to M12, it is however slightly brighter and more concentrated. Can be seen with the naked eye on dark nights. It lies close to the orange star 30 Ophiuchi (spectral type K4, magnitude 5), and so if you locate this star then by using averted vision M10 should be easily seen. With apertures of 20 cm and more, the stars are easily resolved right to the cluster's centre. Under medium aperture and magnification, several coloured components have been reported: a pale blue tinted outer region surrounding a very faint pink area, with a yellow star at the cluster's centre.

Messier 62	NGC 6266	17h 01.2m	–30° 07′	June 6
6.7 m	⊕ 14′		IV	easy

A very nice cluster, visible in binoculars as a small hazy patch of light set in a wonderful star field. Owing to its irregular shape, it bears a cometary appearance, which is apparent even in small telescopes. Has a very interesting structure where concentric rings of stars have been reported by several observers, along with a coloured sheen to its centre, described as both pale red and yellow!

Messier 19	NGC 6273	17h 02.6m	–26° 16′	June 7
6.7 m	⊕ 13.5′		VIII	easy

A splendid albeit faint cluster when viewed through a telescope, it nevertheless can be glimpsed with binoculars, where its egg shape is very apparent. Although a challenge to resolve, it is nevertheless a colourful object, reported as having both faint orange and faint blue stars, while the overall colour of the cluster is a creamy white. Some amateurs also claim that a few faint dark patches mottle the cluster; perhaps this is interstellar dust between us and the cluster.

Messier 9	NGC 6333	17h 19.2m	–18° 31′	June 11
7.6 m	⊕ 9.3′		VII	easy

Visible in binoculars, this is a small cluster, with a brighter core. The cluster is one of the nearest to the centre of our Galaxy, and is in a region conspicuous for its dark nebulae, including Barnard 64; it may be that the entire region is swathed in interstellar dust, which gives rise to the cluster's dim appearance. It lies about 19,000 l.y. away.

Messier 22	NGC 6656	18h 36.4m	–23° 54′	June 30
5.1 m	⊕ 24′		VII	easy

Wonderful, a truly spectacular globular cluster, visible under perfect conditions to the naked eye. Low-power eyepieces will show a hazy spot of light, while high power will resolve a few stars. A 15 cm telescope will give an amazing view of minute bright stars evenly spaced over a huge area. Often passed over by northern hemisphere observers owing to its low declination. Only 10,000 l.y. away, nearly twice as close as M13.

Messier 55	NGC 6809	19h 40.0m	–30° 58′	July 17
6.3 m	⊕ 19′		XI	easy

See July.

Messier 5	NGC 5904	15h 18.6m	+02° 05′	May 11
5.7 m	⊕ 17.4′		V	easy

See May.

Messier 80	NGC 6093	16h 17.0m	–22° 59′	May 26
7.3 m	⊕ 8.9′		II	easy

See May.

Messier 4	NGC 6121	16h 23.6m	–26° 32′	May 28
5.8 m	⊕ 26.3′		IX	easy

See May.

Messier 68	NGC 4590	12h 39.5m	–26° 45′	April 1
7.7 m	⊕ 12′		X	easy

See April.

Caldwell 80	NGC 5139	13h 26.8m	–47° 29′	April 13
3.5 m	⊕ 36′		VIII	easy

See April.

Messier 3	NGC 5272	13h 42.2m	+28° 23′	April 17
5.9 m	⊕ 16′		VI	easy

See April.

Messier 92	NGC 6341	17h 17.1m	+43° 08′	June 10
6.4 m	⊕ 11′		IV	medium

A beautiful cluster, often overshadowed by its more illustrious neighbour, M13. It is a somewhat difficult object to locate, but once found is truly spectacular. It can be glimpsed with the naked eye. In binoculars it will appear as a hazy small patch, but in 20 cm telescopes its true beauty becomes apparent with a bright, strongly concentrated core. It also has several very distinct dark lanes running across the face of the cluster. A very old cluster, 25,000 l.y. distant.

Messier 14	NGC 6402	17h 37.6m	−03° 15′	June 16
7.6 m	⊕ 11.7′		VIII	medium

Located in an empty part of the sky, it is brighter and larger than is usual for a globular. Though visible only in binoculars as a small patch of light, and not resolved even in a small telescope (<15 cm), it is nevertheless worth searching for. It shows a delicate structure with a lot of detail, much of which will be obscured if seen from an urban location. It has a pale yellow tint, and some observers report seeing a definite stellar core, which has a striking orange colour. But this feature is seen only with telescopes of aperture 15 cm and greater and using a high magnification.

Herschel 147	NGC 6304	17h 14.5m	−29° 28′	June 18
8.4 m	⊕ 6.9′		VI	medium

This is a small but bright cluster with only a few resolvable stars near its edge. Nevertheless it will be a challenge to locate with binoculars.

Herschel 49[4]	NGC 6522	18h 03.6m	−30° 02′	June 22
8.6 m	⊕ 5.6′		VI	medium

With telescopes of aperture 20 cm this cluster will appear with a bright core, but an unresolved halo. A difficult object to locate with binoculars.

Herschel 12	NGC 6553	18h 09.3m	−25° 54′	June 24
8.1 m	⊕ 8.1′		XI	medium

Not easily visible in binoculars (although it would prove an observational challenge to locate), it is a fairly evenly bright cluster, with no perceptible increase towards the core.

Messier 28	NGC 6626	18h 24.5m	−24° 52′	June 27
6.8 m	⊕ 11.2′		IV	medium

Only seen as a small patch of faint light in binoculars, this is an impressive cluster in telescopes. With an aperture of 15 cm it shows a bright core with a few resolvable stars at the halo's rim. With a larger aperture the cluster becomes increasingly resolvable and presents a spectacular sight. It lies at a distance of about 22,000 l.y. . Well worth seeking out for large-aperture telescope owners, as it is a lost gem.

[4] Located within the constellation Sagittarius are countless globular clusters; only the brightest are listed here.

Messier 69	NGC 6637	$18^h 31.4^m$	–32° 21′	June 29
7.6 m	⊕ 7.1′		V	medium

Visible as just a hazy spot in binoculars, it appears with a nearly star-like core in telescopes. Large aperture will be needed to resolve any detail, and will show the myriad dark patches located within the cluster.

Messier 70	NGC 6681	$18^h 42.2^m$	–32° 18′	July 2
8.0 m	⊕ 7.8′		V	medium

See July.

Messier 54	NGC 6715	$18^h 55.1^m$	–30° 29′	July 5
7.6 m	⊕ 9.1′		III	medium

See July.

–	NGC 6760	$19^h 11.2^m$	+01° 02′	July 9
9.1 m	⊕ 6.5′		IX	medium

See July.

Messier 56	NGC 6779	$19^h 16.6^m$	+30° 11′	July 11
8.3 m	⊕ 7′		X	medium

See July.

Messier 71	NGC 6838	$19^h 53.8^m$	+18° 47′	July 20
8.0 m	⊕ 7.2′		–	medium

See July.

Messier 75	NGC 6864	$20^h 06.1^m$	–21° 55′	July 23
8.5 m	⊕ 6′		I	medium

See July.

Caldwell 47	NGC 6934	$20^h 34.2^m$	+07° 24′	July 30
8.9 m	⊕ 5.9′		VIII	medium

See July.

Messier 107	NGC 6171	$16^h 32.5^m$	–13° 03′	May 30
8.1 m	⊕ 10′		X	medium

See May.

Messier 53	NGC 5024	$13^h 12.9^m$	+18° 10′	April 9
7.5 m	⊕ 12′		V	medium

See April.

Herschel 9	NGC 5466	14^h 05.5^m	+28° 32′	April 23
9.0 m	⊕ 11′		XII	medium

See April.

Herschel 50	NGC 6229	16^h 47.0^m	+47° 32′	June 3
9.4 m	⊕ 4.5′		IV	difficult

A difficult object to locate, and even with 20 cm telescopes will appear unresolved. Large-aperture telescopes, however, will show structure and detail within the cluster.

Herschel 200	NGC 6528	18^h 04.8^m	–30° 03′	June 22
9.5 m	⊕ 3.7′		V	difficult

Even in large telescope of aperture 35 cm, this cluster is unresolved. It will just appear as a faint glow with a slightly brighter centre. A good challenge for large-aperture telescopes.

Caldwell 66	NGC 5694	14^h 39.6^m	–26° 32′	May 1
10.2 m	⊕ 3.6′		VII	difficult

See May.

Herschel 19	NGC 5897	15^h 17.4^m	–21° 01′	May 11
8.6 m	⊕ 12.5′		XI	difficult

See May.

Herschel 7	NGC 5053	13^h 16.4^m	+17° 42′	April 10
10.9 m	⊕ 10′		XI	difficult

See April.

July

Messier 55	NGC 6809	19^h 40.0^m	–30° 58′	July 17
6.3 m	⊕ 19′		XI	easy

A lovely cluster, easily seen in binoculars, and just visible with the naked eye. Small-aperture telescopes (15 cm) show a bright, easily resolved cluster with a nice concentrated halo. Because it is very open, a lot of detail can be seen such as star arcs and dark lanes, even with quite small telescopes. With a larger aperture, hundreds of stars are seen.

Messier 15	NGC 7078	21^h 30.0^m	+12° 10′	August 13
6.4 m	⊕ 12′		IV	easy

See August.

Messier 2	NGC 7089	21^h 33.5^m	–00° 49′	August 14
6.4 m	⊕ 13′		II	easy

See August.

Messier 30	NGC 7099	21ʰ 40.4ᵐ	–23° 11′	August 16
7.3 m	⊕ 11′		V	easy
See August.				

Messier 13	NGC 6205	16ʰ 41.7ᵐ	+36° 28′	June 1
5.7 m	⊕ 16.5′		V	easy
See June.				

Messier 12	NGC 6218	16ʰ 47.2ᵐ	–01° 57′	June 3
6.8 m	⊕ 14.5′		IX	easy
See June.				

Messier 10	NGC 6254	16ʰ 57.1ᵐ	–04° 06′	June 5
6.6 m	⊕ 15′		VII	easy
See June.				

Messier 62	NGC 6266	17ʰ 01.2ᵐ	–30° 07′	June 6
6.7 m	⊕ 14′		IV	easy
See June.				

Messier 19	NGC 6273	17ʰ 02.6ᵐ	–26° 16′	June 7
6.7 m	⊕ 13.5′		VIII	easy
See June.				

Messier 9	NGC 6333	17ʰ 19.2ᵐ	–18° 31′	June 11
7.6 m	⊕ 9.3′		VII	easy
See June.				

Messier 22	NGC 6656	18ʰ 36.4ᵐ	–23° 54′	June 30
5.1 m	⊕ 24′		VII	easy
See June.				

Messier 5	NGC 5904	15ʰ 18.6ᵐ	+02° 05′	May 11
5.7 m	⊕ 17.4′		V	easy
See May.				

Messier 80	NGC 6093	16ʰ 17.0ᵐ	–22° 59′	May 26
7.3 m	⊕ 8.9′		II	easy
See May.				

Messier 4	NGC 6121	16ʰ 23.6ᵐ	–26° 32′	May 28
5.8 m	⊕ 26.3′		IX	easy
See May.				

Messier 70	NGC 6681	18ʰ 42.2ᵐ	–32° 18′	July 2
8.0 m	⊕ 7.8′		V	medium

A faint binocular object that is a twin of M69. Best viewed with a large aperture, as with a small telescope, it is often mistaken for a galaxy. It lies at a distance of 35,000 l.y.

Messier 54	NGC 6715	18ʰ 55.1ᵐ	–30° 29′	July 5
7.6 m	⊕ 9.1′		III	medium

With telescopic apertures smaller than 35 cm the cluster remains unresolved, and will show only a larger view similar to that seen in binoculars – a faint hazy patch of light. It has a colourful aspect – a pale blue outer region and pale yellow inner core. Recent research has found that the cluster was originally related to the Sagittarius Dwarf Galaxy, but that the gravitational attraction of our Galaxy has pulled the globular from its parent. Among the globular clusters in the Messier catalogue it is one of the densest as well as being the most distant.

–	NGC 6760	19ʰ 11.2ᵐ	+01° 02′	July 9
9.1 m	⊕ 6.5′		IX	medium

A faint, symmetrical cluster with a just perceptible brighter core. High-power binoculars should be able to locate this cluster, and even with a small telescope it should present no problems. But a knowledge of the use of setting circles would be useful, as would a computer-controlled telescope.

Messier 56	NGC 6779	19ʰ 16.6ᵐ	+30° 11′	July 11
8.3 m	⊕ 7′		X	medium

It is situated in a rich star field and in small instruments will appear as a hazy patch with a brighter core. It has often been likened to a comet in its appearance. Resolution of the cluster will need at least a 20 cm aperture telescope, and increasing magnification will show further detail.

Messier 71	NGC 6838	19ʰ 53.8ᵐ	+18° 47′	July 20
8.0 m	⊕ 7.2′		–	medium

A rich and compressed cluster that only shows a very faint glow in binoculars. Located in a glittering star field. Up until recently there was some debate as to whether this was a globular or open cluster. The consensus now is that it is a very young globular cluster, only 13,000 l.y. away. What makes this globular so nice is that the central stars can be resolved all the way to the core, which is rare among globular clusters.

Messier 75	NGC 6864	20ʰ 06.1ᵐ	–21° 55′	July 23
8.5 m	⊕ 6′		I	medium

A difficult object to locate with binoculars as it is so small and faint. Even then it will only appear as a hazy spot (like so many others). Will show a bright core and a few resolved stars in the halo with 25 cm aperture telescopes. One of the most distant globulars in Messier's catalogue, at 60,000 l.y.

Caldwell 47	NGC 6934	20h 34.2m	+07° 24′	July 30
8.9 m	⊕ 5.9′		VIII	medium

A difficult object for binoculars, appearing as a tiny patch if light. Just resolvable with 10 cm aperture telescopes, as a small, bright and round cluster, with a brighter and condensed centre. Some observers report that the use of averted vision aids in seeing some faint structure within the cluster. It has many blue straggler stars, and was one of the first objects to be imaged by the Gemini North Telescope, which resolved its core.

Messier 72	NGC 6981	20h 53.5m	–12° 32′	August 4
9.3 m	⊕ 6′		IX	medium

See August.

Caldwell 42	NGC 7006	21h 01.5m	+16° 11′	August 6
10.6 m	⊕ 2.8′		I	medium

See August.

Messier 92	NGC 6341	17h 17.1m	+43° 08′	June 10
6.4 m	⊕ 11′		IV	medium

See June.

Messier 14	NGC 6402	17h 37.6m	–03° 15′	June 16
7.6 m	⊕ 11.7′		VIII	medium

See June.

Herschel 147	NGC 6304	17h 14.5m	–29° 28′	June 18
8.4 m	⊕ 6.9′		VI	medium

See June.

Herschel 49	NGC 6522	18h 03.6m	–30° 02′	June 22
8.6 m	⊕ 5.6′		VI	medium

See June.

Herschel 12	NGC 6553	18h 09.3m	–25° 54′	June 24
8.1 m	⊕ 8.1′		XI	medium

See June.

Messier 28	NGC 6626	18h 24.5m	–24° 52′	June 27
6.8 m	⊕ 11.2′		IV	medium

See June.

Messier 69	NGC 6637	18h 31.4m	–32° 21′	June 29
7.6 m	⊕ 7.1′		V	medium

See June.

Messier 107	NGC 6171	$16^h\ 32.5^m$	−13° 03′	May 30
8.1 m	⊕ 10′		X	medium

See May.

Palomar 12	–	$21^h\ 46.5^m$	−21° 14′	August 18
11.7 m	⊕ 3′		XII	difficult

See August.

Herschel 50	NGC 6229	$16^h\ 47.0^m$	+47° 32′	June 3
9.4 m	⊕ 4.5′		IV	difficult

See June.

Herschel 200	NGC 6528	$18^h\ 04.8^m$	−30° 03′	June 22
9.5 m	⊕ 3.7′		V	difficult

See June.

Caldwell 66	NGC 5694	$14^h\ 39.6^m$	−26° 32′	May 1
10.2 m	⊕ 3.6′		VII	difficult

See May.

Herschel 19	NGC 5897	$15^h\ 17.4^m$	−21° 01′	May 11
8.6 m	⊕ 12.5′		XI	difficult

See May.

Herschel 19	NGC 5897	$15^h\ 17.4^m$	−21° 01′	May 11
8.6 m	⊕ 12.5′		XI	difficult

See May.

August

Messier 15	NGC 7078	$21^h\ 30.0^m$	+12° 10′	August 13
6.4 m	⊕ 12′		IV	easy

An impressive cluster in telescopes, it can be glimpsed with the naked eye. In binoculars it appears as a hazy object with no stars visible. Averted vision will be necessary in order to see the central stars. It does, however, under medium magnification and aperture, show considerable detail such as dark lanes, arcs of stars and a noticeable asymmetry. It is one of the few globulars that have a planetary nebula located within it – Pease-1, which is seen only in apertures of 30 cm and greater. The cluster is also an X-ray source.

Messier 2	NGC 7089	$21^h 33.5^m$	–00° 49′	August 14
6.4 m	⊕ 13′		II	easy

This is a very impressive non-stellar object. It can be seen with the naked eye, although averted vision will be necessary. However, as it is located in a barren area of the sky it can prove difficult to locate. But when found it is a rewarding object, and even in large binoculars its oval shape is apparent. Telescopes will show its bright core, and larger instruments will show several star chains snaking out from the core. Believed to be about 37,000 l.y. away and to contain over 100,000 stars.

Messier 30	NGC 7099	$21^h 40.4^m$	–23° 11′	August 16
7.3 m	⊕ 11′		V	easy

In binoculars it will appear simply as a tiny, round, hazy patch of light, and even in telescopes of aperture 20 to 25 cm will show just a bright, asymmetrical core with an unresolved halo. At its periphery are several looping arcs of stars.

Messier 55	NGC 6809	$19^h 40.0^m$	–30° 58′	July 17
6.3 m	⊕ 19′		XI	easy

See July.

Messier 13	NGC 6205	$16^h 41.7^m$	+36° 28′	June 1
5.7 m	⊕ 16.5′		V	easy

See June.

Messier 12	NGC 6218	$16^h 47.2^m$	–01° 57′	June 3
6.8 m	⊕ 14.5′		IX	easy

See June.

Messier 10	NGC 6254	$16^h 57.1^m$	–04° 06′	June 5
6.6 m	⊕ 15′		VII	easy

See June.

Messier 62	NGC 6266	$17^h 01.2^m$	–30° 07′	June 6
6.7 m	⊕ 14′		IV	easy

See June.

Messier 19	NGC 6273	$17^h 02.6^m$	–26° 16′	June 7
6.7 m	⊕ 13.5′		VIII	easy

See June.

Messier 9	NGC 6333	$17^h 19.2^m$	–18° 31′	June 11
7.6 m	⊕ 9.3′		VII	easy

See June.

Messier 22	NGC 6656	$18^h 36.4^m$	–23° 54′	June 30
5.1 m	⊕ 24′		VII	easy

See June.

Messier 5	NGC 5904	$15^h 18.6^m$	+02° 05′	May 11
5.7 m	⊕ 17.4′		V	easy

See May.

Messier 80	NGC 6093	$16^h 17.0^m$	–22° 59′	May 26
7.3 m	⊕ 8.9′		II	easy

See May.

Messier 80	NGC 6093	$16^h 17.0^m$	–22° 59′	May 26
7.3 m	⊕ 8.9′		II	easy

See May.

Messier 4	NGC 6121	$16^h 23.6^m$	–26° 32′	May 28
5.8 m	⊕ 26.3′		IX	easy

Messier 72	NGC 6981	$20^h 53.5^m$	–12° 32′	August 4
9.3 m	⊕ 6′		IX	medium

At a distance of about 60,000 l.y., this is faintest globular cluster catalogued by Messier. In binoculars it will appear as just a tiny, hazy point of light, but in telescopes or aperture 20 cm and larger its true nature becomes apparent. The use of averted vision may help you to see any detail within the cluster.

Caldwell 42	NGC 7006	$21^h 01.5^m$	+16° 11′	August 6
10.6 m	⊕ 2.8′		I	medium

Another very distant cluster, at 185,000 l.y. from the Solar System. In telescopes it appears as a small unresolved disc, not unlike a planetary nebula. However, even with large binoculars little will be seen unless averted vision is used. A very old cluster located far out in the galactic halo.

Messier 70	NGC 6681	$18^h 42.2^m$	–32° 18′	July 2
8.0 m	⊕ 7.8′		V	medium

See July.

Messier 54	NGC 6715	$18^h 55.1^m$	–30° 29′	July 5
7.6 m	⊕ 9.1′		III	medium

See July.

–	NGC 6760	$19^h 11.2^m$	+01° 02′	July 9
9.1 m	⊕ 6.5′		IX	medium

See July.

Messier 56	NGC 6779	19h 16.6m	+30° 11′	July 11
8.3 m	⊕ 7′		X	medium
See July.				

Messier 71	NGC 6838	19h 53.8m	+18° 47′	July 20
8.0 m	⊕ 7.2′		–	medium
See July.				

Messier 75	NGC 6864	20h 06.1m	–21° 55′	July 23
8.5 m	⊕ 6′		I	medium
See July.				

Caldwell 47	NGC 6934	20h 34.2m	+07° 24′	July 30
8.9 m	⊕ 5.9′		VIII	medium
See July.				

Messier 92	NGC 6341	17h 17.1m	+43° 08′	June 10
6.4 m	⊕ 11′		IV	medium
See June.				

Messier 14	NGC 6402	17h 37.6m	–03° 15′	June 16
37.6 m	⊕ 11.7′		VIII	medium
See June.				

Herschel 147	NGC 6304	17h 14.5m	–29° 28′	June 18
8.4 m	⊕ 6.9′		VI	medium
See June.				

Herschel 49	NGC 6522	18h 03.6m	–30° 02′	June 22
8.6 m	⊕ 5.6′		VI	medium
See June.				

Herschel 12	NGC 6553	18h 09.3m	–25° 54′	June 24
8.1 m	⊕ 8.1′		XI	medium
See June.				

Messier 28	NGC 6626	18h 24.5m	–24° 52′	June 27
6.8 m	⊕ 11.2′		IV	medium
See June.				

Messier 69	NGC 6637	18h 31.4m	–32° 21′	June 29
7.6 m	⊕ 7.1′		V	medium
See June.				

Messier 107	NGC 6171	$16^h 32.5^m$	−13° 03′	May 30
8.1 m	⊕ 10′		X	medium

See May.

Palomar 12	–	$21^h 46.5^m$	−21° 14′	August 18
11.7 m	⊕ 3′		XII	difficult

A challenge for amateurs who possess telescopes of aperture 20 to 25 cm. It will appear as a faint, small patch of nebulosity. Perfect seeing conditions will be needed.

Herschel 50	NGC 6229	$16^h 47.0^m$	+47° 32′	June 3
9.4 m	⊕ 4.5′		IV	difficult

See June.

Herschel 200	NGC 6528	$18^h 04.8^m$	−30° 03′	June 22
9.5 m	⊕ 3.7′		V	difficult

See June.

Caldwell 66	NGC 5694	$14^h 39.6^m$	−26° 32′	May 1
10.2 m	⊕ 3.6′		VII	difficult

See May.

Herschel 19	NGC 5897	$15^h 17.4^m$	−21° 01′	May 11
8.6 m	⊕ 12.5′		XI	difficult

See May.

September

Messier 15	NGC 7078	$21^h 30.0^m$	+12° 10′	August 13
6.4 m	⊕ 12′		IV	easy

See August.

Messier 2	NGC 7089	$21^h 33.5^m$	−00° 49′	August 14
6.4 m	⊕ 13′		II	easy

See August.

Messier 30	NGC 7099	$21^h 40.4^m$	−23° 11′	August 16
7.3 m	⊕ 11′		V	easy

See August.

Messier 55	NGC 6809	$19^h 40.0^m$	−30° 58′	July 17
6.3 m	⊕ 19′		XI	easy

See July.

Messier 13	NGC 6205	$16^h\ 41.7^m$	+36° 28′	June 1
5.7 m	⊕ 16.5′		V	easy
See June.				

Messier 12	NGC 6218	$16^h\ 47.2^m$	–01° 57′	June 3
6.8 m	⊕ 14.5′		IX	easy
See June.				

Messier 10	NGC 6254	$16^h\ 57.1^m$	–04° 06′	June 5
6.6 m	⊕ 15′		VII	easy
See June.				

Messier 62	NGC 6266	$17^h\ 01.2^m$	–30° 07′	June 6
6.7 m	⊕ 14′		IV	easy
See June.				

Messier 19	NGC 6273	$17^h\ 02.6^m$	–26° 16′	June 7
6.7 m	⊕ 13.5′		VIII	easy
See June.				

Messier 9	NGC 6333	$17^h\ 19.2^m$	–18° 31′	June 11
7.6 m	⊕ 9.3′		VII	easy
See June.				

Messier 22	NGC 6656	$18^h\ 36.4^m$	–23° 54′	June 30
5.1 m	⊕ 24′		VII	easy
See June.				

Messier 72	NGC 6981	$20^h\ 53.5^m$	–12° 32′	August 4
9.3 m	⊕ 6′		IX	medium
See August.				

Caldwell 42	NGC 7006	$21^h\ 01.5^m$	+16° 11′	August 6
10.6 m	⊕ 2.8′		I	medium
See August.				

Messier 70	NGC 6681	$18^h\ 42.2^m$	–32° 18′	July 2
8.0 m	⊕ 7.8′		V	medium
See July.				

Messier 54	NGC 6715	$18^h\ 55.1^m$	–30° 29′	July 5
7.6 m	⊕ 9.1′		III	medium
See July.				

–	NGC 6760	19^h 11.2^m	+01° 02′	July 9
9.1 m	⊕ 6.5′		IX	medium
See July.				

Messier 56	NGC 6779	19^h 16.6^m	+30° 11′	July 11
38.3 m	⊕ 7′		X	medium
See July.				

Messier 71	NGC 6838	19^h 53.8^m	+18° 47′	July 20
8.0 m	⊕ 7.2′		–	medium
See July.				

Messier 75	NGC 6864	20^h 06.1^m	–21° 55′	July 23
8.5 m	⊕ 6′		I	medium
See July.				

Caldwell 47	NGC 6934	20^h 34.2^m	+07° 24′	July 30
8.9 m	⊕ 5.9′		VIII	medium
See July.				

Messier 92	NGC 6341	17^h 17.1^m	+43° 08′	June 10
6.4 m	⊕ 11′		IV	medium
See June.				

Messier 14	NGC 6402	17^h 37.6^m	–03° 15′	June 16
7.6 m	⊕ 11.7′		VIII	medium
See June.				

Herschel 147	NGC 6304	17^h 14.5^m	–29° 28′	June 18
8.4 m	⊕ 6.9′		VI	medium
See June.				

Herschel 49	NGC 6522	18^h 03.6^m	–30° 02′	June 22
8.6 m	⊕ 5.6′		VI	medium
See June.				

Herschel 12	NGC 6553	18^h 09.3^m	–25° 54′	June 24
8.1 m	⊕ 8.1′		XI	medium
See June.				

Messier 28	NGC 6626	18^h 24.5^m	–24° 52′	June 27
6.8 m	⊕ 11.2′		IV	medium
See June.				

Messier 69	NGC 6637	$18^h 31.4^m$	–32° 21′	June 29
7.6 m	⊕ 7.1′		V	medium

See June.

Palomar 12	–	$21^h 46.5^m$	–21° 14′	August 18
11.7 m	⊕ 3′		XII	difficult

See August.

Herschel 50	NGC 6229	$16^h 47.0^m$	+47° 32′	June 3
9.4 m	⊕ 4.5′		IV	difficult

See June.

Herschel 200	NGC 6528	$18^h 04.8^m$	–30° 03′	June 22
9.5 m	⊕ 3.7′		V	difficult

See June.

October

Messier 15	NGC 7078	$21^h 30.0^m$	+12° 10′	August 13
6.4 m	⊕ 12′		IV	easy

See August.

Messier 2	NGC 7089	$21^h 33.5^m$	–00° 49′	August 14
6.4 m	⊕ 13′		II	easy

See August.

Messier 30	NGC 7099	$21^h 40.4^m$	–23° 11′	August 16
7.3 m	≈ 11¢		V	easy

See August.

Messier 55	NGC 6809	$19^h 40.0^m$	–30° 58′	July 17
6.3 m	⊕ 19′		XI	easy

See July.

Messier 13	NGC 6205	$16^h 41.7^m$	+36° 28′	June 1
5.7 m	⊕ 16.5′		V	easy

See June.

Messier 12	NGC 6218	$16^h 47.2^m$	–01° 57′	June 3
6.8 m	⊕ 14.5′		IX	easy

See June.

Messier 10	NGC 6254	$16^h\ 57.1^m$	–04° 06′	June 5
6.6 m	⊕ 15′		VII	easy
See June.				

Messier 62	NGC 6266	$17^h\ 01.2^m$	–30° 07′	June 6
6.7 m	⊕ 14′		IV	easy
See June.				

Messier 19	NGC 6273	$17^h\ 02.6^m$	–26° 16′	June 7
6.7 m	⊕ 13.5′		VIII	easy
See June.				

Messier 9	NGC 6333	$17^h\ 19.2^m$	–18° 31′	June 11
7.6 m	⊕ 9.3′		VII	easy
See June.				

Messier 22	NGC 6656	$18^h\ 36.4^m$	–23° 54′	June 30
5.1 m	⊕ 24′		VII	easy
See June.				

Messier 79	NGC 1904	$05^h\ 24.5^m$	–24° 33′	December 12
7.8 m	⊕ 8.5′		V	medium
See December.				

Messier 72	NGC 6981	$20^h\ 53.5^m$	–12° 32′	August 4
9.3 m	⊕ 6′		IX	medium
See August.				

Caldwell 42	NGC 7006	$21^h\ 01.5^m$	+16° 11′	August 6
10.6 m	⊕ 2.8′		I	medium
See August.				

Messier 70	NGC 6681	$18^h\ 42.2^m$	–32° 18′	July 2
8.0 m	⊕ 7.8′		V	medium
See July.				

Messier 54	NGC 6715	$18^h\ 55.1^m$	–30° 29′	July 5
7.6 m	⊕ 9.1′		III	medium
See July.				

–	NGC 6760	$19^h\ 11.2^m$	+01° 02′	July 9
9.1 m	⊕ 6.5′		IX	medium
See July.				

Messier 56	NGC 6779	19^h 16.6^m	+30° 11′	July 11
8.3 m	⊕ 7′		X	medium
See July.				

Messier 71	NGC 6838	19^h 53.8^m	+18° 47′	July 20
8.0 m	⊕ 7.2′		–	medium
See July.				

Messier 75	NGC 6864	20^h 06.1^m	–21° 55′	July 23
8.5 m	⊕ 6′		I	medium
See July.				

Caldwell 47	NGC 6934	20^h 34.2^m	+07° 24′	July 30
8.9 m	⊕ 5.9′		VIII	medium
See July.				

Messier 92	NGC 6341	17^h 17.1^m	+43° 08′	June 10
6.4 m	⊕ 11′		IV	medium
See June.				

Messier 14	NGC 6402	17^h 37.6^m	–03° 15′	June 16
7.6 m	⊕ 11.7′		VIII	medium
See June.				

Herschel 147	NGC 6304	17^h 14.5^m	–29° 28′	June 18
8.4 m	⊕ 6.9′		VI	medium
See June.				

Herschel 49	NGC 6522	18^h 03.6^m	–30° 02′	June 22
8.6 m	⊕ 5.6′		VI	medium
See June.				

Herschel 12	NGC 6553	18^h 09.3^m	–25° 54′	June 24
8.1 m	⊕ 8.1′		XI	medium
See June.				

Messier 28	NGC 6626	18^h 24.5^m	–24° 52′	June 27
6.8 m	⊕ 11.2′		IV	medium
See June.				

Messier 69	NGC 6637	18^h 31.4^m	–32° 21′	June 29
7.6 m	⊕ 7.1′		V	medium
See June.				

Palomar 12	–	$21^h 46.5^m$	–21° 14′	August 18
11.7 m	⊕ 3′		XII	difficult
See August.				

Herschel 50	NGC 6229	$16^h 47.0^m$	+47° 32′	June 3
39.4 m	⊕ 4.5′		IV	difficult
See June.				

Herschel 200	NGC 6528	$18^h 04.8^m$	–30° 03′	June 22
9.5 m	⊕ 3.7′		V	difficult
See June.				

November

Messier 15	NGC 7078	$21^h 30.0^m$	+12° 10′	August 13
6.4 m	⊕ 12′		IV	easy
See August.				

Messier 2	NGC 7089	$21^h 33.5^m$	–00° 49′	August 14
6.4 m	⊕ 13′		II	easy
See August.				

Messier 30	NGC 7099	$21^h 40.4^m$	–23° 11′	August 16
7.3 m	⊕ 11′		V	easy
See August.				

Messier 55	NGC 6809	$19^h 40.0^m$	–30° 58′	July 17
6.3 m	⊕ 19′		XI	easy
See July.				

Messier 79	NGC 1904	$05^h 24.5^m$	–24° 33′	December 12
7.8 m	⊕ 8.5′		V	medium
A fine cluster, best appreciated with a telescope because binoculars cannot resolve it. Small telescopes of, say, 10 cm aperture can resolve the core, but it will be a challenge because you will need a high magnification, dark skies and averted vision. A perfect test for you and your telescope optics. Telescopes of aperture 40 cm and more will resolve the core with no difficulty. Located at a distance of 41,000 l.y., it is the sole globular cluster for northern hemisphere observers in the winter.				

Messier 72	NGC 6981	$20^h 53.5^m$	–12° 32′	August 4
9.3 m	⊕ 6′		IX	medium
See August.				

Caldwell 42	NGC 7006	$21^h 01.5^m$	+16° 11′	August 6
10.6 m	⊕ 2.8′		I	medium
See August.				

Messier 70	NGC 6681	$18^h 42.2^m$	–32° 18′	July 2
8.0 m	⊕ 7.8′		V	medium
See July.				

Messier 54	NGC 6715	$18^h 55.1^m$	–30° 29′	July 5
7.6 m	⊕ 9.1′		III	medium
See July.				

–	NGC 6760	$19^h 11.2^m$	+01° 02′	July 9
9.1 m	⊕ 6.5′		IX	medium
See July.				

Messier 56	NGC 6779	$19^h 16.6^m$	+30° 11′	July 11
8.3 m	⊕ 7′		X	medium
See July.				

Messier 56	NGC 6779	$19^h 16.6^m$	+30° 11′	July 11
8.3 m	⊕ 7′		X	medium
See July.				

Messier 71	NGC 6838	$19^h 53.8^m$	+18° 47′	July 20
8.0 m	⊕ 7.2′		–	medium
See July.				

Messier 75	NGC 6864	$20^h 06.1^m$	–21° 55′	July 23
8.5 m	⊕ 6′		I	medium
See July.				

Caldwell 47	NGC 6934	$20^h 34.2^m$	+07° 24′	July 30
8.9 m	⊕ 5.9′		VIII	medium
See July.				

Palomar 12	–	$21^h 46.5^m$	–21° 14′	August 18
11.7 m	⊕ 3′		XII	difficult
See August.				

December

Messier 15	NGC 7078	$21^h 30.0^m$	+12° 10′	August 13
6.4 m	⊕ 12′		IV	easy
See August.				

Messier 2	NGC 7089	$21^h 33.5^m$	–00° 49′	August 14
6.4 m	⊕ 13′		II	easy
See August.				

Messier 30	NGC 7099	$21^h 40.4^m$	–23° 11′	August 16
7.3 m	⊕ 11′		V	easy
See August.				

Messier 55	NGC 6809	$19^h 40.0^m$	–30° 58′	July 17
6.3 m	⊕ 19′		XI	easy
See July.				

Messier 79	NGC 1904	$05^h 24.5^m$	–24° 33′	December 12
7.8 m	⊕ 8.5′		V	medium
See December.				

Messier 72	NGC 6981	$20^h 53.5^m$	–12° 32′	August 4
9.3 m	⊕ 6′		IX	medium
See August.				

Caldwell 42	NGC 7006	$21^h 01.5^m$	+16° 11′	August 6
10.6 m	⊕ 2.8′		I	medium
See August.				

Messier 70	NGC 6681	$18^h 42.2^m$	–32° 18′	July 2
8.0 m	⊕ 7.8′		V	medium
See July.				

Messier 54	NGC 6715	$18^h 55.1^m$	–30° 29′	July 5
7.6 m	⊕ 9.1′		III	medium
See July.				

–	NGC 6760	$19^h 11.2^m$	+01° 02′	July 9
9.1 m	⊕ 6.5′		IX	medium
See July.				

Messier 56	NGC 6779	$19^h 16.6^m$	+30° 11′	July 11
8.3 m	⊕ 7′		X	medium
See July.				

Messier 71	NGC 6838	$19^h 53.8^m$	+18° 47′	July 20
8.0 m	⊕ 7.2′		–	medium
See July.				

Messier 75	NGC 6864	$20^h 06.1^m$	–21° 55′	July 23
8.5 m	⊕ 6′		I	medium
See July.				

Caldwell 47	NGC 6934	20h 34.2m	+07° 24′	July 30
8.9 m	⊕ 5.9′		VIII	medium
See July.				

Caldwell 25	NGC 2419	07h 38.1m	+38° 53′	January 15
10.4 m	⊕ 4.1′		II	difficult
See January.				

Palomar 12	–	21h 46.5m	–21° 14′	August 18
11.7 m	⊕ 3′		XII	difficult
See August.				

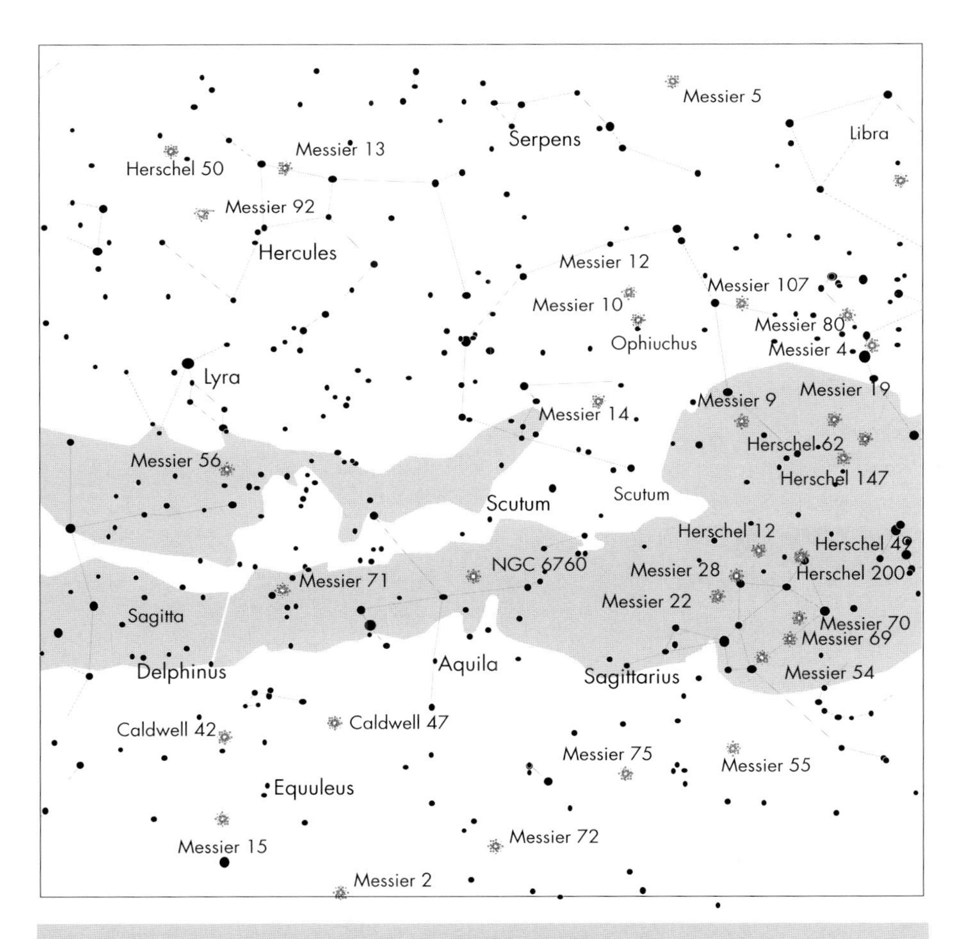

Figure 3.11. Globular clusters 1.

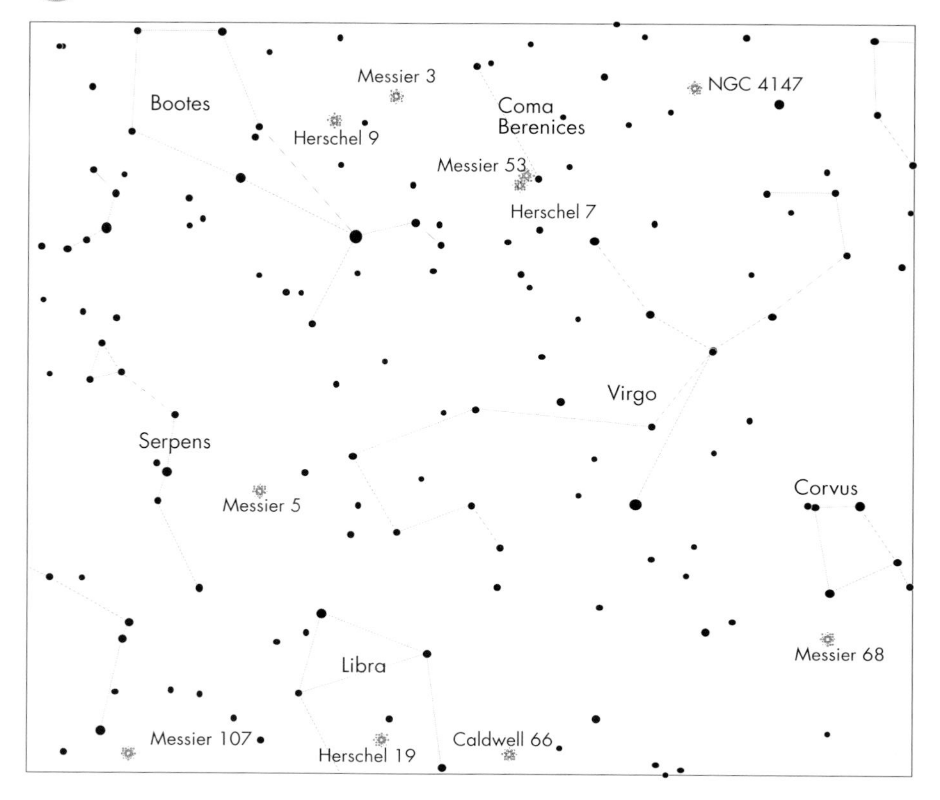

Figure 3.12. Globular clusters 2.

3.4 Stellar Associations and Streams

The previous sections discussed those groups of stars which can easily be recognised as either an open cluster, or a globular cluster, but there exists another type of grouping of stars, that is much more ephemeral and spread over a very large region of the sky. In fact, after reading this section you may think that stellar associations are not a further classification of cluster at all, but something very different!

A stellar association is a loosely bound group of very young stars. They may still be swathed in the dust and gas cloud they formed within and star formation will still be occurring within the cloud. Where they differ from other clusters is in the fact that they are enormous, covering both a sizable angular area of the night sky, and at the same time encompass a comparably large volume in space. As an illustration of this huge size, the Scorpius-Centaurus Association is around 700 by 760 l.y. in extent, and covers about 80°.

There are three types of Stellar association:

OB associations, containing very luminous O- and B- type main sequence, giant and super-giant stars.

B associations, containing only B-type main sequence and giant stars but with an absence of O-type stars. These associations are just older versions of the OB association, and thus the faster evolving O-type stars have been lost to the group as supernovae.

T Associations, are groupings of T Tauri type stars. These are irregular variable stars that are still contracting and evolving toward being A-, F- and G- type main sequence stars. As they are still in their infancy, more often than not they will be shrouded in dark dust clouds, and those that are visible will be embedded in small reflection and emission nebulae (see Chapter 4).

The OB associations are truly enormous objects, often covering many hundreds of light years. This is a consequence of the fact that massive O- and B-type stars can only be formed within the huge giant molecular clouds which are themselves hundreds of light years across. On the other hand, the T associations are much smaller affairs, perhaps only a few light years in diameter. In some cases, the T association is itself located within or near an OB association.

The lifetime of an association is comparatively short. The very luminous O-type stars are soon lost to the group as supernovae, and, as usual, the ever pervasive gravitational effects of the Galaxy soon disrupt the association. The coherence and identity of the group can only exist for as long as the brighter components stay in the same general area of a spiral arm, as well as having a similar space motion through the Galaxy. As time passes, the B-type stars will disappear through stellar evolution, and the remaining A-type and later stars will now be spread over an enormous volume of space, and the only common factor amongst them will be their motion through space. The association is now called a *stellar stream*. An example of such a stream and one which often surprises the amateur (it did me!) is the Ursa Major Stream. This is an enormous group of stars, with the five central stars of Ursa Major[5] (The Plough) being its most concentrated and brightest members. Furthermore, the stream is also known as the Sirius Supercluster after its brightest member. The Sun actually lies within this stream [more information about this fascinating stream can be found below].

Although there are over 70 stellar associations and streams known to exist, only a handful are visible using the naked eye or amateur telescopes., nevertheless, they are wonderful objects due to the fact that they cover an appreciable area in the sky and are composed of dozens of stars of naked eye visibility. In fact they may be amongst the few deep-sky objects that can be observed without any optical aid. Even if the observing conditions for deep sky work are less than favourable, it should be possible to see these amazing objects.

As stellar associations cover such a large area of the sky, it is difficult, and even pointless to specify a specific set of coordinates for a particular association. Thus, any details listed below will refer to the association as a whole and not just to any one specific star, unless otherwise stated.

[5]There is some debate as to whether these five stars are in fact the central stars of an open cluster. If so, it is the nearest to the Solar System at a distance of 75 l.y.

The Orion Association, 1600 l.y.

This association includes most of the stars in the constellation down to 3.5 magnitude, except for γ Orionis and π^3 Orionis. Also included are several 4th, 5th, and 6th magnitude stars. The wonderful nebula M42 is also part of this spectacular association. Several other nebulae[6] [including dark, reflection and emission nebulae] are all located within a vast Giant Molecular Cloud, which is the birthplace of all the O-and B-type supergiant, giant and main sequence stars in Orion. The association is believed to be 800 l.y. across and 1000 l.y. deep. By looking at this association, you are in fact looking deep into our own spiral arm, which incidentally is called the Cygnus-Carina Arm.

The Scorpius-Centaurus Association, 550 l.y.

A much older, but closer association than the Orion association. It includes most of the stars of 1st, 2nd and 3rd magnitude in Scorpius down through Lupus and Centaurus to Crux. Classed as a B-type association because it lacks O-type stars, its angular size on the sky is around 80°. It is estimated to be 750 × 300 l.y. in size and 400 l.y. deep, with the centre of the association midway between α Lupi and ζ Centauri. Its elongated shape is thought to be the result of rotational stresses induced by its rotation about the Galactic centre. Bright stars in this association include θ Ophiuchi, β, ν, δ and σ Scorpii, α, γ Lupi, ε, δ, μ and ε Centauri, and β Crucis.

The Zeta Persei Association, 300 l.y.

Also known as Per OB2, this association includes ζ and ξ Persei, as well as 40, 42 and o Persei. The California nebula, NGC 1499, is also within this association.

The Ursa Major Stream, 75 l.y.

As briefly mentioned earlier in this section, this stream includes the five central stars of the Plough. It is spread over a vast area of the sky, approximately 24°, and is around 20 × 30 l.y. in extent. It includes as members, Sirius [α Canis Majoris], α Coronae Borealis, δ Leonis, β Eridani, δ Aquarii and β Serpentis. Due to the predominance of A1 and A0 stars within the association, its age has been estimated at 300 million years.

The Hyades Stream

There is some evidence (although it is not fully agreed upon) that the Ursa Major stream is itself within a much older and larger stream. This older component includes M44, Praesepe in Cancer and the Hyades in Taurus, with these two open clusters being the core of a very large, but loose grouping of stars. Included within this are Capella [α Aurigae],

---

[6]These nebulae are described in Chapter 4

α Canum Venaticorum,[7] δ Cassiopeiae and λ Ursae Majoris. The stream extends to over 200 l.y. beyond the Hyades star cluster, and 300 l.y. behind the Sun. Thus, the Sun is believed to lie within this stream.[8]

The Alpha Persei Stream, 40 l.y.

Also known as Melotte 20, this is a group of about 100 stars including, α Persei, ψ Persei, 29 and 34 Persei. The stars δ and ϵ Persei are believed to be amongst its most outlying members, as they also share the same space motion as the main groups of stars. The inner region of the stream is measured to be over 33 l.y. in length; the distance between 29 to ψ Persei.

---

[7]Capella and α Canum Venaticorum are also thought to be members of the even larger Taurus Stream, which has a motion through space similar to the Hyades, and thus may be related.
[8]The bright stars that extend from Perseus, Taurus and Orion, and down to Centaurus and Scorpius, including the Orion and Scorpius-Centaurus associations, lie at an angle of about 1 °5 to the Milky Way, and thus to the equatorial plane of the Galaxy. This group or band of stars is often called *Gould's Belt*.

This glorious image of the Crab Nebula was obtained by Hubble's Wide Field and Planetary Camera 2. Images taken with five different colour filters were used build this new false-colour picture.

The image shows ragged shreds of gas that are expanding away from the explosion site at over 3 million miles per hour. The core of the star has survived the explosion as a "pulsar", visible in the Hubble image as the lower of the two moderately bright stars to the upper left centre. The pulsar is a neutron star, spinning on its axis 30 times a second. It heats its surroundings, creating the ghostly diffuse bluish-green glowing gas cloud in its vicinity (including the blue arc just to its right). The colourful network of filaments is material from its outer layers, expelled during the explosion.

The picture is deceptive in that the filaments appear to be close to the pulsar. In reality, the yellowish-green filaments toward the bottom of the image are closer to us, and approaching at some 300 miles per second. The orange and pink filaments toward the top of the picture include material behind the pulsar, rushing away from us at a similar speed.

NASA/Hubble Heritage Team (STScI/AURA)/W.P. Blair (JHU)

Chapter 4 Nebulae

4.1 Introduction

Of all the images you can see in those large-format "coffee table" astronomy books, or posted on the Internet, perhaps those of nebulae are the most spectacular. With their panoramas of glowing red, green and yellow clouds of gas, and impenetrable blankets of black dust lanes, often intertwined and seemingly swirling around each other, they give a wonderfully vivid impression of the amazing mechanisms at work in the universe.

Nebulae are actually very disparate in nature, even though many of them have a rather similar appearance. They are associated with the areas of star formation, cover several aspects of a star's life, and end with the process of star death. This chapter will cover the four main types of nebula – emission, reflecting, dark and planetary – but it includes supernova remnants as well.

However, as this book is a guide for amateur astronomers, and thus seeks to give a true and honest description of what can be seen, I have to begin by saying that the glorious pictures I referred to in the first paragraph give a completely false impression of what one can see with the naked eye, or indeed with optical equipment. It isn't that I am trying to be negative and depressing, but it is a fact that nebulae are very faint and elusive objects. Those wonderful images, which may have been obtained with the world's largest telescopes (including, quite often, the Hubble Space Telescope) and using the most sophisticated techniques, are long-exposure images made under perfect conditions, when there is sufficient time for the faint image to be built up, usually over several hours. The result is a bright and detailed picture.

The naked eye works in a different way from a photographic plate or CCD camera, and the image is seen "instantly" (more as it is by a video camera). This is the reason we don't see images comparable to those in books. Also, add to this the fact that good observing conditions are crucial in order to observe nebulae (and the largest telescopes are located at the best possible places for seeing, including earth-orbit) and it comes as no surprise that nebulae can often disappoint the inexperienced observer.

Having now completely dampened your spirits, let me lift them by saying there are many nebulae that can be seen with moderate instruments, and several that can be glimpsed with the naked eye or binoculars. Indeed, many of the brighter or more distinctive nebulae can be easily recognised, and a few show faint traces or tints of subtle colour.

As with most astronomical objects (except perhaps star clusters) the bigger the aperture, the more detail that will be seen, and telescopes of aperture 20 cm and greater will show hundreds of different objects: wispy emission nebulae, blue–green discs of planetary nebula, the barely discernible, breath-like reflecting nebulae, difficult and mysterious dark nebulae, and the rare and awe-inspiring supernovae remnants.

With the advent in recent years of the availability of specialised filters to improve visual contrast (once these were the province only of professional astronomers), even more nebulae become available to the amateur, and there are a handful that are among the most spectacular and beautiful objects in the night sky!

On crisp winter nights, or warm and still summer evenings, searching for and observing nebulae is a wonderful and rewarding pastime.

4.2 Emission Nebulae

This type of nebula is probably the easiest to observe, owing to its brightness and size. Emission nebulae are actually clouds of gas – mainly hydrogen, but there can be small amounts of oxygen and nitrogen present. Within the clouds are very hot O- and B-type stars which produce immense amounts of ultraviolet radiation. Usually, these very luminous stars are actually born within and from the material of the clouds, and so many emission nebulae are "stellar nurseries". Radiation from the stars causes the (usually) hydrogen gas to undergo a process called *fluorescence*, and it is this which is responsible for the glow observed from the clouds of gas. A full explanation requires a reasonable knowledge of astrophysics but I'll attempt a simple explanation by considerably simplifying things ...

The energy provided by ultraviolet radiation from the young and hot stars ionises the hydrogen. In other words energy, in this case in the form of ultra-violet radiation, is absorbed by the atom, and transferred to an electron which is sitting comfortably in what is called an energy level or orbital shell.[1] The electron, having gained extra energy, can leave the energy level it is in, and in some instances actually break free from the atom. This process when an atom loses an electron is called *ionisation*.

If electrons are broken free from their atoms, the hydrogen cloud contains some hydrogen atoms without electrons, and a corresponding number of free electrons. Eventually[2] the electrons recombine with the atoms, but an electron can't just settle down back to the state it originally had before it absorbed the extra energy – it has to lose the extra energy that the ultraviolet imparted. For this to happen, the electron moves down the atomic energy levels until it reaches its original level, losing energy as it goes. In hydrogen (the most common gas in the nebula, you remember), an electron moving down from the third energy level to the second emits a photon of light at 656.3 nanometres.

This is the origin of the famous "hydrogen alpha line", usually written as *H-alpha*. It is a lovely red colour and is responsible for all the pink and red glowing gas clouds seen in the images referred to at the start of this chapter. Unfortunately the red glow is usually too weak to be seen at the eyepiece.

---

[1]Our simple model of an atom has a central nucleus with electrons orbiting around it, somewhat like planets orbiting a sun. Not all orbits are allowed by quantum mechanics: in order to move up to higher energy levels, electrons need a very specific amount energy. Too much or too little and an electron will not move.

[2]The time spent before recombining is very short, millionths of seconds, but also depends on how much radiation is present and the density of the gas cloud.

When electrons move down from other energy levels within the atom, other specific wavelengths of light are emitted. For instance, when an electron moves from the second level to the first, it emits a photon in the ultraviolet part of the spectrum – this particular wavelength is called the *Lyman alpha line* of hydrogen.

It is this process of atoms absorbing radiation to ionise a gas, with electrons subsequently cascading down the energy levels of an atom, that is responsible for nearly all the light we see from emission nebulae.

If a gas cloud is particularly dense, oxygen in it may be ionised and the resulting recombination of the electron and atom produces the doubly ionised lines, at wavelengths of 495.9 nm and 500.7 nm. These lines are a rich blue–green colour, and under good seeing and with clean optics this colour can be glimpsed in the Orion Nebula, M42.

Emission nebula are sometimes called HII regions, pronounced "aitch two". This astrophysical term refers to hydrogen which has lost one electron by ionisation. The term HI, or "aitch one", refers to hydrogen which is unaffected by any radiation, that is, neutral hydrogen. The doubly ionised oxygen line mentioned above is termed OIII ("oh three"); the "doubly" means that *two* of the outermost electrons have been lost from the atom by ionisation.[3]

Physically, the shape of an emission nebula is dependent on several factors: the amount of radiation available, the density of the gas cloud and the amount of gas available for ionisation. In the case where there is a significant amount of radiation, coupled with a small and low-density cloud, then most probably all of the cloud will be ionised, and thus the resulting HII region will be of an irregular shape – just the shape of the cloud itself. However, if the cloud of gas is large and dense, then the radiation can penetrate only a certain distance before it is used up – that is, there is only a fixed amount of radiation available for ionisation. In this case, the HII region will be a sphere,[4] often surrounded by the remaining gas cloud, which is not fluorescing. Many of the emission regions that are irregular in shape include M42, the Orion Nebula, M8, the Lagoon Nebula, and M17 in Sagittarius. Those that exhibit a circular shape, and thus are in fact spherical, are M20, the Trifid Nebula, and NGC 2237, the Rosette Nebula, to name only two.

After a suitable period of time, usually several million years, the group of young O- and B-type stars located at the centre of the nebulae will be producing so much radiation that they can in affect sweep away the residual gas and dust clouds that surround them. This produces a "bubble" of clear space surrounding the cluster of stars. Several emission regions show this, for example, NGC 6276 and M78, which show the star cluster residing in a circular clear area within the larger emission nebula

Many emission nebulae are faint and have a low surface brightness, making them not exactly difficult objects to observe, but rather featureless and indistinct, though in some instances the brighter nebulae do show several easily seen features. These nebulae will be described in detail in the following list. It goes without saying that clean optics (for maximum contrast) and excellent seeing conditions should be a priorities when observing these nebulae, and dark adaption and averted vision are required techniques.

There exist several classification schemes for emission nebula, but only one is used here. It is a measure of the visibility of the nebula as seen on the photographic plates from the Palomar Observatory Sky Survey (POSS). The photographic brightness is assigned a value from 1 through to 6; those nebulae rated at 1 are just barely detectable on the plate, while those quoted at a value of 6 are easily seen on the photographic plate. In the context of this

---

[3]In some astrophysical contexts, such as in the centre of quasars, conditions exist which can give rise to terms such as Fe23. The amount of radiation is so phenomenal that the atom of iron (Fe) has been ionised to such an extent that it has lost 22 of its electrons!

[4]This is often called the Stromgren sphere, named after the astronomer Bengt Stromgren, who did some pioneering work on HII regions.

book, it is just the measure of the difficulty (or ease) of observation, and is given the symbol ●. The size of an object is also given in arc seconds and is indicated by the symbol ⊕. Where a value of ⊕ is given as $x|y$, then the object is approximately x arc seconds long by y arc minutes wide. The remaining usual nomenclature applies. Finally, two very simple star maps showing the positions of the emission nebulae are at the end of this section.

January–February–March

Messier 42	NGC 1976	$05^h 35.4^m$	–05° 27′	December 15
● 1–5	⊕ 65|60′			easy
See October–November–December.				

Messier 43	NGC 1982	$05^h 35.6^m$	–05° 16′	December 15
● 1–5	⊕ 20|15′			easy
See October–November–December.				

Caldwell 49	NGC 2237–39	$06^h 32.3^m$	+05° 03′	December 29
● 1–5	⊕ 80|60′			easy
See October–November–December.				

Gum 4	NGC 2359	$07^h 18.6^m$	–13° 12′	January 10
● 2–5	⊕ 9|6′			moderate
Also known as the Duck Nebula. This is a bright emission nebula easily seen in telescopes of aperture 20 cm. It consists of two patches of nebulosity, with the northern patch being the larger and less dense. Using an OIII filter will greatly improve the appearance of the emission nebula, showing the delicate filamentary nature.				

Herschel 261	NGC 1931	$05^h 31.4^m$	+34° 15′	December 14
● 1–5	⊕ 4|4′			moderate
See October–November–December.				

-	NGC 1999	$05^h 36.5^m$	–06° 42′	December 15
● 1–5	⊕ 2|2′			moderate
See October–November–December.				

-	NGC 2024	$05^h 40.7^m$	02° 27′	December 16
● 2–5	⊕ 30|30′			moderate
See October–November–December.				

Messier 78	NGC 2068	$05^h 46.7^m$	00° 03′	December 17
● 1–5	⊕ 8|6′			moderate
See October–November–December.				

Caldwell 46	NGC 2261	$06^h 39.2^m$	+08° 44′	December 31
● 1–5	⊕ 3.5|1.5′			moderate
See October–November–December.				

April–May–June

Messier 20	NGC 6514	18h 02.3m	–23° 02′	June 22
● 1–5	⊕ 20I20′			easy

Also known as the Trifid Nebula. This emission nebula can be glimpsed as a small hazy patch of nebulosity, and in fact is difficult to locate on warm summer evenings unless the skies are very transparent. With aperture around 15 cm, the nebula is easy to see, along with its famous three dark lanes which give it its name. They radiate outwards from the central object, an O8-type star which is the power source for the nebula. The northern nebulosity is in fact a reflection nebula, and thus harder to observe. In large telescopes the nebula is wonderful and repays long and careful observation.

Messier 8	NGC 6523	18h 03.8m	–24° 23′	June 22
● 1–5	⊕ 45I30′			easy

Also known as the Lagoon Nebula. Visible to the naked eye on summer evenings, this is thought by many to be the premier emission nebula of the summer sky. Binoculars will show a vast expanse of glowing green–blue gas split by a very prominent dark lane. Using a light filters and telescopes of aperture 30 cm will show much intricate and delicate detail, including many dark bands. The Lagoon Nebula is located in the Sagittarius–Carina Spiral Arm of our Galaxy, at a distance of 5400 l.y.

Messier 17	NGC 6618	18h 20.8m	–16° 11′	June 27
● 1–5	⊕ 20I16′			easy

Also known as the Swan or Omega Nebula. This is a magnificent object in binoculars, and is perhaps a rival to the Orion Nebula, M42, for the summer sky. Not often observed by amateurs, which is a pity as it offers much. With telescopes the detail of the nebula becomes apparent, and with the addition of a light filter it can in some instances surpass M42. Certainly, it has many more dark and light patches than its winter cousin, although it definitely needs an OIII filter for the regions to be fully appreciated. Another celestial object that warrants slow and careful study.

Messier 16	IC 4703	18h 18.6m	–13° 58′	June 26
● 1–5	⊕ 35I30′			moderate

Also known as the Star Queen or Eagle Nebula. A famous though not often observed nebula. Although it can be glimpsed in binoculars, and will appear as a hazy patch with the naked eye, telescopic observation is needed to see any detail. As is usual, the use of a filter enhances the visibility. The “Black Pillar” and associated nebulosity are difficult to see, even though they are portrayed in many beautiful photographs. (A prime example of astronomical imagery fooling the amateur into thinking that these justifiably impressive objects can easily be seen through a telescope.) Nevertheless, they can be spotted by an astute observer under near-perfect conditions.

Gum 4	NGC 2359	07h 18.6m	–13° 12′	January 10
● 2–5	⊕ 9I6′			moderate

See January–February–March.

July–August–September

Messier 20	NGC 6514	18h 02.3m	–23° 02′	June 22
● 1–5	⊕ 20|20′			easy

See April–May–June.

Messier 8	NGC 6523	18h 03.8m	–24° 23′	June 22
● 1–5	45|30′			easy

See April–May–June.

Messier 17	NGC 6618	18h 20.8m	–16° 11′	June 27
● 1–5	⊕ 20|16′			easy

See April–May–June.

–	IC 5067–70	20h 50.8m	+44° 21′	August 4
● 1–5	⊕ 60|50′			easy

Also known as the Pelican Nebula. This nebula, close to the North America Nebula (see the entry below), has been reported to be visible to the naked eye. It is easily glimpsed in binoculars as a triangular faint hazy patch of light. It can be seen best with averted vision, and the use of light filters.

Caldwell 20	NGC 7000	20h 58.8m	+44° 12′	August 6
● 1–5	⊕ 120|100′			easy

Also known as the North America Nebula. A famous emission nebula, visible on dark nights to the naked eye. Located just west of Deneb, it is magnificent in binoculars, melding as it does into the stunning star fields of Cygnus. Providing you know where, and what to look for, the nebula is visible to the naked eye. With small- and large-aperture telescopes details within the nebula become visible, though several amateurs have reported that increasing aperture decreases the nebula's impact. The dark nebula lying between it and the Pelican nebula is responsible for their characteristic shape. Until recently, Deneb was thought to be the star responsible for providing the energy to make the nebula glow, but recent research points to several unseen stars being the power sources.

–	IC 1396	21h 39.1m	+57° 30′	August 16
● 3–5	⊕ 170|40′			easy ©

One of the few emission nebula visible to the naked eye (under perfect seeing of course!), and easily spotted in binoculars. It is an enormous patch of nebulosity, over 3°, spreading south of the orange star Mu (μ) Cephei. Any telescope will lessen the impact of the nebula but the use of filters will help to locate knots and patches of brighter nebulosity and dark dust lanes. Dark adaption and averted vision will all enhance the observation of this giant emission nebula.

Messier 16	IC 4703	$18^h 18.6^m$	−13° 58′	June 26
● 1–5	⊕ 35|30′			moderate
See April–May–June.				

–	NGC 604	$01^h 33.9^m$	+30° 39′	October 14
● 3–5	⊕ 60|35′			moderate
See October–November–December.				

Herschell 144	NGC 6857	$20^h 01.9^m$	+33° 31′	July 22
● 1	⊕ 1|1′			difficult
This is a very faint and small emission nebula, which requires telescopes of at least 20 cm aperture. It could easily be mistaken for a planetary nebula.				

Caldwell 27	NGC 6888	$20^h 12.0^m$	+38° 21′	July 25
● 1–5	⊕ 20|10′			difficult
Also known as the Crescent Nebula. Although visible in binoculars, a dark location and a light filter will make its location much easier. With good conditions, the emission nebula will live up to its name, having an oval shape with a gap in the ring on its south-eastern side. The nebula is known as a *stellar wind bubble*, and is the result of a fast-moving stellar wind from a Wolf–Rayet star which is sweeping up all the material that it had previously ejected during its red giant stage.				

Caldwell 19	IC 5146	$21^h 53.4^m$	+47° 16′	August 19
● 3–5	⊕ 12|12′			difficult
Also known as the Cocoon Nebula. A very difficult nebula to find and observe. It has a low surface brightness and appears as nothing more than a hazy amorphous glow surrounding a couple of 9th-magnitude stars. The dark nebula Barnard 168 (which the Cocoon lies at the end of) is surprisingly easy to find, and thus can act as a pointer to the more elusive emission nebula. The whole area is a vast stellar nursery and recent infrared research indicates the presence of many new and proto-stars within the nebula itself.				

Caldwell 11	NGC 7635	$23^h 20.7^m$	+61° 12′	September 11
● 1–5	⊕ 16|9′			difficult ©
Also known as the Bubble Nebula. This is a very faint nebula, even in telescopes of aperture 20 cm. The use of averted vision will help in its detection. An 8th-magnitude star within the emission nebula and a nearby 7th-magnitude star hinder in its detection owing to their combined glare. Research suggests that a strong stellar wind from a star pushes material out – the "Bubble" – and also heats up a nearby molecular cloud, which in turn ionises the "Bubble". It really does bear a striking resemblance to a soap bubble.				

October–November–December

Messier 42	NGC 1976	05h 35.4m	–05° 27′	December 15
● 1–5	⊕ 65∣60′			easy

Also known as the Orion Nebula. Incredible! The premier emission nebula and one of the most magnificent objects in the entire sky. Visible to the naked eye as a barely resolved patch of light, it shows detail from the smallest aperture upwards. It is really one of those objects where words cannot describe the view seen. In binoculars its pearly glow will show structure and detail, and in telescopes of aperture 10 cm the whole field will be filled. The entire nebulosity is glowing owing to the light (and thus energy) provided by the famous Trapezium stars located within it. What is also readily seen along with the glowing nebula are the dark, apparently empty and starless regions. These are still part of the huge complex of dust and gas, but are not glowing by the process of fluorescence – instead they are vast clouds of obscuring dust. The emission nebula is one of the few that shows definite colour. Many observers report seeing a greenish glow, along with pale grey and blue. The British amateur astronomer Don Tinkler has this to say about M42: "The size of M42 always amazes me, under dark skies it seems endless, with no edge or boundary. A wonderful nebula and a celestial showpiece." Many amateurs state that with very large apertures of 35 cm a pinkish glow can be seen. Located within the nebula are the famous Kleinmann–Low Sources and the Becklin–Neugebauer Object, which are believed to be dust-enshrouded young stars. The whole nebula complex is a vast stellar nursery. M42 is at a distance of 1700 l.y., and about 40 l.y. in diameter. Try to spend a long time observing this object – you will benefit from it, and many observers just let the nebula drift into the field of view. Truly wonderful.

Messier 43	NGC 1982	05h 35.6m	–05° 16′	December 15
● 1–5	⊕ 20∣15′			easy

Visible in binoculars, this emission nebula is part of the M42 complex, and some observers find it difficult to discriminate between them. Visible to the north of M42, it takes magnification well, and will show many intricate details.

Caldwell 49	NGC 2237–39	06h 32.3m	+05° 03′	December 29
● 1–5	⊕ 80∣60′			easy

Also known as the Rosette Nebula. This giant emission nebula has the dubious reputation of being very difficult to observe. But this is wrong – on clear nights it can be seen with binoculars. It is over 1° in diameter, and thus covers an area of sky four times larger than a full Moon! With a large aperture and light filters the complexity of the nebula becomes readily apparent, and under perfect seeing conditions dark dust lanes can be glimpsed. The brightest parts of the emission nebula have their own NGC numbers: 2237, 2238, 2239 and 2246. It is a young nebula, perhaps only half a million years old, and star formation may still be occurring within it. Photographs show that the central area contains the star cluster NGC 2244, along with the "empty" cavity caused by the hot young stars blowing the dust and gas away. Also known as the Rosette Molecular Complex (RMC).

–	NGC 1499	04h 00.7m	+36° 37′	November 21
● 1–5	⊕ 160I50′			easy/moderate

Also known as the California Nebula. This emission nebula presents a paradox. Some observers state that it can be glimpsed with the naked eye, others that binoculars are needed. The combined light from the emission nebula results in a magnitude of 6, but the surface brightness falls to around the 14th magnitude when observed through a telescope. Most observers agree, however, that the use of filters is necessary, especially from an urban location and when the seeing is not ideal. Clean optics are also a must to locate this nebula. Glimpsed as a faint patch in binoculars, with telescopes of aperture 20 cm, the emission nebula is seen to be nearly 3° long. Whatever optical instrument is used, it will remain faint and elusive.

–	NGC 604	01h 33.9m	+30° 39′	October 14
● 3–5	⊕ 60I35′			moderate

This may come as quite a surprise to many observers, but this is possibly the brightest emission nebula that can be glimpsed which is actually in another Galaxy. It resides in M33, in Triangulum. It appears as a faint hazy glow some 10′ northeast of M33's core. Owing to M33's low surface brightness (which often makes it a difficult object to find), the emission nebula may be visible while the galaxy isn't! It is estimated to be about 1000 times bigger than the Orion Nebula.

Herschel 261	NGC 1931	05h 31.4m	+34° 15′	December 14
● 1–5	⊕ 4I4′			moderate

Visible in telescopes of aperture 20 cm, the emission nebula appears as a very small faint hazy patch surrounding a triangle of stars. Larger apertures will show a slight brightening of the emission nebula at its centre.

–	NGC 1999	05h 36.5m	–06° 42′	December 15
● 1–5	⊕ 2I2′			moderate

In telescopes of aperture 20 cm, this small but bright emission nebula resembles a planetary nebula, and even has a star in its central region of magnitude 9.4. It lies about 1° south of M42.

–	NGC 2024	05h 40.7m	02° 27′	December 16
● 2–5	⊕ 30I30			moderate

This difficult nebula lies next to the famous star Zeta Orionis, which is unfortunate as the glare from the star makes observation difficult. It can however be glimpsed in binoculars as an unevenly shaped hazy and faint patch to the east of the star, providing the star is placed out of the field of view. With large telescopes and filters the emission nebula is a striking object, and has a shape reminiscent of a maple-leaf.

Messier 78	NGC 2068	05h 46.7m	00° 03′	December 17
● 1–5	⊕ 8I6′			moderate

A bright but small emission nebula that can be seen in binoculars. It has a fan shape, whereas some observers liken it to a comet. There are two 10th-magnitude stars located within the nebula which can give the false impression of two cometary nuclei. With a large-aperture telescope and high magnification, some very faint detail can be glimpsed along the eastern edge of the nebulosity, but excellent seeing will be needed in order to observe this.

Caldwell 46	NGC 2261	06h 39.2m	+08° 44′	December 31
● 1–5	⊕ 3.5I1.5′			moderate

Also known as Hubble's Variable Nebula. Easily seen in small telescopes of 10 cm as a small, comet-like nebula that can be seen from the suburbs. Larger apertures just amplify what is seen with little detail visible. What we are observing is the result of a very young and hot star clearing away the debris from which it was formed. The star R Monocerotis (buried within the nebula and thus invisible to us) emits material from its polar regions, and we see the north polar emissions, with the southern emission blocked from view by an accretion disc. The variability of the nebula, reported by Edwin Hubble in 1916, is due to a shadowing effect caused by clouds of dust drifting near the stars. It was also the first object to be officially photographed with the 200 inch Hale Telescope.

Gum 4	NGC 2359	07h 18.6m	−13° 12′	January 10
● 2–5	⊕ 9I6′			moderate

See January–February–March.

Herschell 258	NGC 1491	04h 03.4m	+51° 19′	November 21
● 1–5	⊕ 25I25′			difficult ©

Another faint and elusive emission nebula, with a telescope of aperture 25 cm, it appears as a hazy fan-shaped glow. The use of filters enhances the emission nebula quite well.

-	IC 417	05h 28.1m	+34° 26′	December 13
● 2–5	⊕ 13I10′			difficult

In order to observe this emission nebula the use of averted vision is required. It appears as a very faint ghostly haze with a few faint stars located within it. Filters are again useful with the emission nebula, as are perfect seeing conditions.

-	NGC 1554–55	04h 21.8m	+19° 32′	November 26
● 2–5	⊕ 1I7′ variable			very difficult

Also known as Hind's Variable Nebula. This famous but incredibly faint emission nebula is located to the west of the also famous stars T Tauri, the prototype for a class of variable star. The nebula has been much brighter in the past, but now is an exceedingly difficult object to locate. With large aperture, it will appear as a small faint hazy patch. When (and if!) located, it does bear higher magnification well. It may become brighter in the future, so is well worth looking for in the hope that it makes an unexpected reappearance.

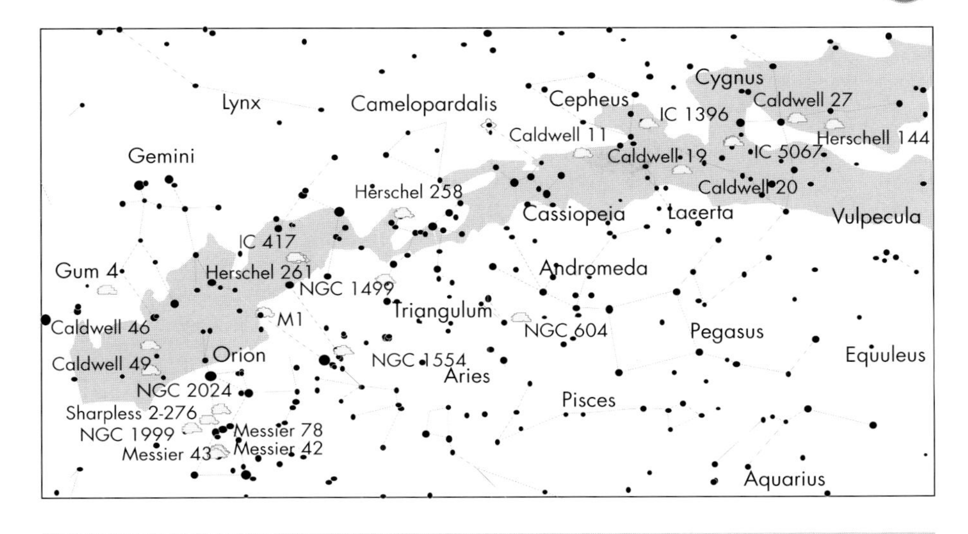

Figure 4.1. Emission nebulae 1.

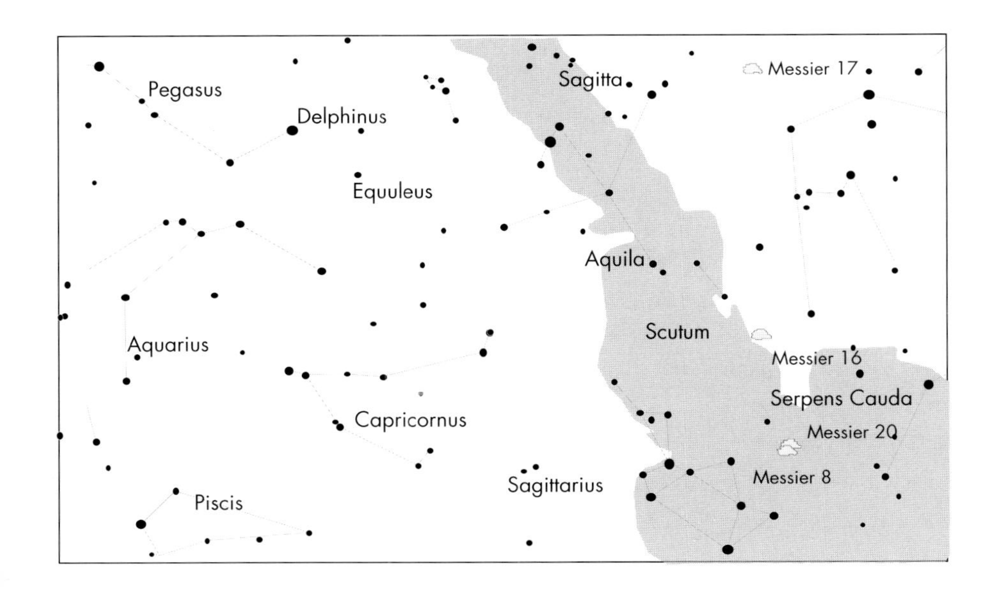

Figure 4.2. Emission nebulae 2.

4.3 Reflection Nebulae

As the name suggests, these nebula shine by the light reflected from stars within the nebula, or perhaps close-by. Like the emission nebula, these vast clouds consist of both gas and dust. The term "dust" is used throughout this book and in many other texts, but it isn't the dust we are familiar with on Earth. It is something very different indeed. In space, conditions exist which allow for the formation of very small particles (or *grains*) of graphite, metals, silicates and ices. They are roughly 0.0001 cm in size, and usually spindle-shaped. All these grains are embedded within a much larger cloud of molecular hydrogen. One of the characteristics of particles, or grains, that are so small (in proportion to the wavelength of light) is their property of selectively scattering light of a particular wavelength. If a beam of white light shines upon a cloud containing the grains, the blue light is scattered in all directions, a phenomena similar to that seen in the Earth's sky, which gives it its blue colour. This is the reason why reflection nebulae appear so blue on photographs – it is just the blue wavelengths of the light from (usually) hot blue stars nearby. To be scientifically correct, these nebulae should be called "scattering nebulae" instead of reflection nebulae, but the name has stuck. An interesting property of the scattered light is that the scattering process itself polarises the light, which is useful in the studies of grain composition and structure.

Several reflection nebulae reside within the same gas clouds as emission nebula; the Trifid Nebula is a perfect example. The inner parts of the nebula are glowing with the tell-tale pink colour, indicative of the ionisation process responsible for the emission, whereas away from the centre the edge material is definitely blue, thus signposting the scattering nature of the nebula. Visually, reflection nebulae are very faint objects having a low surface brightness, so they are not easy targets. Most require large-aperture telescopes with moderate magnification in order to be seen, but a feware visible in binoculars and small telescopes. There is also considerable debate concerning one particular nebula, which some amateurs claim can actually be glimpsed with the naked eye! As is to be expected, excellent seeing conditions are necessary, and very dark skies. Note that many of the larger emission nebulae have reflection nebulae associated with them, especially around their peripheries. The constellations of Sagittarius and Orion abound in combined bright emission and faint reflection nebulae. A simple star map of the reflection nebulae is at the end of this section.

January–February–March–April

–	NGC 1333	$03^h\ 29.3^m$	+31° 25′	November 13
● 3–5	⊕ 613			moderate
See September–October–November–December.				

–	NGC 1435	$03^h\ 46.1^m$	+23° 47′	November 17
● 2–5	⊕ 30130			moderate
See September–October–November–December.				

Caldwell 31	IC 405	$05^h\ 16.2^m$	+34° 16′	December 10
● 2–5	⊕ 30119			difficult
See September–October–November–December.				

–	NGC 1973–75–77	$05^h 35.1^m$	–04° 44′	December 14
● 1–5	⊕ 515			difficult

See September–October–November–December.

May–June–July–August

Caldwell 4	NGC 7023	$21^h 00.5^m$	+68° 10′	August 6
● 1–5	⊕ 18118			easy ©

Though small, this is a very nice, easy to observe reflection nebula. It has a star cluster at its centre which can hinder observation. However, what makes the reflection nebula easy to detect is its location. It is surrounded by a larger area of dark nebulosity, probably part of the same nebula complex. The contrast between the background stars, the dark nebula and the reflection nebula makes for a very interesting region.

–	NGC 1333	$03^h 29.3^m$	+31° 25′	November 13
● 3–5	⊕ 613			moderate

See September–October–November–December.

–	NGC 1435	$03^h 46.1^m$	+23° 47′	November 17
● 2–5	⊕ 30130			moderate

See September–October–November–December.

September–October–November–December

–	NGC 1333	$03^h 29.3^m$	+31° 25′	November 13
● 3–5	⊕ 613			moderate

This is a nice, easily seen reflection nebula, and appears as an elongated hazy patch. Larger aperture telescopes will show some detail along with two fainter dark nebulae Barnard 1 and 2, lying toward that north and south of the reflection nebula.

–	NGC 1435	$03^h 46.1^m$	+23° 47′	November 17
● 2–5	⊕ 30130			moderate

Also known as Tempel's Nebula. This faint patch of reflection nebula is located within the most famous star cluster in the sky. The nebula surrounds the star Merope, one of the brighter members of the Pleiades, and under perfect conditions can be glimpsed with binoculars. It is a comet-shaped cloud and was described by W. Tempel in 1859 as resembling "a breath on a mirror". Several other members of the cluster are also enshrouded by nebulosity, but these require exceptionally clear nights, and, incidentally, clean optics, as even the slightest smear on, say, a pair of binoculars, will reduce the chances to nil. To my knowledge, this reflection nebula has never been seen from an urban location in the UK. Perhaps, though, the news has just not reached me yet.

Caldwell 31	IC 405	05ʰ 16.2ᵐ	+34° 16′	December 10
● 2–5	⊕ 30119			difficult

Also known as the Flaming Star Nebula. A very hard reflection nebula to observe. It is actually several nebula including IC 405, 410 and 417, plus the variable star AE Aurigae. Narrow-band filters are justified with this reflection nebula, as they will highlight the various components. A challenge to the observer.

–	NGC 1973–75–77	05ʰ 35.1ᵐ	–04° 44′	December 14
● 1–5	⊕ 515			difficult

The location of these omission nebulae so close to M42 has meant that they are often neglected. They lie between M42 on their south and the cluster NGC 11981 to their north. They are also difficult to see because the glare from the star 42 Orionis tends to make observation difficult.

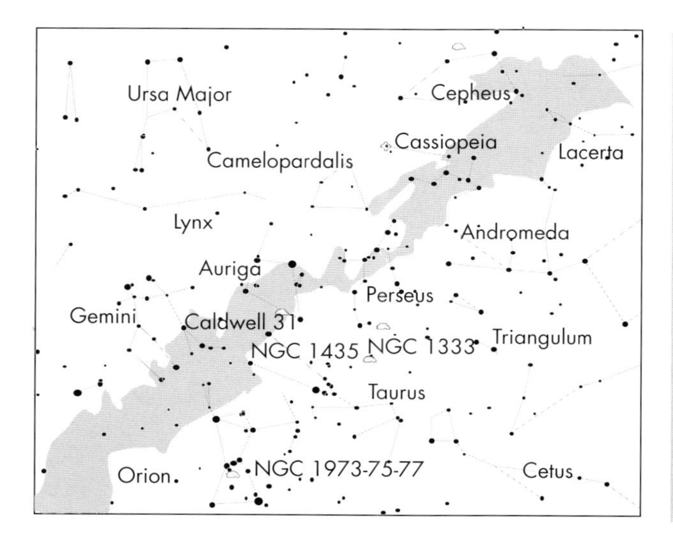

Figure 4.3. Reflection nebulae.

4.4 Dark Nebulae

Dark nebulae, by their very nature, are different in one major respect, from all the others. They do not shine. In fact, when you observe them, you are actually not seeing them by any light-emitting process, but rather because of their light-blocking ability. They are vast clouds of dust grains, of the same type as described in the section on reflection nebulae, seen against other nebulae or the star fields of the Milky Way. The nebulae appear dark because of their vast size, and so are very effective at scattering all of the light with the result that hardly any reaches the naked eye. The process of scattering the light is so effective, that, for instance, visible light emitted from the centre of our Galaxy is extinguished nearly 100 per cent by the dust clouds between us and the centre. This is the reason why it is still a mystery as to what the central region looks like in *visible light*. Don't be confused, however, by thinking that these clouds of dust grains are very dense objects. They are not.

Most of the material in the cloud is molecular hydrogen (along with carbon monoxide which is responsible for their radio emission), and the resulting density is low. There is also some evidence that the dust grains present in the clouds have properties different from those dust grains in the interstellar medium.

Many dark nebulae are actually interacting with their environs, as witnessed by the spectacular images taken by the Hubble Space Telescope of M16 in Serpens, which show dust clouds containing dense regions, or globules, that are resisting the radiation pressure from close, hot young stars, with the result that many of the globules have long tails of material trailing from them. The area near the Horsehead Nebula in Orion is also famous for its image of the radiation from the supergiant stars of Orion's Belt impacting on the dark clouds to either side of the Horsehead, with the result that material is ionised and streaming from the surface of the cloud.

Most of the dark clouds listed in the text will have vastly different shapes, for several reasons. It may be that the cloud would have originally been spherical in shape, and thus present a circular image to us, but hot stars in its environment will have disrupted this by radiation pressure and stellar winds. Shock fronts from nearby supernovae can also have an impact. The gravitational effects from other clouds, stars and even that of the Milky Way itself will all have a role to play in determining the shape of the cloud. It is also thought that magnetic fields may have some limited affect. As many of theses dark clouds are part of a much larger star-forming region, the new stars will themselves influence and alter its shape.

The "opacity" of a dark nebula is a measure of how opaque the cloud is to light, and thus how dark it will appear. There is a rough classification system that can be used; a value of 1 for a dark nebula would indicate that it only very slightly attenuates the starlight from the background Milky Way, while at the other end of the scale a value of 6 would mean that the cloud is nearly black, and is given the symbol ◆. Observing dark nebulae can be a very frustrating pastime. The best advice is to always use the lowest magnification possible. This will enhance the contrast between the dark nebulae and that of the background star field. If a high magnification is used, the contrast will be lost and you will only see the area surrounding the dark nebula, and not the nebula itself. Of course there are exceptions to this rule; a high magnification can be employed on many of the dark nebulae in Sagittarius, but try a low magnification first in order to locate the object, then proceed to higher ones. Dark skies are a must with these objects, as even a hint of light pollution will make their detection an impossible task. Two simple star maps which show the locations of the dark nebulae are at the end of the section.

January–February–March–April

Barnard 33	05^h 40.9^m	–02° 28′	December 16
◆ 4	⊕ 6I4′		difficult
See September–October–November–December.			

Barnard 228	15^h 45.5^m	–34° 24′	May 18
◆ 6	⊕ 240I20′		easy
See May–June–July–August.			

May–June–July–August

Barnard 228	15^h 45.5^m	–34° 24′	May 18
◆ 6	⊕ 240120′		easy

This is a long band of dark nebula, easily spotted in binoculars. It lies halfway between Psi (ψ) and Chi (χ) Lupi. Best seen in low-power, large-aperture binoculars, as it stands out well against the rich background star field.

Barnard 59, 65–7	LDN 1773	17^h 21.0^m	–27° 23′	June 11
◆ 6	⊕ 300160′			easy

Also known as the Pipe Nebula (Stem), and Lynds Dark Nebula 1773. This is a large dark nebula visible to the naked eye. It is conspicuous because it stands against a star-studded field. Best viewed with lower-power binoculars. With the unaided eye, it appears as a straight line, but under magnification its many variations can be glimpsed.

Barnard 78	LDN 42	17^h 33.0^m	–26° 30′	June 14
◆ 5	⊕ 2001150′			easy

Also known as the Pipe Nebula (Bowl). Part of the same dark nebula as above, the bowl appears as a jagged formation, covering over 9°. The whole region is studded with dark nebula, and is thought to be a part of the same complex as that which encompasses Rho (ρ) Ophiuchi and Antares, which are over 700 l.y. away from it.

Barnard 86	LDN 93	18^h 03.0^m	–27° 53′	June 22
◆ 5	⊕ 6′			easy

Also known as the Ink Spot. Located within the Great Sagittarius Star Cloud, this is a near-perfect example of a dark nebula, appearing as a completely opaque blot against the background stars.

Barnard 87, 65–7	LDN 1771	18^h 04.3^m	–32° 30′	June 22
◆ 4	⊕ 12′			easy

Also known as the Parrot Nebula. Although not a very distinct nebula, it stands out because of its location within a stunning background of stars. Visible in binoculars as a small circular dark patch, it is best seen in small telescope of around 10 to 15 cm.

Barnard 103	LDN 497	18^h 39.4^m	–06° 41′	July 1
◆ 6	⊕ 40115′			easy

Easily seen at the north-east edge of the famous Scutum Star Cloud. It is a curved dark line. Can be glimpsed in binoculars, but best seen at apertures of around 10 to 15 cm.

Barnard 110–1	18^h 50.1^m	–04° 48′	July 4
◆ 6	⊕ 11′		easy

An easily seen complex of dark nebulae that can be seen in binoculars. The contrast between the background star clouds and the darkness of the nebulae is immediately seen.

Barnard 142-3	19^h 41.0^m	+10° 31′	July 17
◆ 6	⊕ 45′		easy
This is an easily seen pair of dark nebulae, visible in binoculars. It appears as a cloud with two "horns" extending towards the west. The nebula contrasts very easily with the background Milky Way and so is a fine object. With a rich field telescope and large binoculars, the dark nebula actually appears to be floating against the star field.			

Barnard 145	20^h 02.8^m	+37° 40′	July 22
◆ 4	⊕ 35I8′		easy
Visible in binoculars, it is a triangular dust cloud that stands out well against the impressive star field. As it is not completely opaque to starlight, several faint stars can be seen shining through it.			

Barnard 343	20^h 13.5^m	+40° 16′	July 25
◆ 5	⊕ 13I6′		easy
Easily seen as a "hole" in the background Milky Way, this is an oval dark nebula, which although glimpsed in binoculars is at its best in telescopes.			

Lynds 906	20^h 40.0^m	+42° 00′	August 1
◆ 5	⊕ – –		easy
Also known as the Northern Coalsack. This is probably the largest dark nebulosity of the northern sky. It is an immense region, easily visible on clear moonless nights just south of Deneb. It lies just at the northern boundary of the Great Rift, a collection of several dark nebulae which bisects the Milky Way. The Rift is of course part of a spiral arm of the Galaxy, so features prominent on photographs of other galaxies such as NGC 891 in Andromeda.			

Barnard 352	20^h 57.1^m	+45° 54′	August 5
◆ 5	⊕ 20I10′		moderate
Visible in binoculars, this is part of the much more famous North American Nebula, though this dark part is located to the north. It is a well-defined triangular dark nebula.			

Barnard 33	05^h 40.9^m	−02° 28′	December 16
◆ 4	⊕ 6I4′		difficult
See September–October–November–December.			

December

Barnard 103	LDN 497	18^h 39.4^m	−06° 41′	July 1
◆ 6	⊕ 40I15′			easy
See May–June–July–August.				

Barnard 110-1	18^h 50.1^m	−04° 48′	July 4
◆ 6	⊕ 11′		easy
See May–June–July–August.			

Barnard 142–3	$19^h\,41.0^m$	+10° 31′	July 17
◆ 6	⊕ 45′		easy

See May–June–July–August.

Barnard 145	$20^h\,02.8^m$	+37° 40′	July 22
◆ 4	⊕ 35|18′		easy

See May–June–July–August.

Barnard 343	$20^h\,13.5^m$	+40° 16′	July 25
◆ 5	⊕ 13|6′		easy

See May–June–July–August.

Lynds 906	$20^h\,40.0^m$	+42° 00′	August 1
◆ 5	⊕ – –		easy

See May–June–July–August.

Barnard 352	$20^h\,57.1^m$	+45° 54′	August 5
◆ 5	⊕ 20|10′		moderate

See May–June–July–August.

Barnard 33	$05^h\,40.9^m$	–02° 28′	December 16
◆ 4	⊕ 6|4′		difficult

Also known as the Horsehead Nebula. Often photographed, but very rarely observed, this famous dark nebula is very difficult to see. It is a small dark nebula which is seen in silhouette against the dim glow of the emission nebula IC 434. Both are very faint and will need perfect seeing conditions. Such is the elusiveness of this object that even telescopes of 40 cm are not guaranteed a view. Dark adaption and averted vision, along with the judicious use of filters, may result in its detection. Nevertheless, have a go!

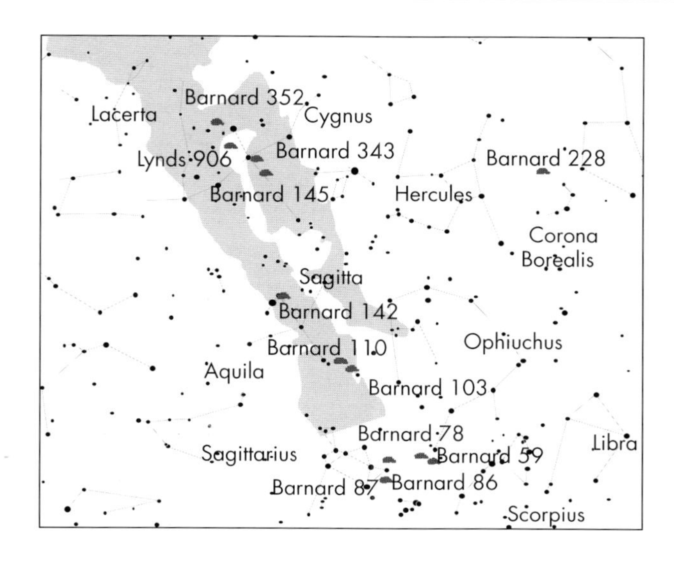

Figure 4.4. Dark nebulae 1.

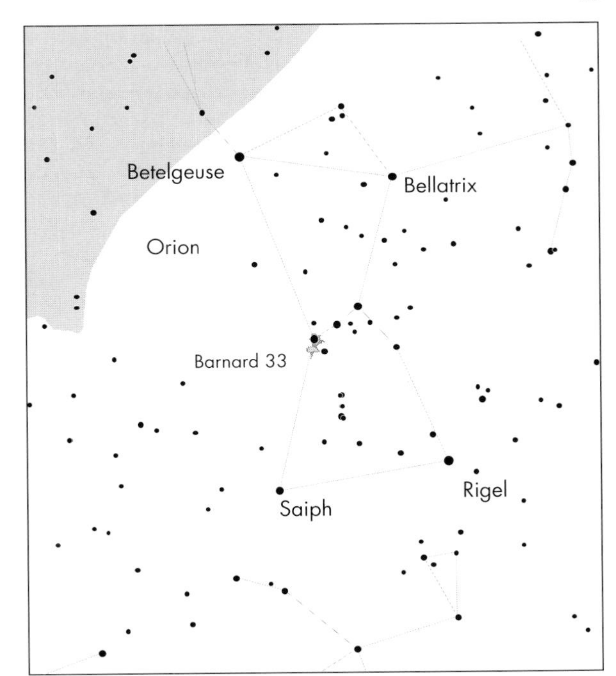

Figure 4.5.
Dark nebulae 2.

4.5 Planetary Nebulae

Planetary nebulae[5] are some of the most interesting objects in the sky and have a lot to offer the amateur. They range across the whole of the observational spectrum: some are easy to find in binoculars, while others will require large aperture, patience, and even maybe specialised filters in order for them to be distinguished from the background star fields. These small shells of gas, once the atmosphere of stars, come in a variety of shapes, sizes and brightness. Many have a hot central star within the nebula, which is visible in amateur equipment, and is the power source, providing the energy for the gas to glow.

The origin of a planetary nebula is very different from that of any nebula mentioned previously. In stars with approximately solar mass (and perhaps just a little more), a point will arrive when the nuclear reactions in the core have transmuted all the hydrogen and helium into oxygen and carbon. The star is by this time is extremely old, and will have become a red giant and a variable star.[6] The variability of the star is of a class designated a pulsating variable. However, the core never proceeds beyond carbon and oxygen in its nucleosynthesis, because the temperature never rises to a sufficiently hot enough temperature for these reactions to occur. The outer layers of the star pulsate and in fact the material achieves escape velocity and so is able to expand into the space surrounding the stars. The outer layers, now puffed into space, are the planetary nebulae we observe.

---

[5]The name "planetary nebula" was first applied to such an object by Herschell, who thought that it looked like Jupiter when seen in a telescope.
[6]These stars are described in Chapter 2.

Several nebulae have a multiple shell appearance, and this is thought due to the red giant experiencing several periods of pulsation where the material escapes from the star. The strong stellar winds and magnetic fields of the star are also thought responsible for the many observed exotic shapes of the nebulae. Planetary nebulae are only a fleeting feature in our Galaxy. After only a few tens of thousands of years, a nebula will have dissipated into interstellar space, and so no longer exist. However, this aspect of a stars evolution is apparently very common, and there are over 1400 planetary nebulae in our part of the Galaxy alone!

Visually, a nebula is one of the few deep sky objects that actually appears coloured. Around 90 per cent of its light comes from the doubly ionised oxygen line, OIII, at wavelengths 495.9 nm and 500.7 nm. This is a very characteristic blue–green colour, and, it so happens, the colour at which the dark-adapted eye is at its most sensitive. The specialised light filters are also extremely useful for observing planetaries as they isolate the OIII light in particular, increasing the contrast between the nebula and the sky background, thus markedly improving the nebula's visibility.

Such is the variety of shapes and sizes that there is something to offer every type of observer. Some planetaries are so tiny that even at high magnification, using large aperture telescopes, the nebula will still appear starlike. Others are much larger, for instance the Helix Nebula, Caldwell 63, is half the size of the full Moon, but can be observed only with low magnification, and perhaps only in binoculars, as any higher magnification will lower its contrast to such an extent that it will simply disappear from view. Many exhibit a bipolar shape, such as the Dumbbell Nebula, M27, in Vulpecula. Others, though, show ring shapes, such as the ever popular M57 in Lyra.

An interesting minor point is the possibility of observing the central stars of the nebulae. These are very small, subdwarf and dwarf stars. They are similar to main sequence stars of type O and B, but, as they are running down their nuclear reactions, or in some cases, no longer producing energy by nuclear reactions, are consequently fainter and smaller. These two characteristics make observation very difficult. The brightest central star is possibly that of NGC 1514 in Taurus, at 9.4 magnitude, but the majority are at magnitude 10 or fainter.

The *Vorontsoz-Vellyaminov classification system* can be used to describe the appearance of a planetary. Although it is of limited application, it will be used here.

Planetary Nebula Morphology Types

1 Starlike
2 Smooth disclike appearance
   a. bright towards centre
   b. uniform brightness
   c. possible, faint ring structure
3 Irregular disclike appearance
   a. irregular brightness distribution
   b. possible, faint ring structure
4 Definite ring structure
5 Irregular shape
6 Unclassified shape – can be a combination of two classifications e.g. 4 + 3 (ring and irregular disc)

The usual information is given for each object, with the addition of morphology class [ ] and central star brightness [ ]. In addition, the magnitude quoted is the magnitude of the planetary nebula as if it where a point source. This last parameter can often be confusing, so even if a nebula has a quoted magnitude of, say, 8, it may be much fainter than this, and consequently, hard to find.

January

Caldwell 39	NGC 2392	$07^h\ 29.2^m$	+20° 55′	January 12
8.6 m	⊕ 15′	3b + 3b	9.8	easy

Also known as the Eskimo Nebula. This is a small but famous planetary nebula which can be seen as a pale blue dot in a telescope of 10 cm, although it can be glimpsed in binoculars as the apparent southern half of a double star. Higher magnification will resolve the central star and the beginnings of its characteristic "Eskimo" face. With aperture of 20 cm, the blue disc becomes apparent. The shell, ring and halo structure will need apertures of 40 cm in order to become easily resolvable. Research indicates that we are seeing the planetary nebula pole-on, although this is by no means certain. Its distance is also in doubt, with values ranging from 1600 to 7500 l.y.

Caldwell 59	NGC 3242	$10^h\ 24.8^m$	–18° 38′	February 26
8.4 m	⊕ 16′	4 + 3b	12.1	easy

See February.

Herschel 64	NGC 2440	$07^h\ 41.9^m$	–18° 13′	January 16
9.4 m	⊕ 14ǀ32′	5 + 3	19	moderate

With a telescope of aperture 25 cm, this planetary nebula will appear as an oval shaped disc of hazy light, and may show a faint green tint. Large telescopes will show several distinct patched of light and dark area within the planetary nebula.

PK196–10.1	NGC 2022	$05^h\ 42.1^m$	+09° 05′	December 16
12 m	⊕ 18′	4 + 2	15	moderate

See December.

PK221–12.1	IC 2165	$06^h\ 21.7^m$	–12° 59′	December 26
12.9 m	⊕ 4.0′	3b	15	difficult

See December.

February

Caldwell 59	NGC 3242	10^h 24.8^m	−18° 38′	February 26
8.4 m	⊕ 16′	4 + 3b	12.1	easy

Also known as the Ghost of Jupiter. One of the brighter planetary nebula, and the brightest in the spring sky for northern observers, this is a fine sight in small telescopes. Visible in binoculars as a tiny blue disc. With an aperture of 10 cm, the blue colour becomes more pronounced along with its disc, which is approximately the same size as that of Jupiter in a similar aperture. It is bright, but with a faint hazy edge. In larger telescopes the disc becomes oval in appearance, with barely perceptible extensions to either side. Photographs, (and keen-eyed observers who possess large telescopes) suggest the planetary nebula has a strong resemblance to a human eye. The central star has a reported temperature of about 100,000 K.

Caldwell 39	NGC 2392	07^h 29.2^m	+20° 55′	January 12
8.6 m	⊕ 15′	3b + 3b	9.8	easy

See January.

Herschel 64	NGC 2440	07^h 41.9^m	−18° 13′	January 16
9.4 m	⊕ 14 I 32′	5 + 3	19	moderate

See January.

Messier 97	NGC 3587	11^h 14.8^m	+55° 01′	March 11
9.9 m	⊕ 194′	3a	16	moderate ©

See March.

Sharpless 2–290	PK219–31.1	08^h 54.2^m	+08° 55′	February 3
12.0 m	⊕ 980′	3a	15.5	difficult

Although this is an extremely large planetary nebula, it is very difficult to locate. This is one planetary nebula where the use of a light filter is required or it proves pointless to try to observe. However, with a filter, the giant circular shape of the planetary nebula becomes obvious.

PK123+34.1	IC 3568	12^h 32.9^m	+82° 33′	March 30
10.6 m	⊕ 6.0′	2 +2a	11.5	difficult ©

See March.

March

Caldwell 39	NGC 2392	07^h 29.2^m	+20° 55′	January 12
8.6 m	⊕ 15′	3b + 3b	9.8	easy

See January.

Caldwell 59	NGC 3242	10^h 24.8^m	−18° 38′	February 26
8.4 m	⊕ 16′	4 + 3b	12.1	easy

See February.

Messier 97	NGC 3587	11^h 14.8^m	+55° 01′	March 11
9.9 m	⊕ 194′	3a	16	moderate ©

Also known as the Owl Nebula. Not visible in binoculars owing to its low surface brightness; apertures of at least 20 cm will be needed to glimpse the "eyes" of the nebula. At about 10 cm aperture, the planetary nebula will appear as a very pale blue tinted circular disc, although the topic of colour in regard to this particular planetary nebula is in question. Larger aperture will show more detail and the central star can be glimpsed if seeing conditions permit.

PK123+34.1	IC 3568	12^h 32.9^m	+82° 33′	March 30
10.6 m	⊕ 6.0′	2 +2a	11.5	difficult ©

A planetary nebula that will appear almost star-like unless a high magnification is used. Indeed, it is often mistaken for a star, and thus overlooked. Telescopes of aperture 20 cm will show the disc, but larger apertures resolve the lovely blue–green colour of the planetary nebula. The central star can be seen to be slightly off-centre.

April

Caldwell 59	NGC 3242	10^h 24.8^m	–18° 38′	February 26
8.4 m	⊕ 16′	4 + 3b	12.1	easy

See February.

Messier 97	NGC 3587	11^h 14.8^m	+55° 01′	March 11
9.9 m	⊕ 194′	3a	16	moderate ©

See March.

PK123+34.1	IC 3568	12^h 32.9^m	+82° 33′	March 30
10.6 m	⊕ 6.0′	2 +2a	11.5	difficult ©

See March.

Caldwell 6	NGC 6543	17^h 58.6^m	+66° 38′	June 21
8.3 m	⊕ 18|350′	3a + 2	11	easy ©

See June.

PK64 + 5.1		19^h 34.8^m	+30° 31′	June 30
10 m	⊕ 7.5′	4	10.0	difficult

See June.

May

Caldwell 6	NGC 6543	17^h 58.6^m	+66° 38′	June 21
8.3 m	⊕ 18|350′	3a + 2	11	easy ©

See June.

Messier 57	NGC 6720	$18^h\,53.6^m$	+33° 02′	July 5
8.8 m	⊕ 71′	4 +3	15.3	easy
See July.				

Caldwell 15	NGC 6826	$19^h\,44.8^m$	+50° 31′	July 18
8.8 m	⊕ 25′	3a + 2	11	easy ©
See July.				

Messier 27	NGC 6853	$19^h\,59.6^m$	+22° 43′	July 22
7.3 m	⊕ 348′	3 + 2	13.8	easy
See July.				

PK54–12.1	NGC 6891	$20^h\,15.2^m$	+12° 42′	July 26
10.5 m	⊕ 14′	2a + 2b	12.5	easy
See July.				

Messier 97	NGC 3587	$11^h\,14.8^m$	+55° 01′	March 11
9.9 m	⊕ 194′	3a	16	moderate ©
See March.				

PK33–2.1	NGC 6741	$19^h\,02.6^m$	–00° 27′	July 7
11.4 m	⊕ 6′	4	17.5	moderate
See July.				

Herschel 743	NGC 6781	$19^h\,18.4^m$	+06° 33′	July 11
11.4 m	⊕ 110′	3b + 3	16.2	moderate
See July.				

Herschel 16	NGC 6905	$20^h\,22.4^m$	+20° 05′	July 27
11.1 m	⊕ 40′	3 +3	15.5	moderate
See July.				

PK123+34.1	IC 3568	$12^h\,32.9^m$	+82° 33′	March 30
10.6 m	⊕ 6.0′	2 +2a	11.5	difficult ©
See March.				

PK64 + 5.1		$19^h\,34.8^m$	+30° 31′	June 30
10 m	⊕ 7.5′	4	10.0	difficult
See June.				

June

Caldwell 6	NGC 6543	17h 58.6m	+66° 38′	June 21
8.3 m	⊕ 18∣350′	3a + 2	11	easy ©

Also known as the Cat's Eye Nebula. Seen as a bright oval planetary nebula with a fine blue-green colour, this is one of the planetary nebula that became famous after the HST published its image. Visible even in a telescope of 10 cm, but a large telescope (20 cm) will show some faint structure, while to observe the central star requires 40 cm aperture. The incredibly beautiful and complex structure is thought to be the result of a binary system, with the central star classified as a Wolf–Rayet.

Messier 57	NGC 6720	18h 53.6m	+33° 02′	July 5
8.8 m	⊕ 71′	4 +3	15.3	easy

See July.

Caldwell 15	NGC 6826	19h 44.8m	+50° 31′	July 18
8.8 m	⊕ 25′	3a + 2	11	easy ©

See July.

Messier 27	NGC 6853	19h 59.6m	+22° 43′	July 22
7.3 m	⊕ 348′	3 + 2	13.8	easy

See July.

PK54–12.1	NGC 6891	20h 15.2m	+12° 42′	July 26
10.5 m	⊕ 14′	2a + 2b	12.5	easy

PK33–2.1	NGC 6741	19h 02.6m	–00° 27′	July 7
11.4 m	⊕ 6′	4	17.5	moderate

See July.

Herschel 743	NGC 6781	19h 18.4m	+06° 33′	July 11
11.4 m	⊕ 110′	3b + 3	16.2	moderate

See July.

Herschel 16	NGC 6905	20h 22.4m	+20° 05′	July 27
11.1 m	⊕ 40′	3 +3	15.5	moderate

See July.

Caldwell 69	NGC 6302	17ʰ 13.7ᵐ	–37° 06′	June 10
9.6 m	⊕ 50′	6	21	difficult

Also known as the Bug Nebula. Visible in nearly all sizes of telescope; with a large aperture, a distinct brightening at the centre will be seen, which many incorrectly assume is the central star. In fact it is just the bright central region. A dust lane runs across the planetary nebula and obscures the central star from view. It is class of nebula called *bipolar*, which to some will, resemble an extended butterfly. The star responsible for the nebula has a temperature of 380,000 °K. Unfortunately, the nebula is not visible from the UK.

PK64 + 5.1		19ʰ 34.8ᵐ	+30° 31′.	June 30
10 m	⊕ 7.5′	4	10.0	difficult

This is a very small planetary nebula, which even with apertures of 20 cm and greater will require a high magnification. What makes it even more difficult to locate is the multitude of stars in the background. However, a pointer to the planetary nebula is the star responsible for it – Campbell's Hydrogen Star, which has a lovely orange colour.

July

Messier 57	NGC 6720	18ʰ 53.6ᵐ	+33° 02′	July 5
8.8 m	⊕ 71′	4 +3	15.3	easy

Also known as the Ring Nebula. The most famous of all planetary nebula, and, surprisingly – and pleasantly – visible in binoculars. However, it will not be resolved into the famous "smoke-ring" shape seen so often in colour photographs; it will, rather, resemble an out-of-focus star. It is just resolved in telescopes of about 10 cm aperture, and at 20 cm the classic smoke-ring shape becomes apparent. At high magnification (and larger aperture), the Ring Nebula is truly spectacular. The inner region will be seen to be faintly hazy, but large aperture and perfect conditions will be needed to see the central star. Does the planetary nebula appear perfectly circular, or is it slightly oval?

Caldwell 15	NGC 6826	19ʰ 44.8ᵐ	+50° 31′	July 18
8.8 m	⊕ 25′	3a + 2	11	easy ©

Also known as the Blinking Planetary. A difficult planetary nebula to locate, but well worth the effort. The blinking effect is due solely to the physiological structure of the eye. If you stare at the central star long enough, the planetary nebula will fade from view. At this point should you move the eye away from the star, and the planetary nebula will "blink" back into view at the periphery of your vision. Visually, it is a nice blue–green disc, which will take high magnification well. Although not visible in amateur telescopes, the planetary nebula is made up of two components – an inner region consisting of a bright shell and two *ansae*, and a halo which is delicate in structure with a bright shell.

Messier 27	NGC 6853	19^h 59.6^m	+22° 43′	July 22
7.3 m	⊕ 348′	3 + 2	13.8	easy

Also known as the Dumbbell Nebula. This famous planetary nebula can be seen in small binoculars as a box-shaped hazy patch, and many amateurs rate this as the sky's premier planetary nebula. In apertures of 20 cm, the classic dumbbell shape is apparent, with the brighter parts appearing as wedge shapes which spread out to the north and south of the planetary nebula's centre. The central star can be glimpsed at this aperture. With perfect observing conditions, a faint glow can be seen in its outer parts. A wonderful object.

PK54–12.1	NGC 6891	20^h 15.2^m	+12° 42′	July 26
10.5 m	⊕ 14′	2a + 2b	12.5	easy

A pleasant planetary nebula which can be seen in apertures of 10 cm. It is a small but bright disc, which will appear like a star in low magnification. Using larger apertures, the central star will become visible along with the colour of the blue–green disc.

Caldwell 6	NGC 6543	17^h 58.6^m	+66° 38′	June 21
8.3 m	⊕ 18 I 350′	3a + 2	11	easy ©

See June.

PK33–2.1	NGC 6741	19^h 02.6^m	–00° 27′	July 7
11.4 m	⊕ 6′	4	17.5	moderate

This is a small but nicely coloured planetary nebula. A small-aperture telescope of about 20 cm is needed to resolve the planetary nebula into a disc, but it will show the blue–green colour. Larger apertures will just make the image bright but with no increase in resolution of detail.

Herschel 743	NGC 6781	19^h 18.4^m	+06° 33′	July 11
11.4 m	⊕ 110′	3b + 3	16.2	moderate

This is an easily located planetary nebula, large, circular and bright. Under excellent seeing and using averted vision using dark adaption, a darkening of its centre will be revealed along with the fainter part of its northern periphery. Large-aperture instruments will show far more detail, including the halo. Not visible in binoculars.

Herschel 16	NGC 6905	20^h 22.4^m	+20° 05′	July 27
11.1 m	⊕ 40′	3 +3	15.5	moderate

Also known as the Blue Flash Nebula. The true nature of this planetary nebula only becomes apparent at apertures of at least 20 cm, when the lovely blue colour is seen. The colour and brightness also increase towards its centre. The central star can be seen only under good seeing conditions. Larger aperture will show that the planetary nebula has a mottled, unevenly bright disc.

Caldwell 55	NGC 7009	$21^h\ 04.2^m$	−11° 22′	August 7
8.3 m	⊕ 25′	4 + 6	12.78	moderate
See August.				

PK89+0.1	NGC 7026	$21^h\ 06.3^m$	+47° 51′	August 7
10.9 m	⊕ 21′	3a	14.9	moderate
See August.				

Caldwell 69	NGC 6302	$17^h\ 13.7^m$	−37° 06′	June 10
9.6 m	⊕ 50′	6	21	difficult
See June.				

PK64 + 5.1		$19^h\ 34.8^m$	+30° 31′	June 30
10 m	⊕ 7.5′	4	10.0	difficult
See June.				

Caldwell 63	NGC 7293	$22^h\ 29.6^m$	−20° 48′	August 29
6.3 m	⊕ 770′	4 + 3	13.5	difficult
See August.				

August

Messier 57	NGC 6720	$18^h\ 53.6^m$	+33° 02′	July 5
8.8 m	⊕ 71′	4 +3	15.3	easy
See July.				

Caldwell 15	NGC 6826	$19^h\ 44.8^m$	+50° 31′	July 18
8.8 m	⊕ 25′	3a + 2	11	easy ©
See July.				

Messier 27	NGC 6853	$19^h\ 59.6^m$	+22° 43′	July 22
7.3 m	⊕ 348′	3 + 2	13.8	easy
See July.				

PK54−12.1	NGC 6891	$20^h\ 15.2^m$	+12° 42′	July 26
10.5 m	⊕ 14′	2a + 2b	12.5	easy
See July.				

Caldwell 55	NGC 7009	$21^h 04.2^m$	–11° 22′	August 7
8.3 m	⊕ 25′	4 + 6	12.78	moderate

Also known as the Saturn Nebula. Although it can be glimpsed in small aperture, a telescope of at least 25 cm is needed to see the striking morphology of the planetary nebula which gives it its name. There are extensions, or ansae, on either side of the disc, along an east–west direction, which can be seen under perfect seeing. The disc is a nice blue–green and the central star may be glimpsed. High magnification is also justified in this case. Recent theory predicts a companion to the central star, which may be the cause of the peculiar shape.

PK89+0.1	NGC 7026	$21^h 06.3^m$	+47° 51′	August 7
10.9 m	⊕ 21′	3a	14.9	moderate

A nicely coloured bluish-green planetary nebula, which has a slightly brighter centre. It will be stellar-like in binoculars, and 20 cm aperture telescopes will be need to resolve it into a disc.

PK33–2.1	NGC 6741	$19^h 02.6^m$	–00° 27′	July 7
11.4 m	⊕ 6′	4	17.5	moderate

See July.

Herschel 743	NGC 6781	$19^h 18.4^m$	+06° 33′	July 11
11.4 m	⊕ 110′	3b + 3	16.2	moderate

See July.

Herschel 16	NGC 6905	$20^h 22.4^m$	+20° 05′	July 27
11.1 m	⊕ 40′	3 +3	15.5	moderate

See July.

Caldwell 63	NGC 7293	$22^h 29.6^m$	–20° 48′	August 29
6.3 m	⊕ 770′	4 + 3	13.5	difficult

Also known as the Helix Nebula. Thought to be the closest planetary nebula to the Earth, at about 450 l.y., it has an angular size of over $^1/_4$° – half that of the full Moon. However, it has a very low surface brightness and is thus notoriously difficult to locate. With binoculars, the planetary nebula appears as a ghostly image. With an aperture of 10 cm, low magnification is necessary, and averted vision is useful in order to glimpse the central star. The use of an OIII filter will drastically improve the image.

September

Messier 57	NGC 6720	$18^h 53.6^m$	+33° 02′	July 5
8.8 m	⊕ 71′	4 +3	15.3	easy

See July.

Caldwell 15	NGC 6826	$19^h 44.8^m$	+50° 31′	July 18
8.8 m	⊕ 25′	3a + 2	11	easy ©

See July.

Messier 27	NGC 6853	$19^h 59.6^m$	+22° 43′	July 22
7.3 m	⊕ 348′	3 + 2	13.8	easy

See July.

PK54–12.1	NGC 6891	$20^h 15.2^m$	+12° 42′	July 26
10.5 m	⊕ 14′	2a + 2b	12.5	easy

See July.

Caldwell 22	NGC 7662	$23^h 25.9^m$	+42° 33′	September 12
8.6 m	⊕ 12′	4 + 3	13.2	moderate

Also known as the Blue Snowball. This nice planetary nebula is visible in binoculars owing to its striking blue colour, but will only appear stellar-like. In telescopes of 20 cm, the disc is seen, along with some ring structure. With larger aperture, subtle colour variations appear – blue–green shading. Research indicates that the planetary nebula has a structure similar to that seen in the striking HST image of the Helix Nebula, showing fast low-ionisation emission regions (Fliers). These are clumps of above-average-density gas ejected from the central star before it formed the planetary nebula.

Caldwell 2	NGC 40	$00^h 13.0^m$	+72° 32′	September 24
12.4 m	⊕ 36′	3b + 3	11.6	moderate ©

This is a spectacular object, often overlooked. Appearing as a star in binoculars, it needs an aperture of at least 20 cm for its planetary nebula nature to become apparent. Bright and oval, it has brighter regions still at its west and east sections and has a lighter northern area, but this latter feature is seen only under perfect seeing conditions.

Caldwell 55	NGC 7009	$21^h 04.2^m$	−11° 22′	August 7
8.3 m	⊕ 25′	4 + 6	12.78	moderate

See August.

PK89+0.1	NGC 7026	$21^h 06.3^m$	+47° 51′	August 7
10.9 m	⊕ 21′	3a	14.9	moderate

See August.

Caldwell 63	NGC 7293	$22^h 29.6^m$	−20° 48′	August 29
6.3 m	⊕ 770′	4 + 3	13.5	difficult

See August.

October

Messier 57	NGC 6720	$18^h 53.6^m$	+33° 02′	July 5
8.8 m	⊕ 71′	4 +3	15.3	easy

See July.

Caldwell 15	NGC 6826	$19^h 44.8^m$	+50° 31′	July 18
8.8 m	⊕ 25′	3a + 2	11	easy ©

See July.

Messier 27	NGC 6853	$19^h 59.6^m$	+22° 43′	July 22
7.3 m	⊕ 348′	3 + 2	13.8	easy

See July.

PK54–12.1	NGC 6891	$20^h 15.2^m$	+12° 42′	July 26
10.5 m	⊕ 14′	2a + 2b	12.5	easy

See July.

Caldwell 56	NGC 246	$00^h 47.0^m$	–11° 53′	October 2
8.5 m	⊕ 225′	3b	11	moderate

A nice planetary nebula that is large and bright and shows a distinct circular appearance. An aperture of at least 20 cm is needed for its true nature to become apparent. With larger apertures the mottling appearance is easily seen, with bright and dark areas making up the characteristic shape of this planetary nebula. Its central star is very strange, believed to be one of the hottest stars known, with a temperature of at least 135,000 °K. It is also thought to be a binary star, which may account for its peculiar shape.

Messier 76	NGC 650–1	$01^h 42.4^m$	+51° 34′	October 16
10.1 m	⊕ 65′	3 + 6	15.9	moderate ©

Also known as the Little Dumbbell Nebula. This is a small planetary nebula that shows a definite non-symmetrical shape. In small telescopes of aperture 10 cm, and using averted vision, two distinct "nodes" or protuberances can be seen. With apertures of around 30 cm, the planetary nebula will appear as two bright but small discs which are in contact. Even larger telescopes will show considerably more detail.

Caldwell 22	NGC 7662	$23^h 25.9^m$	+42° 33′	September 12
8.6 m	⊕ 12′	4 + 3	13.2	moderate

See September.

Caldwell 2	NGC 40	$00^h 13.0^m$	+72° 32′	September 24
12.4 m	⊕ 36′	3b + 3	11.6	moderate ©

See September.

Caldwell 55	NGC 7009	$21^h\ 04.2^m$	–11° 22′	August 7
8.3 m	⊕ 25′	4 + 6	12.78	moderate

See August.

PK89+0.1	NGC 7026	$21^h\ 06.3^m$	+47° 51′	August 7
10.9 m	⊕ 21′	3a	14.9	moderate

See August.

November

PK206–40.1	NGC 1535	$04^h\ 14.2^m$	–12° 44′	November 24
9.6 m	⊕ 18′	4 + 2c	12	easy

One of the few planetary nebulae that show a distinct circular appearance in small telescopes. With an aperture of 5 cm and magnification of at least 100×, a small hazy glow will be seen. Under higher apertures, the disc is resolved along with the nice blue colour. Telescopes of 20 cm will easily resolve the subtle hazy outer ring structure, surrounding the bright bluish-green disc. The central star is easily seen at this aperture. A nice but not well known planetary nebula.

PK220–53.1	NGC 1360	$03^h\ 33.3^m$	–25° 51′	November 14
9.4 m	⊕ 400′	3	11	moderate

This is one of those planetary nebulae that gets better with larger apertures. Easily seen in telescopes of 20 cm, it will appear as a large, hazy but faint patch, with a prominent central star. Using a filter improves its visibility. As bigger apertures are used, it becomes more impressive, showing an oval disc. Often overlooked because it is very far south, it nevertheless merits more attention than it receives. This is a planetary nebula that owners of large telescopes should seek out and observe.

Herschel 53	NGC 1501	$04^h\ 07.0^m$	+60° 55′	November 22
11.5 m	⊕ 52′	3	14.5	moderate ©

Has been called the Oyster Nebula. A very nice blue planetary nebula, easily seen in telescopes of 20 cm, and glimpsed in apertures of 10 cm. With larger aperture, some structure can be glimpsed, and many observers liken this planetary nebula to that of the Eskimo Nebula. The central star can be seen if a high magnification is used – 300×.

Caldwell 56	NGC 246	$00^h\ 47.0^m$	–11° 53′	October 2
8.5 m	⊕ 225′	3b	11	moderate

See October.

Messier 76	NGC 650–1	$01^h\ 42.4^m$	+51° 34′	October 16
10.1 m	⊕ 65′	3 + 6	15.9	moderate ©

See October.

Caldwell 22	NGC 7662	$23^h 25.9^m$	+42° 33′	September 12
8.6 m	⊕ 12′	4 + 3	13.2	moderate

See September.

Caldwell 2	NGC 40	$00^h 13.0^m$	+72° 32′	September 24
12.4 m	⊕ 36′	3b + 3	11.6	moderate ©

See September.

PK196–10.1	NGC 2022	$05^h 42.1^m$	+09° 05′	December 16
12 m	⊕ 18′	4 + 2	15	moderate

See December.

December

PK206–40.1	NGC 1535	$04^h 14.2^m$	–12° 44′	November 24
9.6 m	⊕ 18′	4 + 2c	12	easy

See November.

PK196–10.1	NGC 2022	$05^h 42.1^m$	+09° 05′	December 16
12 m	⊕ 18′	4 + 2	15	moderate

This is a very small, faint, grey planetary nebula, but can be glimpsed in telescopes of 20 cm. Using larger aperture will resolve the disc appearance along with a pale greenish tint.

PK220–53.1	NGC 1360	$03^h 33.3^m$	–25° 51′	November 14
9.4 m	⊕ 400′	3	11	moderate

See November.

Herschel 53	NGC 1501	$04^h 07.0^m$	+60° 55′	November 22
11.5 m	⊕ 52′	3	14.5	moderate ©

See November.

Caldwell 56	NGC 246	$00^h 47.0^m$	–11° 53′	October 2
8.5 m	⊕ 225′	3b	11	moderate

See October.

Messier 76	NGC 650–1	$01^h 42.4^m$	+51° 34′	October 16
10.1 m	⊕ 65′	3 + 6	15.9	moderate ©

See October.

PK221–12.1	IC 2165	06h 21.7m	–12° 59′	December 26
12.9 m	⊕ 4.0′	3b	15	difficult
Another planetary nebula that needs a high magnification to resolve its non-stellar properties. With aperture of 20 cm, the small, faintly blue disc can be seen. Using larger aperture will resolve the non-circular shape along with a slight brightening at its centre.				

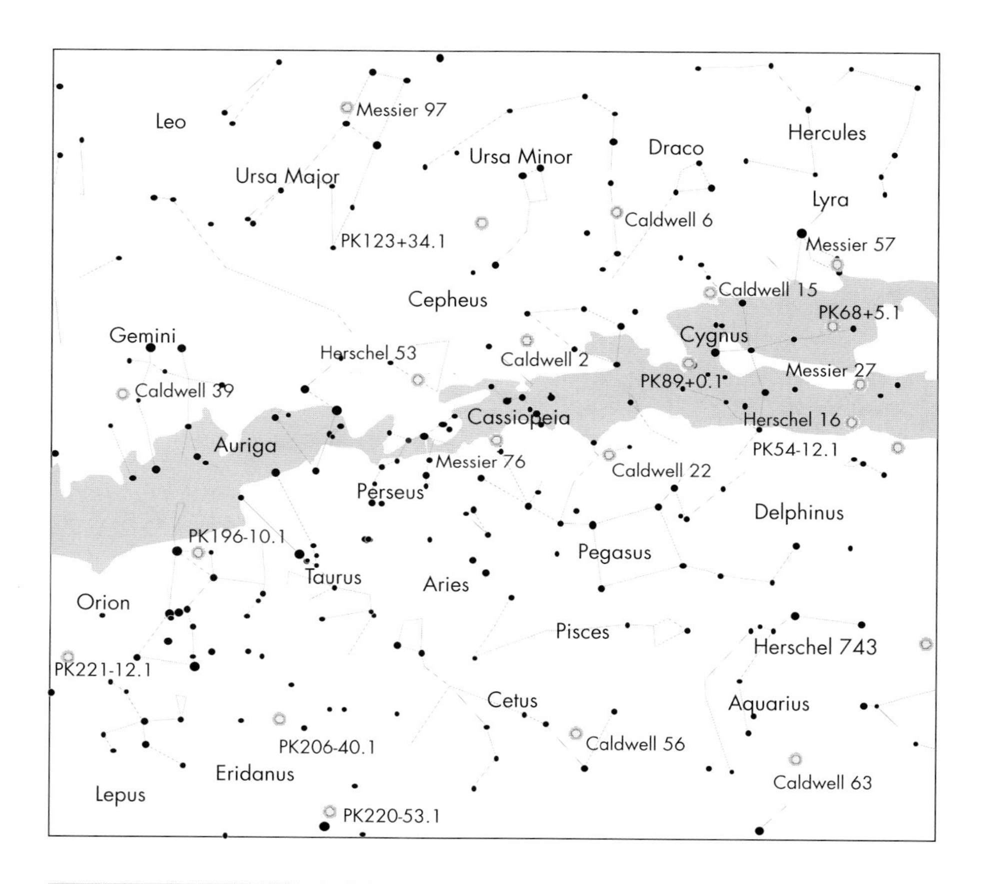

Figure 4.6. Planetary nebulae 1.

4.6 Supernova Remnants

The final topic in this chapter concerns those objects which, although they are few and far between from an amateur's viewpoint, represent the final and spectacularly explosive stage of a supergiant star's life. The finer details that lead up to a supernova are far beyond the scope of this book, but a rough picture can be sketched.

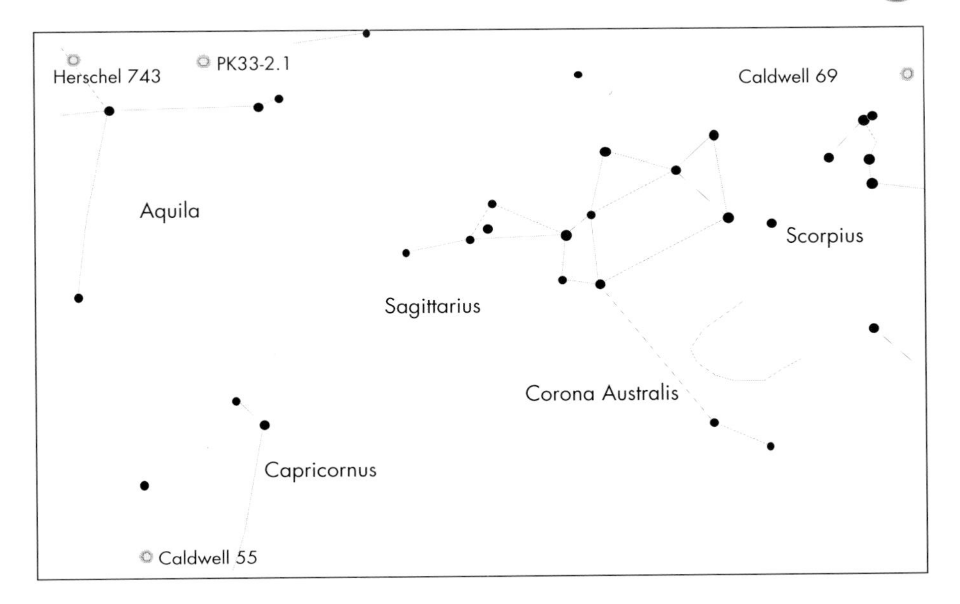

Figure 4.7. Planetary nebulae 2.

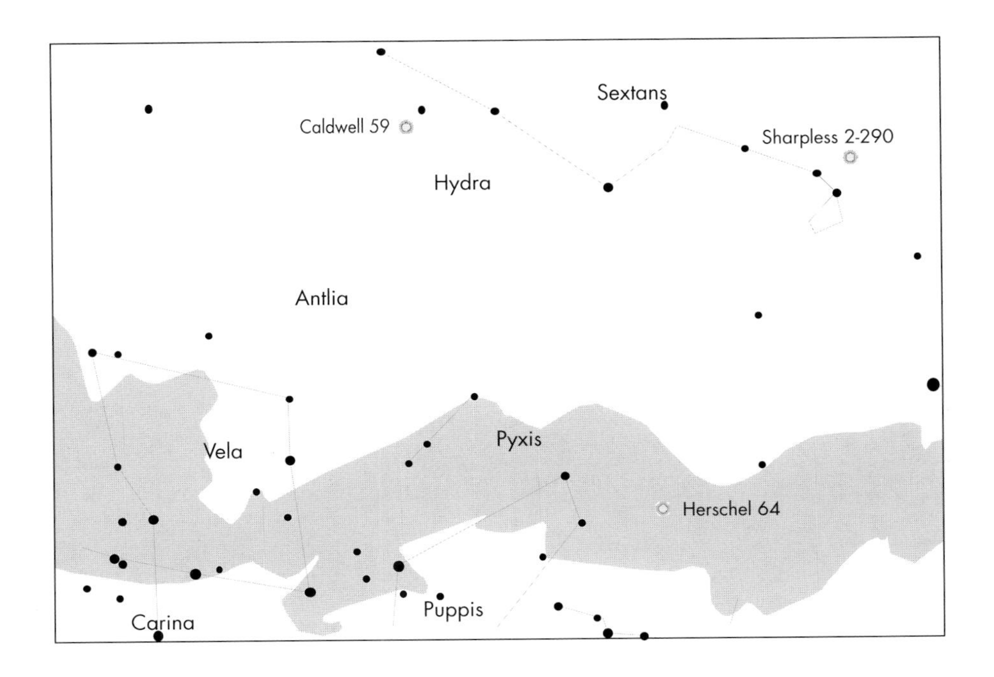

Figure 4.8. Planetary nebulae 3.

Basically, massive stars will, as mentioned earlier in the book, gradually turn all the hydrogen in the core to helium, and with an accompanying increase in temperature, turn this helium into carbon, then oxygen. Providing that the temperature increase needed to allow nucleosynthesis to occur is available, then many elements in the periodic table, up to and including iron, will be produced. The reactions are very complex, but the main outcome of all this will be an increase in the star's size, along with higher core temperatures. However, up until iron, all of the reactions have been *exothermic*, that is, they produce energy, and thus heat. This energy balances the tremendous gravitational force which otherwise would make the star collapse. The star is held in a state of equilibrium – the gravity balanced by the radiation produced in the core. The reactions involving iron however, become *endothermic* – they take in energy. Thus the delicate balancing act is disrupted, gravity wins the day, and the core collapses very rapidly, maybe in a matter of seconds. The net outcome of this dramatic implosion is for a wave of energy and neutrinos to hurl the outer layers of the star into space at phenomenal speeds in the region of 10,000 kilometres per second. The increasing surface area of the star represents an increase in luminosity, and this is what we see as a supernova.

The supernova remnant (usually abbreviated to SNR) represents the debris of the explosion, the layers of the star that have been hurled into space, and the remains of the core which will now be a *neutron star*. From an observational viewpoint however, you are only concerned with the parts that can be observed with amateur instruments – the shell of material flung out during the explosion. The visibility of the remnant actually depends on several factors; it's age, whether there is an energy source to continue making it shine, and the original type of supernova explosion.

As the remnant ages, its velocity will decrease, usually from 10,000 kilometres per second to maybe 200. It will of course fade during this time. A few SNRs have a neutron star at their centre that provides a replenishing source of energy to the far-flung material. The classic archetypal SNR which undergoes this process is the Crab Nebula, M1 in Taurus. What we see is the radiation produced by electrons travelling at velocities near the speed of light as they circle around magnetic fields. This radiation is called synchrotron radiation, and is the pearly, faint glow we observe. Some SNRs glow as the speeding material impacts dust grains and atoms in interstellar space, while others emit radiation as a consequence of the tremendous kinetic energies of the exploding star material.

What this all means to an amateur astronomer is that, unfortunately, only a handful of SNRs are observable. The good news is that most of them can be glimpsed with binoculars. A simple star map showing the location of the SNRs is at the end of the section.

January–February–March–April

Messier 1	NGC 1952	$05^h\ 34.5^m$	+22° 01′	December 14
● 1–5	⊕ 6I4′			easy
See September–October–November–December.				

–	IC 2118	$05^h\ 06.9^m$	–07° 13′	December 7
● 3–5	⊕ 180I60			moderate
See September–October–November–December.				

Sharpless 2-276	$05^h\ 56.0^m$	–02° 00′	December 20
● 5–6	⊕ 600′		difficult
See September–October–November–December.			

May–June–July–August

Caldwell 34	NGC 6960	20ʰ 45.7ᵐ	+30° 43′	August 2
● 3–5	⊕ 70 16′			moderate

Also known as the Veil Nebula (Western Section). This is the western portion of the Great Cygnus Loop, which is the remnant of a supernova that occurred about 30,000 years ago. It is easy to locate because it is close to the star 52 Cygni, though the glare from this star makes it difficult to see. Dark skies are needed and a light filter makes a vast difference. Positioning the telescope so that 52 Cygni is out of the field of view also helps. The nebulosity we observe is the result of the shockwave from the supernova explosion impacting on the much denser interstellar medium. The actual remains of the star have not been detected.

Caldwell 33	NGC 6992–5	20ʰ 56.4ᵐ	+31° 43′	August 5
● 2–5	⊕ 60 18′			moderate

Also known as the Veil Nebula (Eastern Section). A spectacular object when viewed under good conditions. It is the only part of the Loop that can be seen in binoculars, and has been described as looking like a fish-hook. It takes large aperture and high magnification well, and 40 cm telescopes will show the southern knot. Using such a telescope, it becomes apparent why the nebula has been the named the Filamentary Nebula, as lacy and delicate strands will be seen. However, there is a down side: it is notoriously difficult to find. Patience, clear skies and a good star atlas will help. A showpiece of the summer sky (when you have finally found it).

-	NGC 6974–79	20ʰ 51.0ᵐ	+32° 09′	August 4
● 2–5	⊕ 7 13′			difficult

Also known as the Veil Nebula (Central Section). This part of the Great Cygnus Loop is difficult to see, but the use of light filters makes it easier to locate and observe. It appears as a triangular hazy patch of light. A very transparent sky is need to glimpse this.

September–October–November–December

-	IC 2118	05ʰ 06.9ᵐ	–07° 13′	December 7
● 3–5	⊕ 180 160			moderate

Also known as the Witch Head Nebula. This is a very faint patch of nebulosity which apparently is the last of a very old supernova remnant. It resembles a long ribbon of material, which can be glimpsed with binoculars. It is glowing by reflecting the light of nearby Rigel. Very rarely mentioned in observing guides, it deserves more attention because it is one of a very select club of observable supernova remnants.

Messier 1	NGC 1952	05ʰ 34.5ᵐ	+22° 01′	December 14
● 1–5	⊕ 6 14′			easy

Also known as the Crab Nebula. The most famous supernova remnant in the sky, it can be glimpsed in binoculars as an oval light of plain appearance. With telescopes of aperture 20 cm it becomes a ghostly patch of grey light. Larger aperture will show some faint mottled structure. In all apertures (except very large – 40 cm) it will remain uniform in appearance. In 1968, in its centre was discovered the Crab Pulsar, the source of the energy responsible for the pearly glow observed, a rapidly rotating neutron star which has also been optically detected.

The Crab Nebula is a type of supernova remnant called a *plerion*, which, however, is far from common among supernova remnants.

Sharpless 2–276	05^h 56.0^m	–02° 00′	December 20
● 5–6	⊕ 600′		difficult

Also known as Barnard's Loop. Often mentioned in books, but very rarely observed, this is a huge arcing loop of gas located to the east of the constellation Orion. It encloses both the sword and belt of Orion, and if it were a complete circle it would be about 10° in diameter. The eastern part of the loop is well defined, but the western part is exceedingly difficult to locate, and has never to my knowledge been seen visually, only being observed by the use of photography or using a CCD. Impossible to see through a telescope, recent rumours have emerged that it has been glimpsed by a select few, by using either an OIII filter or an ultra-high-contrast filter. Needless to say, perfect conditions and very dark skies will greatly heighten the chances of it being seen. Possibly the greatest observing challenge to the naked-eye observer. It seems to be the remains of a very old supernova.

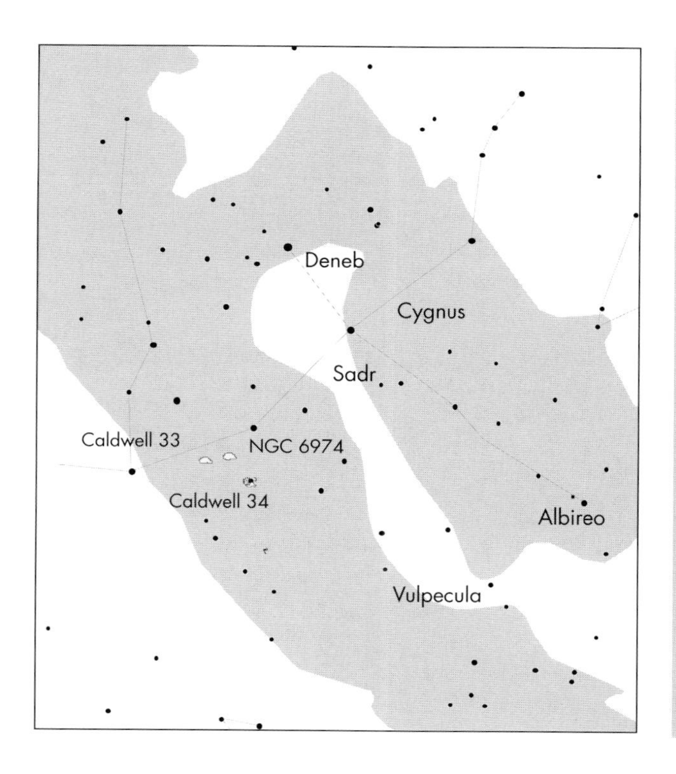

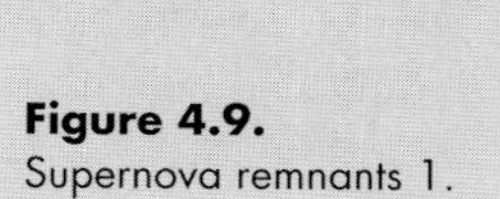
Figure 4.9. Supernova remnants 1.

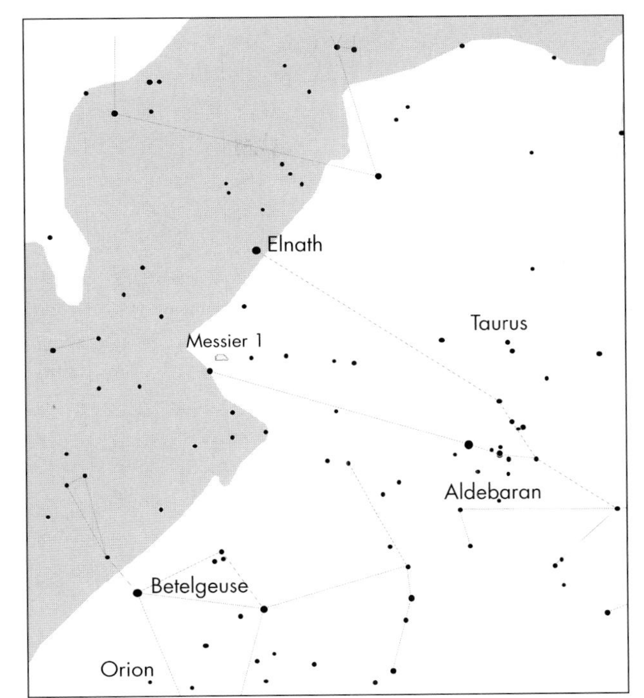

Figure 4.10.
Supernova remnants 2.

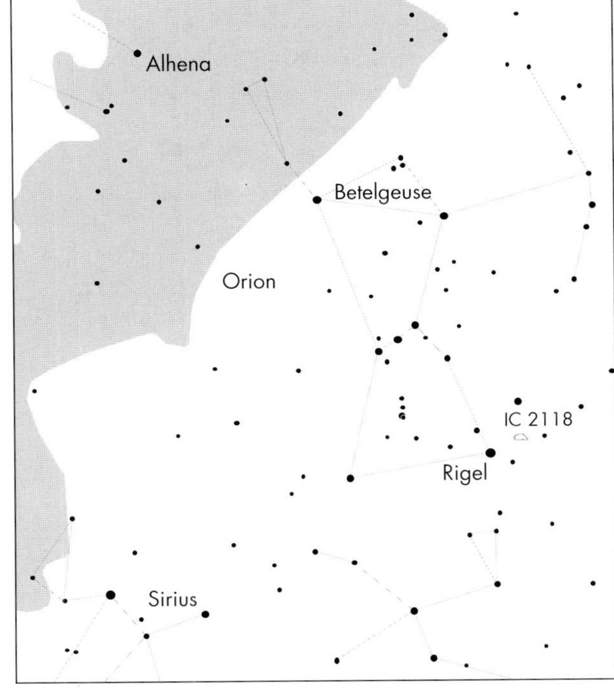

Figure 4.11.
Supernova remnants 3.

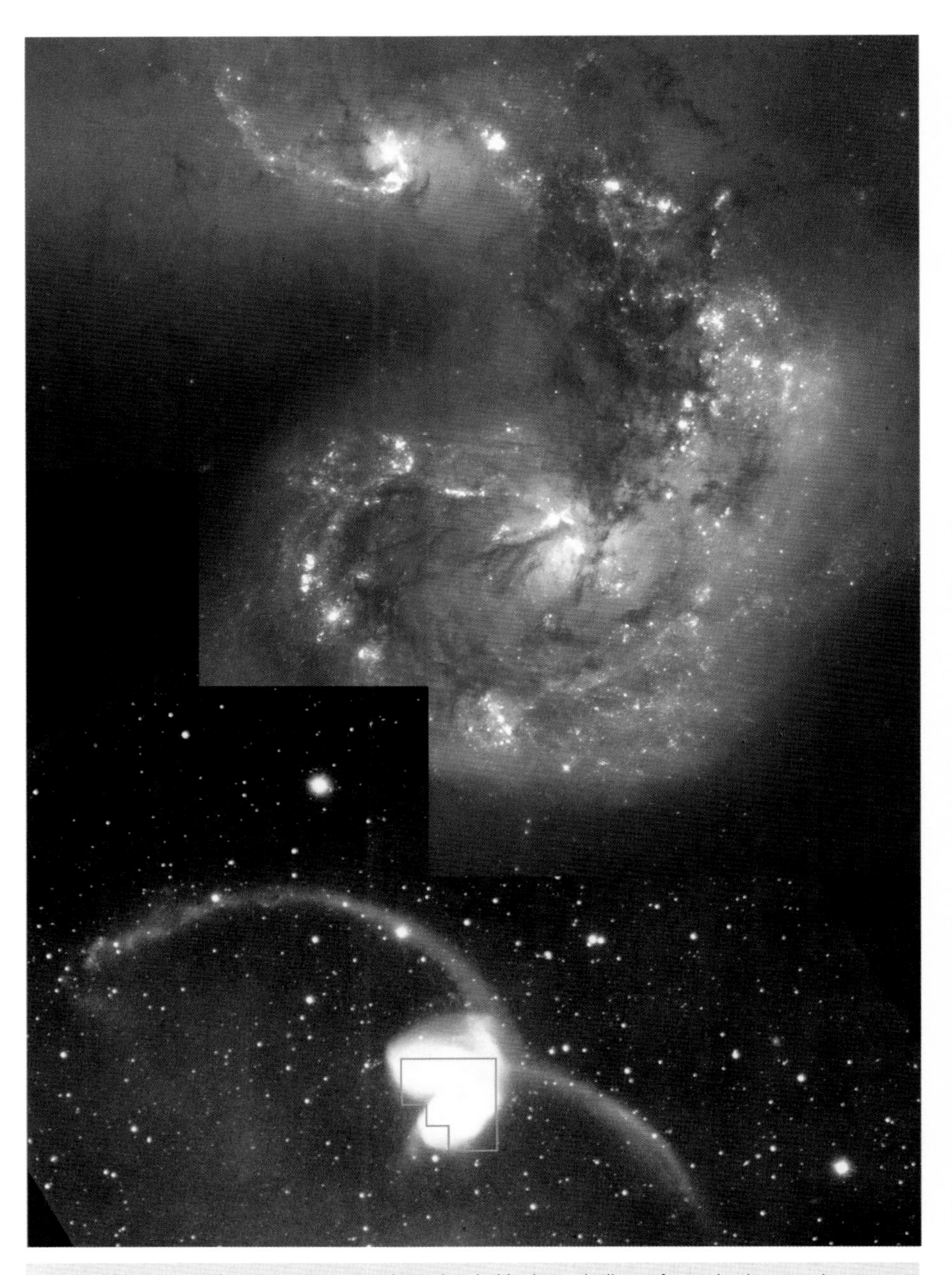

This Hubble Space Telescope image provides a detailed look at a brilliant "fireworks show" at the centre of a collision between two galaxies. The HST has uncovered over a thousand bright, young star clusters bursting to life as a result of the head-on wreck.

Lower image: a ground-based telescopic view of the Antennae galaxies (NGC 4038/4039) – named after a pair of long "antennae" of luminous matter caused by the gravitational tidal forces of their encounter, and resembling an insect's antennae. The galaxies are located 63 million light years away in the southern constellation Corvus.

Upper image: HST image shows the respective cores of the twin galaxies as orange blobs, above and below the middle of the picture, crisscrossed by filaments of dark dust. A wide band of chaotic dust, called the "overlap region", stretches between the cores of the two galaxies. The sweeping spiral-like patterns, traced by bright blue star clusters, show the result of the firestorm of star-birth activity that was triggered by the collision.

This natural-colour image is a composite of four separately filtered images taken with the Wide Field Planetary Camera 2.

Brad Whitmore (STScI)/NASA

5.1 Introduction

This penultimate chapter deals with those objects which every amateur astronomer not only knows about but has usually seen at least a handful of – galaxies.[1]

For the majority of amateurs, however, galaxies remain faint and elusive objects and I am probably correct in saying that only perhaps 10 to 15 galaxies are ever observed by 99 per cent of amateur astronomers. But with the proper optical system and under optimum seeing conditions (and a copy of this book), many more than that are within reach of even small telescopes or binoculars. A few are visible to the naked eye.

Galaxies are vast, immense collections of stars. Indeed they are the source of all stars, because stars are not born outside galaxies. The number of stars in galaxies varies considerably. In some giant galaxies, there may be a trillion (1×10^{12}) stars – a number that staggers the mind. In small dwarf galaxies, such as Leo I, there may be only a few hundred thousand.

Galaxies come in a variety of shapes and sizes, but most can be grouped into a few distinct classifications. When astronomers began studying the galaxies, the characteristic that immediately became apparent was their shape, or morphology. Broadly speaking, they can be classified into three major categories:

Spiral galaxies appear as flat white discs with yellowish bulges at their centres. The disc regions are occupied by dust and cool gas, interspersed with hotter ionised gas, as is the case in the Milky Way. Their most obvious characteristic is the beautiful spiral arms.

Elliptical galaxies are somewhat redder, more rounded in appearance, like a football. Compared with spiral galaxies, ellipticals contain far less cool gas and dust, but very much more hot ionised gas.

Those galaxies that appear neither disc-like nor rounded are classified as *irregular galaxies*.

---

[1]The Milky Way galaxy is often referred to as the "Galaxy", with a capital letter, whereas any other is simply a "galaxy".

The classification system is further sub-divided and specialised to take account of, for instance, the brightness of the nuclear region (the tight compact central region of the galaxy), the tightness of the spiral arms, etc.

5.2 Galaxy Structure

Before going any further it will I think be helpful to describe in a little more detail the structure of a galaxy as this will, albeit in a small way, provide some insight into why galaxies appear the way they do. The books mentioned in the appendix will have a much more detailed coverage of this topic, along with discussions on the origin and formation of galaxies.

Spiral galaxies have a thin *disc* extending outward from a central *bulge*. The bulge merges smoothly into what is called the *halo*, which can extend to a radius in excess of 100,000 l.y. Both the bulge and halo make up what is called the *spheroidal component*. There are no clear boundaries as to what divides this component up into its constituent parts, but a ball-park figure, often used, is that stars within 10,000 l.y. of the centre can be considered to be bulge stars, whereas those outside this radius are members of the halo.

The *disc component* of a spiral galaxy cuts through both the halo and the bulge, and can, in a large spiral galaxy such as the Milky Way, extend 50,000 l.y. from the centre. The disc area of all spirals contains a mixture of gas and dust, called the *interstellar medium*, but the amounts and proportions of the gas, whether atomic, ionised or molecular, will be different from galaxy to galaxy.

The stars contained within a spiral galaxy can also be classified depending on where they reside. Those that lie in the disc region are called *Population I* stars, and are often young, hot and blue stars: those in the bulge region are old, red giant stars, called *Population II* stars. This is why photographs often show the spiral arms coloured blue, owing to the Population I stars, with the bulge coloured orange because of the Population II ones. The spiral arms may also be dotted with pink and red HII regions,[2] areas of star formation. Thus, new stars are usually formed in the spiral arms of galaxies, seldom in the bulge.

Some spiral galaxies exhibit a straight bar of stars which cuts across the centre, with spiral arms curling away from the ends of the bars. Galaxies with these features are known as *barred spirals*. Galaxies which possess discs but not spiral arms are called *lenticular* galaxies, because they look lens-shaped when seen edge-on.

About 75 per cent of large galaxies in the observable universe are apparently spiral or lenticular. Some spiral galaxies that can be found in a loose collection of other spirals – this is known as a *group* – spread over several million light years. Our Galaxy, the Milky Way, is a member of the Local Group.

Elliptical galaxies differ significantly from spirals in that they do not have a significant disc component. Therefore an elliptical has only the spheroidal component. The interstellar medium is also different from that in spirals; it is a mixture of low-density, hot, X-ray-emitting gas, with little cool gas or dust. Contrary to what you may read in some books, ellipticals do possess a little gas and dust, and some have a small gaseous disc at their centre which is believed to be the remains of spiral galaxies which the elliptical has consumed.

---

[2]See Chapter 4 for a description of HII regions.

The stars in the spheroidal population of elliptical galaxies give a clue to possible star formation, if any. Such stars are orange and red, with an absence of blue stars, indicating that they are old, and that star formation occurred a long time ago.

Elliptical galaxies are often found in large clusters of galaxies, usually located near their centre. They make up about 15 per cent of the large galaxies found outside clusters, but about 50 per cent of the large galaxies within a cluster. Very small galaxies, called *dwarf elliptical galaxies*, are often found accompanying large spiral ones. A perfect example of such an arrangement, and one that is visible to the amateur, is the Great Andromeda Galaxy, M31, which is a classic spiral galaxy, and its attendants, M32 and M110, both dwarf ellipticals.

Several galaxies can be observed that do not belong to either the spiral or the elliptical galaxy category. These are irregular galaxies, and in fact more or less include all those galaxies which do not easily fall into the two previous classes. They include small galaxies such as the Magellanic Clouds,[3] and those galaxies which are peculiar owing to tidal interactions. These systems of galaxies are usually white and dusty, as spirals are, though there the resemblance ends. Deep imaging has shown that the more distant galaxies are irregular, which indicates that this type of galaxy was more common when the universe was much younger.

5.3 Hubble Classification of Galaxies

The famous American astronomer Edwin Hubble was the first person to put the many disparate types of galaxies into some sort of order. The *Hubble Classification*, as it is now known, is used as a means of categorising any galaxy. Further amendments have of course been added, particularly by the astronomer de Vaucouleurs.

Basically the classification is as follows. An upper-case letter followed by either a number or a lower-case letter is assigned to the galaxy in question, and this identifies its morphology. For instance, in the case of an elliptical galaxy, the letter E is used followed by a number. The larger the number, the flatter the galaxy. An E0 galaxy is round, whereas an E7 galaxy is very elongated. There exists a subgroup for the ellipticals with the following nomenclature: D signifies a diffuse halo, c is a supergiant galaxy, and d represents a dwarf galaxy.

A spiral galaxy is assigned the letter S, but can also be assigned SA, to signify that it is an ordinary spiral, or SB, where the B indicates it has a bar. It is then followed by a lower-case letter: a, b, c or d. Intermediate classes also exist, for example ab, bc, cd, dm and m. The lower-case letter a to d indicates the size of the bulge region, the dustiness of the disc, and the tightness of the spiral arms, while m denotes a stage where the spiral shape is barely discernible. An Sa galaxy will usually have a large bulge, with a modest amount of dust, and tightly wound arms, whereas an Sd galaxy will have a small bulge and very loosely wound arms. An SBc galaxy, for instance, will have a bar and also a small bulge.

[3]Recent research has shown that the Large Magellanic Cloud, often classified as irregular, is in fact a spiral galaxy, even though it bears little resemblance to the classic spiral shape.

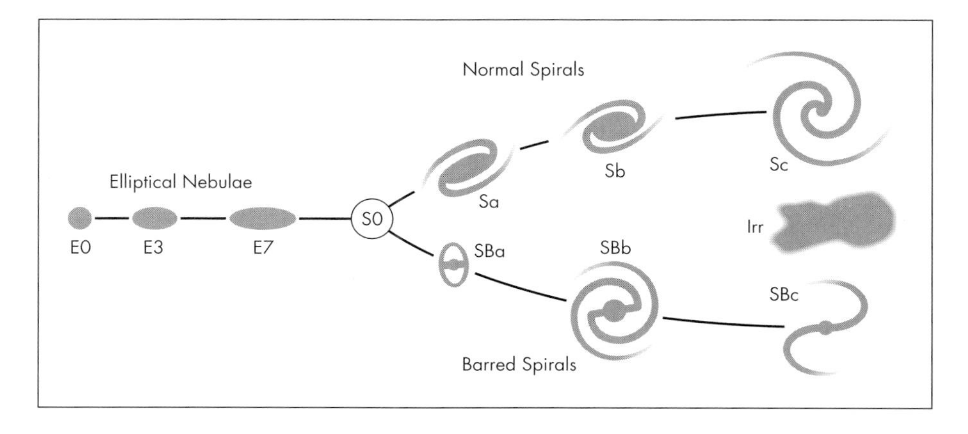

Figure 5.1. Hubble's tuning fork diagram showing the main galaxy types.

There is also a classification for galaxies intermediate between spirals and ellipticals, the lenticular galaxies, classified as SO, while SAO is for those that are ordinary, and SBO for those that are barred. In addition, for galaxies intermediate between type S and SB, there is the classification SAB. Both lenticular and spiral galaxies can be surrounded by an outer ring, or perhaps the spiral arms will nearly close upon themselves, thus forming a pseudo-ring. These new features are classified as R and R′ respectively.

Finally, there are classifications for those galaxies that do not easily fall into any of the above three! These include Pec, for peculiar galaxies that may have a distorted form. Some galaxies have an irregular morphology, and are classed as Irr. They can also be further classified as unstructured, IA, and barred, IB. Dwarf galaxies are classified as d. The difference between a galaxy classified as Pec and one classified as Irr can be very small, but it appears that a peculiar galaxy is one that may have suffered considerable tidal distortion for the passage of another galaxy nearby.

The Hubble classification system can be represented by a diagram (Fig. 5.1); note, however, that the diagram, and the classification generally, do not represent an evolutionary sequence. Galaxies do not start as ellipticals and then progress to be spirals, though there is some evidence that the reverse is the truth.

The classification system can be confusing (an understatement!), and the system and descriptions outlined above are by no means complete – there are further subdivisions to all the classes. But don't let that worry you – the complete system is only of relevance to those astrophysicists who study galaxies; to the observer, the important aspect is whether the galaxy is a spiral, and, if so, whether it is barred, or elliptical. In those few galaxies where the spiral or elliptical structure is very apparent, the subdivisions of, say, E1, E2 and Sa, Sb, SBa, etc., will be useful. Like most things in observational astronomy, it will all become easier with use.

There are some times when the classification system may be of no apparent use at all, namely when a galaxy is inclined to our line of sight. For instance, a galaxy such as M83, which is a nice spiral galaxy, is face-on to us, and thus the classic spiral shape is very apparent. However, a galaxy such as NGC 891, which is classified as a spiral, will just appear as a thin streak of light, because the galaxy is edge-on to us, and the spiral shape will not be visible. However, finding galaxies which present an edge-on or nearly edge-on perspective adds another element to the pleasure of locating and observing these faint objects.

5.4 Observing Galaxies

To the amateur astronomer, observing galaxies can present something of a dilemma. In astronomy magazines and books, you are bombarded with images of galaxies, their spiral arms resplendent and multi-coloured, speckled throughout with distinct pink HII regions. However, when you look at that same galaxy through your telescope, all you can see is a pale tiny blob!

It is true to say that in nearly every case, and from an urban location, the galaxy will be faint and indistinct. Only with the largest telescopes and the darkest possible skies can any real structure be seen. But take heart! Even with the naked eye, you will be astonished at what you can actually see with practice and from the right location. I recall that during a visit to the wilder parts of Turkey, on several occasions under utterly dark skies, I was able to see M31 in Andromeda and M33 in Triangulum in such amazing detail that even today the memory takes my breath away. Using just my naked eyes, I was able to trace M31 to nearly $2^1/_2°$ across the sky, and M33 was a huge amorphous glow. Also, had I only known enough to look, I would have been able to see several other galaxies with the naked eye as well, but I was under the common misapprehension that the naked-eye limit is about 6th magnitude, whereas I now know that, with extremely dark skies and light-adapted vision, magnitude 8 is more like the limit.

The purpose of this anecdote is to remind you that, in order to see faint galaxies and the detail therein, dark skies are indispensable. With such dark skies, and armed only with a pair of binoculars, many galaxies will be within reach. If you have a telescope that number increases dramatically.

As usual, dark skies, dark-adapted vision, and averted vision will all help in tracking down and seeing galaxies. Clean optics will also greatly aid you in your observations. Dust and smears of grease will reduce by a surprising amount the light that reaches your eyes, and in particular will reduce the contrast.

Generally, those galaxies which have a brightness greater than 13th magnitude are usually visible in telescopes of aperture 15 cm, and those of aperture 30 cm will see down to about 14.5 magnitude. There will of course be galaxies which will have much brighter magnitudes than these, and so will be visible in much smaller instruments. In some cases only the brightest part of a galaxy will be visible – perhaps its core (nuclear region), with the spiral part unobservable.

To be able to trace out the finer details of the spiral arms of galaxies, and to locate the bulge area, faint halo and HII regions, you will invariably require a large-aperture telescope. But if the purpose of your observing is just to locate these elusive objects, and to be amazed that the light that is entering your eye may have began its journey over 100 million years ago, then there are a plethora of galaxies awaiting your visit.[4] That is the real purpose of this chapter.

The usual nomenclature applies in the following descriptions, but with the following changes. Galaxies are *extended objects*, which means that they cover an appreciable part of the sky: in some cases a few degrees, in others only a few arc minutes. The light from the galaxy is therefore "spread out" and thus the quoted magnitude will be the magnitude of the galaxy were it the "size" of a star; this magnitude is often termed the *integrated magnitude*. This can cause confusion, as a galaxy with, say, a magnitude of 8, will appear fainter than an 8th-magnitude star, and in some cases, where possible, the surface brightness of a galaxy will be given. This will give a better idea of what the overall magnitude of the galaxy

[4]Of course, I don't really have to mention that if you have a medium-to-large-aperture telescope, then the number of galaxies visible to you is vast, and the detail you will be able to see will astound you!

will be. For instance, Messier 64, the Black Eye Galaxy, has a magnitude of 8.5, whereas its surface brightness is 12.4. The surface brightness will be given in italics after the quoted magnitude, and to use the above example of M64 will appear like this: 8.5 m [*12.4 m*].

Following on from the previous paragraph, the designation "easy", "moderate" or "difficult" takes into account not only the brightness of the galaxy, but also the area of the sky the galaxy spans. Thus, a galaxy may be bright, with – say – a magnitude of 8, which under normal circumstances would be visible in binoculars and designated as "easy"; but if it covers a significant amount of the sky (and thus its surface brightness is low, making it more difficult to observe) I designate it "moderate".

In addition, spiral galaxies can exhibit a variety of views, depending on their inclination to the Solar System. Some will appear face-on, others at a slight angle, and a few completely edge-on. As an indicator of inclination, the following symbols will be used.

Face-on:

Slight inclination:

Edge-on:

Finally, the Hubble classification of galaxies I outlined earlier will also be used.

5.5 Spiral Galaxies

January

Caldwell 7	NGC 2403	07h 36.9m	+65° 35′	January 14
8.5 m [*13.9 m*]	17.8′ l 10.0′		SAB(s)cd	easy ©

This is one of the brightest galaxies that was somehow missed from the Messier catalogue, and is often left out of an observer's schedule. In binoculars it appears as a large, oval hazy patch with a brighter central region. With averted vision, and an aperture of about 20 cm, faint hints of a spiral arm will become apparent. Larger apertures will of course present even further detail. It is not a member of the Local Group of Galaxies[5] but believed to be a member of the M81–M82 group. It was the first galaxy outside the Local Group to have cepheid[6] variable stars discovered within it, and the current estimate of its distance is 11.5 million l.y.

–	NGC 2683	08h 52.7m	+33° 25′	February 3
9.7 m [*12.9 m*]	9.3′ l 2.5′		SA(rs)b	easy

See February.

Messier 81	NGC 3031	09h 55.6m	+69° 04′	February 18
6.9 m [*13.0 m*]	26′ l 14′		SA(s)ab	easy ©

See February.

[5]The Local Group is a cluster of several galaxies, including the Milky Way. It consists of M31, M33, M110, M32, The Large and Small Magellanic Clouds and about 25 other dwarf galaxies, including Leo I and II , And I and II, the Draco, Carina, Sextans and Phoenix dwarfs.

[6]Cepheid variables are used as *standard candles*, for measuring distances to other extra-galactic objects.

Messier 96	NGC 3368	$10^h 46.8^m$	+11° 49′	March 3
9.2 m [*12.9 m*]	7.6′ l 5.2′		SAB(rs)ab	easy
See March.				

Messier 65	NGC 3623	$11^h 18.9^m$	+13° 05′	March 12
9.3 m [*12.4 m*]	9.8′ l 2.9′		SAB(rs)a	easy
See March.				

Messier 66	NGC 3627	$11^h 20.2^m$	+12° 59′	March 12
9.0 m [*12.5 m*]	9.1′ l 4.2′		SA(s)b	easy
See March.				

Messier 106	NGC 4258	$12^h 19.0^m$	+47° 18′	March 27
8.3 m [*13.8 m*]	18.6′ l 7.2′		SAB(s)bc	easy
See March.				

Messier 88	NGC 4501	$12^h 32.0^m$	+14° 25′	March 30
9.6 m [*12.6 m*]	6.9′ l 3.7′		SA(rs)b	easy
See March.				

Caldwell 38	NGC 4565	$12^h 36.3^m$	+25° 59′	March 31
9.6 m [*13.5 m*]	15.5′ l 1.9′		SA(s)bsp	easy
See March.				

Messier 90	NGC 4569	$12^h 36.8^m$	+13° 10′	March 31
9.5 m [*13.6 m*]	9.5′ l 4.4′		SAB(rs)ab	easy
See March.				

Messier 77	NGC 1068	$02^h 42.7^m$	–00° 01′	November 1
8.9 m [*13.2 m*]	7.1′ l 6.0′		(R)SA(rs)b	easy
See November.				

Caldwell 48	NGC 2775	$09^h 10.3^m$	+07° 02′	February 7
10.1 m [*13.1 m*]	5.0′ l 4.0′		SA(R)ab	moderate
See February.				

Messier 98	NGC 4192	$12^h 13.8^m$	+14° 54′	March 25
10.1 m [*13.2 m*]	9.8′ l 2.8′		SAB(s)ab	moderate
See February.				

Messier 99	NGC 4254	$12^h 18.8^m$	+14° 25′	March 27
9.9 m [*13.0 m*]	5.4′ l 4.7′		SA(s)c	moderate
See February.				

Messier 61	NGC 4303	$12^h 21.9^m$	+04° 28′	March 28
9.6 m [*13.4 m*]	6.5′ l 5.8′		SAB(rs)bc	moderate

See March.

Messier 100	NGC 4321	$12^h 22.9^m$	+15° 49′	March 28
9.3 m [*13.3 m*]	7.4′ l 6.3′		SAB(rs)bc	moderate

See March.

Cadwell 36	NGC 4559	$12^h 35.9^m$	+27° 57′	March 31
9.8 m [*13.9 m*]	13.0′ l 5.23′		SAB(rs)cd	moderate

See March.

Cadwell 40	NGC 3626	$11^h 20.1^m$	+18° 21′	March 12
11.0 m [*12.8 m*]	2.8′ l 2.0′		(R)SA(rs)	difficult

See March.

Caldwell 26	NGC 4244	$12^h 17.5^m$	+37° 48′	March 26
10.4 m [*14.0 m*]	18.5′ l 2.3′		SA(s)cd:sp	difficult

See March.

Caldwell 5	IC 342	$03^h 46.8^m$	+68° 05′	November 17
9.2 m [*15.2 m*]	17.8′ l 17.4′		SAB(rs)cd	difficult ©

See November.

February

–	NGC 2683	$08^h 52.7^m$	+33° 25′	February
39.7 m [*12.9 m*]	9.3′ l 2.5′		SA(rs)b	easy

Easily seen with binoculars, providing viewing conditions are at optimum. This cigar-shaped galaxy is inclined nearly edge-on to our line of sight. What is nice about it is the way that the nucleus clearly stands out from the thin disc. A fine sight in all sizes of optical equipment.

Messier 81	NGC 3031	$09^h 55.6^m$	+69° 04′	February 18
6.9 m [*13.0 m*]	26′ l 14′		SA(s)ab	easy ©

A spectacular object! In binoculars it will show a distinct oval form, and using high-power binoculars the nuclear region will easily stand out from the spiral arms. Using a telescope will show a considerable lot more detail, and it is one of the grandest spiral galaxies on view. With an aperture of about 15 cm, traces of several of the spiral arms will be glimpsed. A real challenge, however, is to see if you can locate this galaxy with the naked eye. Several observers have reported seeing it from dark locations. If you do glimpse it without any optical aid, then you are probably looking at one of the furthest objects[7] that can be seen with the naked eye, lying at a distance of some 4.5 million l.y. M81 is partner galaxy to M82 (see the part of Section 5.8 on irregular galaxies), and both these spectacular objects can be glimpsed in the same field of view.

[7]The galaxy M83 lies at the same distance, and has reportedly been seen with the naked eye.

Caldwell 7	NGC 2403	07h 36.9m	+65° 35′	January 14
8.5 m [*13.9 m*]	17.8′ I 10.0′		SAB(s)cd	easy ©

See January.

Messier 96	NGC 3368	10h 46.8m	+11° 49′	March 3
9.2 m [*12.9 m*]	7.6′ I 5.2′		SAB(rs)ab	easy

See March.

Messier 65	NGC 3623	11h 18.9m	+13° 05′	March 12
9.3 m [*12.4 m*]	9.8′ I 2.9′		SAB(rs)a	easy

See March.

Messier 66	NGC 3627	11h 20.2m	+12° 59′	March 12
9.0 m [*12.5 m*]	9.1′ I 4.2′		SA(s)b	easy

See March.

Messier 106	NGC 4258	12h 19.0m	+47° 18′	March 27
8.3 m [*13.8 m*]	18.6′ I 7.2′		SAB(s)bc	easy

See March.

Messier 88	NGC 4501	12h 32.0m	+14° 25′	March 30
9.6 m [*12.6 m*]	6.9′ I 3.7′		SA(rs)b	easy

See March.

Caldwell 38	NGC 4565	12h 36.3m	+25° 59′	March 31
9.6 m [*13.5 m*]	15.5′ I 1.9′		SA(s)bsp	easy

See March.

Messier 90	NGC 4569	12h 36.8m	+13° 10′	March 31
9.5 m [*13.6 m*]	9.5′ I 4.4′		SAB(rs)ab	easy

See March.

Caldwell 48	NGC 2775	09h 10.3m	+07° 02′	February 7
10.1 m [*13.1 m*]	5.0′ I 4.0′		SA(R)ab	moderate

A difficult object for binoculars, this galaxy really requires a telescope. With an aperture of about 20 cm, the galaxy will show itself as a large blob. Detail within the object is conspicuously absent, but a brighter core and fainter outer region can be resolved. The absence of detail (that is, spiral arm dust and gas) has been attributed to an early era of star formation, which used up all the material. The evidence for this was found in the galaxy's spectrum, which lacked emission lines – these lines are usually caused by star-forming regions in and around spiral arms.

Messier 98	NGC 4192	12h 13.8m	+14° 54′	March 25
10.1 m [*13.2 m*]	9.8′ I 2.8′		SAB(s)ab	moderate

See March.

Messier 99	NGC 4254	12h 18.8m	+14° 25′	March 27
9.9 m [13.0 m]	5.4′ l 4.7′		SA(s)c	moderate
See March.				

Messier 61	NGC 4303	12h 21.9m	+04° 28′	March 28
39.6 m [13.4 m]	6.5′ l 5.8′		SAB(rs)bc	moderate
See March.				

Messier 100	NGC 4321	12h 22.9m	+15° 49′	March 28
9.3 m [13.3 m]	7.4′ l 6.3′		SAB(s)bc	moderate
See March.				

Caldwell 36	NGC 4559	12h 35.9m	+27° 57′	March 31
9.8 m [13.9 m]	13.0′ l 5.2′		SAB(rs)cd	moderate
See March.				

Caldwell 40	NGC 3626	11h 20.1m	+18° 21′	March 12
11.0 m [12.8 m]	2.8′ l 2.0′		(R)SA(rs)	difficult
See March.				

Caldwell 26	NGC 4244	12h 17.5m	+37° 48′	March 26
10.4 m [14.0 m]	18.5′ l 2.3′		SA(s)cd:sp	difficult
See March.				

March

Messier 96	NGC 3368	10h 46.8m	+11° 49′	March 3
9.2 m [12.9 m]	7.6′ l 5.2′		SAB(rs)ab	easy

This faint galaxy can be seen in binoculars as a faint hazy oval patch of light. But what you are observing is in fact just the bright central nucleus of the galaxy, as the spiral arms are too faint to be resolved. Telescopes will bring out further detail, and with good conditions the spiral arm features will be seen. There is some slight controversy over M96, as recent measurements of its distance place it at 38 million l.y., which is 60 per cent greater than the previous value. It forms a nice triangle with two other galaxies, M95 and M105.

Messier 65	NGC 3623	11h 18.9m	+13° 05′	March 12
39.3 m [12.4 m]	9.8′ l 2.9′		SAB(rs)a	easy

Visible in binoculars, this is one half of the most famous galaxy pair in the sky, after M81 and M82. Along with M66, it is a galaxy that shows up quite well with low-power optics. It appears a nice oval patch of light, and with higher magnification both spiral arms and a dust lane can be glimpsed. However, it is difficult to observe, as the brightness of the sky background tends to meld in with the galaxy details. A nice challenge for observers in an urban location.

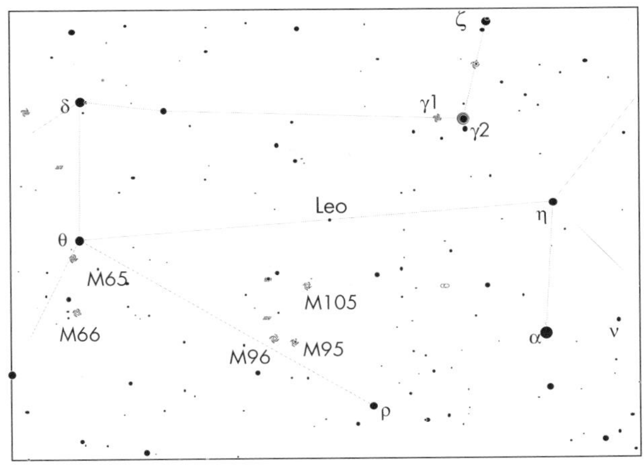

Figure 5.2.

Messier 66	NGC 3627	11^h 20.2^m	+12° 59′	March 12
9.0 m [*12.5 m*]	9.1′l 4.2′		SA(s)b	easy

The other half of the galaxy duo mentioned above. This is bright galaxy, easily seen in binoculars where its distinct elliptical shape and bright centre can be resolved. With telescopes, the oval shape of the nucleus becomes apparent, and with higher magnification a spiral arm and dark patch can be seen. Large-aperture telescopes will show considerable detail consisting of dark and light patches.

Messier 106	NGC 4258	12^h 19.0^m	+47° 18′	March 27
8.3 m [*13.8 m*]	18.6′l 7.2′		SAB(s)bc	easy

The galaxy appears as a large glow when seen in binoculars and has a distinct elliptical shape. Large binoculars reveal the presence of the nucleus. With telescopes of small aperture (10 cm), and low magnification, the spiral arms become apparent, and with higher magnification further detail can be seen. It is a galaxy that takes large aperture and magnification well, and will reveal a surprising amount of detail. The galaxy is nearly face-on to us, but a cloud of gas and dust surrounding the nucleus is apparently edge-on. Furthermore, there is evidence that a black hole resides at the core!

Messier 88	NGC 4501	12^h 32.0^m	+14° 25′	March 30
9.6 m [*12.6 m*]	6.9′l 3.7′		SA(rs)b	easy

A fine galaxy for binoculars, with a bright nucleus and a faint hazy glow surrounding it, caused by the spiral arms. What is a problem, however, is that the galaxy is located in a barren patch of the sky, making location a problem. With a telescope of medium aperture, say 20 cm, structure becomes visible. The spiral arms and nucleus can be seen; but use averted vision.

Caldwell 38	NGC 4565	12ʰ 36.3ᵐ	+25° 59′	March 31
9.6 m [*13.5 m*]	15.5′ l 1.9′		SA(s)bsp	easy

A striking example of an edge-on galaxy. With small binoculars, the classic spindle shape can be seen set against the background stars, and with large ones the central core region can be seen. Using a telescope of 15 cm, the lovely edge-on shape becomes even clearer, along with the star-like nucleus. The dust lane can be seen, but will require apertures of at least 20 cm and more. A very massive galaxy, and thought to be similar to the Milky Way, its dust lane the equivalent of our Great Rift.

Messier 90	NGC 4569	12ʰ 36.8ᵐ	+13° 10′	March 31
9.5 m [*13.6 m*]	9.5′ l 4.4′		SAB(rs)ab	easy

This is an impressive sight in binoculars. If large binoculars are used, and the conditions are right, the star-like nucleus and elliptical glow caused by the spiral arms can be seen. Telescopically, it is equally impressive, with its spiral arms and mottled nucleus resolvable. M90 is one of the largest spiral galaxies in the Virgo Cluster, with an estimated mass of 80 million Suns. It has a measured diameter of 150,000 l.y.

–	NGC 2683	08ʰ 52.7ᵐ	+33° 25′	February
3 9.7 m [*12.9 m*]	9.3′ l 2.5′		SA(rs)b	easy

See February.

Messier 81	NGC 3031	09ʰ 55.6ᵐ	+69° 04′	February 18
6.9 m [*13.0 m*]	26′ l 14′		SA(s)ab	easy ©

See February.

Messier 58	NGC 4579	12ʰ 37.7ᵐ	+11° 49′	April 1
9.6 m [*13.0 m*]	5.9′ l 4.7′		SAB(rs)b	easy

See April.

Messier 104	NGC 4594	12ʰ 40.0ᵐ	–11° 37′	April 1
8.0 m [*11.6 m*]	8.7′ l 3.5′		SA(s)asp	easy

See April.

Messier 94	NGC 4736	12ʰ 50.9ᵐ	+41° 07′	April 4
8.2 m [*13.0 m*]	11.2′ l 9.1′		(R)SA(r)ab	easy

See April.

Messier 64	NGC 4826	12ʰ 56.7ᵐ	+21° 41′	April 5
8.5 m [*12.4 m*]	9.3′ l 5.4′		(R)SA(rs)ab	easy

See April.

Caldwell 29	NGC 5005	13ʰ 10.9ᵐ	+37° 03′	April 9
9.8 m [*12.6 m*]	6.3′ l 3.0′		SAB(rs)bc	easy

See April.

Messier 63	NGC 5055	13h 15.8m	+42° 02′	April 10
8.6 m [*13.6 m*]	12.6′ l 7.2′		SA(rs)bc	easy

See April.

Messier 51	NGC 5194	13h 29.9m	+47° 12′	April 14
8.4 m [*13.1 m*]	11.2′ l 6.9′		SA(s)bcP	easy

See April.

Messier 83	NGC 5236	13h 37.0m	–29° 52′	April 16
7.5 m [*13.2 m*]	12.9′ l 11.5′		SAB(s)c	easy

See April.

Caldwell 7	NGC 2403	07h 36.9m	+65° 35′	January 14
8.5 m [*13.9 m*]	17.8′ l 10.0′		SAB(s)cd	easy ©

See January.

Messier 98	NGC 4192	12h 13.8m	+14° 54′	March 25
10.1 m [*13.2 m*]	9.8′ l 2.8′		SAB(s)ab	moderate

This faint galaxy lies at the edge of the great Coma–Virgo galaxy cluster, an area studded with galaxies, both faint and bright. It is a difficult object to locate, and requires a small telescope of at least 10 cm aperture. It is highly inclined to us, and has a very elongated shape. With excellent seeing conditions, a high power will show spectacular spiral arms and dust lanes, and under very rare conditions the entire halo can be seen to surround the galaxy. An excellent object to observe, but one that requires skill and patience to see any detail in, especially from an urban location.

Messier 99	NGC 4254	12h 18.8m	+14° 25′	March 27
9.9 m [*13.0 m*]	5.4′ l 4.7′		SA(s)c	moderate

A difficult object for binoculars; its circular disc shape will need dark skies to be resolved. With a telescope of aperture 10 cm, the galaxy will remain a hazy round patch but a small nucleus will be seen. Higher magnification with perhaps greater aperture will begin to show two spiral arms. However, M99 is one of those galaxies where the likelihood of seeing detail depends greatly on the seeing conditions. Try observing on a good night and then on an average night, and compare what you see. It lies at a distance of about 55 million l.y. and is one of the galaxies within the Coma–Virgo Cluster.[8]

Messier 61	NGC 4303	12h 21.9m	+04° 28′	March 28
9.6 m [*13.4 m*]	6.5′ l 5.8′		SAB(rs)bc	moderate

A difficult object to locate in binoculars, it will need dark skies, and even then will resemble nothing more than a small faint circular patch of light. For small-telescope users, however, it is a delight, and although it is small and can be a problem to find, is an ideal open-faced spiral galaxy. The use of averted vision is a must for this object, as only then will the nucleus and any spiral arm detail become apparent. A nice addition is the fact that the galaxy is located within the Virgo Cluster of galaxies, and when observing M61 you may notice several faint and indistinct glows in the same field of view, which are at the limit of your vision. These are probably unresolved galaxies!

[8]Sometimes the cluster is just referred to as the Virgo Cluster. For a description of the cluster, see the entry on M87.

Messier 100	NGC 4321	12h 22.9m	+15° 49′	March 28
9.3 m [13.3 m]	7.4′ l 6.3′		SAB(s)bc	moderate

Located very close to M99 (see above), this is a fainter and more elusive target. Although larger than M99 and half a magnitude brighter, it subsequently has a lower surface brightness. Thus excellent seeing conditions may be needed. Telescopically, it will be seen as a brightish patch of hazy light, although some structure may be glimpsed with perfect seeing and high magnification. Research indicates that there has been a recent spate of star formation within M100 with its inner spiral arms rotating in an opposite direction to the spiral arms which lie further out.

Caldwell 36	NGC 4559	12h 35.9m	+27° 57′	March 31
9.8 m [13.9 m]	13.0′ l 5.2′		SAB(rs)cd	moderate

A member of the Virgo Cluster, this is often overlooked. Not really a binocular object, it is perfect for small telescopes. At about 20 cm aperture, the clearly seen oval shape will be resolved, along with a brightening of the nucleus. A just perceptible hint of further detail may be glimpsed. Large aperture shows considerably more detail. There is tentative evidence that a black hole lurks at the core of the galaxy.

Caldwell 45	NGC 5248	13h 37.5m	+08° 53′	April 16
10.3 m [13.8 m]	6.8′ l 5.0′		SAB(rs)bc	moderate

See April.

Messier 101/2[9]	NGC 5457	14h 03.2m	+54° 21′	April 22
7.9 m [14.8 m]	28.8′ l 26.9′		SAB(rs)cd	moderate ©

See April.

Caldwell 40	NGC 3626	11h 20.1m	+18° 21′	March 12
11.0 m [12.8 m]	2.8′ l 2.0′		(R)SA(rs)	difficult

This galaxy is virtually unknown among amateurs. It lies close to several brighter Messier objects, and is often mistaken for NGC 3607. Nevertheless, it is worth searching out for. Visually it is a featureless oval patch of light, and apertures of at least 20 cm are needed to resolve the object. What makes it special, however, is that it is a *multispin* galaxy. This means the molecular and ionised gas is rotating around the galaxy in the opposite direction to that of its stars. The origin of this phenomena is unknown, but one school of thought believes that the galaxy recently collided with and assimilated a huge gas cloud with a mass of about one billion solar masses.

Caldwell 26	NGC 4244	12h 17.5m	+37° 48′	March 26
10.4 m [14.0 m]	18.5′ l 2.3′		SA(s)cd:sp	difficult

One of the most needle-like galaxies that can be seen with amateur instruments This edge-on galaxy really is exceedingly thin, and apertures of at least 20 cm will be need to locate and observe it. It has a faint but easily resolvable star-like nucleus, but detail within the galaxy itself is very rarely seen even with large apertures. It has a Hubble classification similar to that of M33, the Pinwheel Galaxy in Triangulum, and when such a galaxy is seen edge-on, the tiny nucleus and loose, open arms give it its indistinct appearance.

[9]M102 is now believed to be just a duplicate observation of M101.

April

Messier 58	NGC 4579	12h 37.7m	+11° 49′	April 1
9.6 m [13.0 m]	5.9′ l 4.7′		SAB(rs)b	easy

Through binoculars this galaxy will appear as a faint, hazy patch of light with a barely discernible nucleus. You may also glimpse in the same field of view the galaxies M59 and M60. A telescope of about 10 cm aperture will show some structure in the halo, along with faint patches of light and dark. There are some reports that a 20 cm telescope will allow the bar connecting the spiral arms to the nucleus to be resolved. Can you see it? It has about the same mass as the Milky Way and is about 95,000 l.y. in diameter.

Messier 104	NGC 4594	12h 40.0m	−11° 37′	April 1
8.0 m [11.6 m]	8.7′ l 3.5′		SA(s)asp	easy

Also known as the Sombrero Galaxy. This is an extragalactic treasure! A marvellous sight in almost all binoculars and telescopes. With small binoculars it will reveal itself as an oval disc which increases in brightness toward the centre. Using large binoculars will however reveal its true beauty. The dark dust lane that cuts across the galaxy readily becomes apparent. With a telescope even more detail is brought out. With a 10 cm aperture a bright core is seen, along with the long, spindle-like dust lane. With higher magnifications, the spiral arms stand out. Large apertures and higher magnifications will reveal a wealth of detail. It was the first galaxy other than the Milky Way to have its rotation determined. Glorious!

Messier 94	NGC 4736	12h 50.9m	+41° 07′	April 4
8.2 m [13.0 m]	11.2′ l 9.1′		(R)SA(r)ab	easy

This galaxy is visible in binoculars and will appear as a small circular hazy patch with a star-like nucleus. Telescopes will reveal some structure, possibly a faint spiral arm. Several observers have reported that a central ring can be seen near the core, which gives it an appearance very similar to M64, the Black Eye Galaxy. A further elliptical ring has been reported outside the edge of the galaxy. Can you see it? Needless to say, exceptionally dark skies will be need to observe this elusive feature.

Messier 64	NGC 4826	12h 56.7m	+21° 41′	April 5
8.5 m [12.4 m]	9.3′ l 5.4′		(R)SA(rs)ab	easy

Also known as the Black Eye Galaxy. This famous galaxy can be seen in small binoculars as an oval hazy patch with a slightly brighter centre. The feature which gives the galaxy its name has been reported to be visible with large binoculars on very dark nights. Small telescopes show a very bright nucleus encased in a patch of glowing light. The "eye" is a vast dust lane, some 40,000 l.y. in diameter. There is considerable debate as to whether the "eye" can be seen in small instruments. Some observers report that an aperture as small as 6 cm will resolve it, while others claim at least 20 cm is needed. What is agreed upon is that high magnification is necessary. There is also further controversy as to whether the nucleus is star-like or not. It may be that magnification plays an important role here. Observe it and see who you side with!

Caldwell 29	NGC 5005	13h 10.9m	+37° 03′	April 9
9.8 m [*12.6 m*]	6.3′ l 3.0′		SAB(rs)bc	easy

This galaxy is not a binocular object, and with telescopes of about 15 cm aperture will reveal itself as an oval patch with a bright nucleus. The galaxy doesn't have any conspicuous spiral arms so that even with large-aperture telescopes further detail will be sparse, and only some slight irregularity in overall brightness will be resolved. Although it is similar to the Milky Way, what makes this galaxy special is that it is an active galaxy, of a class called a *LINER* (*L*ow *I*onisation *N*uclear *E*mission-line *R*egion). In the centre of the galaxy is some sort of mechanism that gives rise to both the observed spectral lines and a radio source. It may be due to massive stars called *warmers*, or an accretion disc around a black hole.

Messier 63	NGC 5055	13h 15.8m	+42° 02′	April 10
8.6 m [*13.6 m*]	12.6′ l 7.2′		SA(rs)bc	easy

Also known as the Sunflower Galaxy. This is a somewhat difficult object for binoculars, even though it has a fairly bright magnitude. In small binoculars it will just appear as a faint patch of light, but with large binoculars the classic oval shape will become apparent. In telescopes, it reveals a lot of detail, with many faint and detailed spiral arms. A very nice object for large-aperture instruments.

Messier 51	NGC 5194	13h 29.9m	+47° 12′	April 14
8.4 m [*13.1 m*]	11.2′ l 6.9′		SA(s)bcP	easy

Also known as the Whirlpool Galaxy. This famous galaxy is easily visible in binoculars and will appear as a small glowing patch with a bright, star-like nucleus. Many now believe that M51 is the finest example of a face-on spiral galaxy. But what makes it so special is the small, irregular class, satellite galaxy NGC 5195 that is close to it. Deep photographs reveal that the galaxies are physically connected by a bridge of material. Unfortunately this satellite galaxy cannot be seen in most binoculars, and even in giant binoculars it would appear only as a slight bump on the side of M51. With small telescopes (10 cm) not a lot of detail is visible, just perhaps the slightest hint of spiral structure. With an aperture of 25 cm a lot more detail is resolved: spiral arms, structure within the arms, and dark patches. What is a matter of debate however is whether the bridge of material connecting M51 to NGC 5195 can be seen with small telescopes. Some observers claim it can be seen with 10 cm aperture, others that at least 30 cm is needed. What everyone agrees upon is that absolute, perfect transparency is needed, as even the slightest haze or dust in the atmosphere will make observations much more difficult. What can you see?

Messier 83	NGC 5236	13h 37.0m	–29° 52′	April 16
7.5 m [*13.2 m*]	12.9′ l 11.5′		SAB(s)c	easy

Often missed by observers, this is a nice galaxy located within the star fields of Hydra and is a showpiece for small telescopes. In binoculars it will appear as a hazy patch of light with a bright, star-like nucleus. With telescopes much more detail is seen, including spiral arms, dust lanes, bright knots and even detail within the nucleus itself. It is one of those objects that repays several observing sessions. Also, it ties with M81 as being one of the furthest objects visible to the naked eye at about 4.5 million l.y. As it is located so far south, there may be a problem locating it for observers living in the UK.

Messier 96	NGC 3368	$10^h 46.8^m$	+11° 49′	March 3
9.2 m [*12.9 m*]	7.6′ l 5.2′		SAB(rs)ab	easy

See March.

Messier 65	NGC 3623	$11^h 18.9^m$	+13° 05′	March 12
9.3 m [*12.4 m*]	9.8′ l 2.9′		SAB(rs)a	easy

See March.

Messier 66	NGC 3627	$11^h 20.2^m$	+12° 59′	March 12
9.0 m [*12.5 m*]	9.1′ l 4.2′		SA(s)b	easy

See March.

Messier 106	NGC 4258	$12^h 19.0^m$	+47° 18′	March 27
8.3 m [*13.8 m*]	18.6′ l 7.2′		SAB(s)bc	easy

See March.

Messier 88	NGC 4501	$12^h 32.0^m$	+14° 25′	March 30
9.6 m [*12.6 m*]	6.9′ l 3.7′		SA(rs)b	easy

See March.

Caldwell 38	NGC 4565	$12^h 36.3^m$	+25° 59′	March 31
9.6 m [*13.5 m*]	15.5′ l 1.9′		SA(s)bsp	easy

See March.

Messier 90	NGC 4569	$12^h 36.8^m$	+13° 10′	March 31
9.5 m [*13.6 m*]	9.5′ l 4.4′		SAB(rs)ab	easy

See March.

Caldwell 45	NGC 5248	$13^h 37.5^m$	+08° 53′	April 16
10.3 m [*13.8 m*]	6.8′ l 5.0′		SAB(rs)bc	moderate

Barely detectable in binoculars as a tiny patch of glowing light. With telescopes of aperture 15 cm, it will reveal itself as a bright nuclear region surrounded by a barely perceived haze, which is the spiral arms. In order to resolve the spiral structure, large aperture is needed, maybe 40 cm or more.

Messier 101/2[10]	NGC 5457	$14^h 03.2^m$	+54° 21′	April 22
7.9 m [*14.8 m*]	28.8′ l 26.9′		SAB(rs)cd	moderate ©

A difficult object for binoculars because even though its integrated magnitude is high, it has a large surface area, and so the light is spread out, making it surface brightness very low. If glimpsed in binoculars it will appear as a very hazy and faint patch of light, and the nucleus can be seen if averted vision is used. Medium-to-large-aperture telescopes will however bring out a lot of detail. This is a fine face-on spiral galaxy, with two nice spiral arms and a bright star-like nucleus. At a magnitude of 7.9, it should theoretically be detectable with the naked eye. Needless to say, the choice of observing site would be important – say, atop Mauna Kea in Hawaii, along with perfect seeing conditions. If you do glimpse it, you will be seeing an object that lies at a distance of about 17 million l.y. Quite a challenge!

[10]M102 is now believed to be just a duplicate observation of M101.

Messier 98	NGC 4192	$12^h 13.8^m$	+14° 54′	March 25
10.1 m [*13.2 m*]	9.8′ l 2.8′		SAB(s)ab	moderate

See March.

Messier 99	NGC 4254	$12^h 18.8^m$	+14° 25′	March 27
9.9 m [*13.0 m*]	5.4′ l 4.7′		SA(s)c	moderate

See March.

Messier 61	NGC 4303	$12^h 21.9^m$	+04° 28′	March 28
9.6 m [*13.4 m*]	6.5′ l 5.8′		SAB(rs)bc	moderate

See March.

Messier 100	NGC 4321	$12^h 22.9^m$	+15° 49′	March 28
9.3 m [*13.3 m*]	7.4′ l 6.3′		SAB(s)bc	moderate

See March.

Caldwell 36	NGC 4559	$12^h 35.9^m$	+27° 57′	March 31
9.8 m [*13.9 m*]	13.0′ l 5.2′		SAB(rs)cd	moderate

See March.

Caldwell 40	NGC 3626	$11^h 20.1^m$	+18° 21′	March 12
11.0 m [*12.8 m*]	2.8′ l 2.0′		(R)SA(rs)	difficult

See March.

Caldwell 26	NGC 4244	$12^h 17.5^m$	+37° 48′	March 26
10.4 m [*14.0 m*]	18.5′ l 2.3′		SA(s)cd:sp	difficult

See March.

May

Messier 58	NGC 4579	$12^h 37.7^m$	+11° 49′	April 1
9.6 m [*13.0 m*]	5.9′ l 4.7′		SAB(rs)b	easy

See April.

Messier 104	NGC 4594	$12^h 40.0^m$	–11° 37′	April 1
8.0 m [*11.6 m*]	8.7′ l 3.5′		SA(s)asp	easy

See April.

Messier 94	NGC 4736	$12^h 50.9^m$	+41° 07′	April 4
8.2 m [*13.0 m*]	11.2′ l 9.1′		(R)SA(r)ab	easy

See April.

Messier 64	NGC 4826	$12^h 56.7^m$	+21° 41′	April 5
8.5 m [*12.4 m*]	9.3′ l 5.4′		(R)SA(rs)ab	easy
See April.				

Caldwell 29	NGC 5005	$13^h 10.9^m$	+37° 03′	April 9
9.8 m [*12.6 m*]	6.3′ l 3.0′		SAB(rs)bc	easy
See April.				

Messier 63	NGC 5055	$13^h 15.8^m$	+42° 02′	April 10
38.6 m [*13.6 m*]	12.6′ l 7.2′		SA(rs)bc	easy
See April.				

Messier 51	NGC 5194	$13^h 29.9^m$	+47° 12′	April 14
8.4 m [*13.1 m*]	11.2′ l 6.9′		SA(s)bcP	easy
See April.				

Messier 83	NGC 5236	$13^h 37.0^m$	–29° 52′	April 16
7.5 m [*13.2 m*]	12.9′ l 11.5′		SAB(s)c	easy
See April.				

Messier 96	NGC 3368	$10^h 46.8^m$	+11° 49′	March 3
9.2 m [*12.9 m*]	7.6′ l 5.2′		SAB(rs)ab	easy
See March.				

Messier 65	NGC 3623	$11^h 18.9^m$	+13° 05′	March 12
9.3 m [*12.4 m*]	9.8′ l 2.9′		SAB(rs)a	easy
See March.				

Messier 66	NGC 3627	$11^h 20.2^m$	+12° 59′	March 12
9.0 m [*12.5 m*]	9.1′ l 4.2′		SA(s)b	easy
See March.				

Messier 106	NGC 4258	$12^h 19.0^m$	+47° 18′	March 27
8.3 m [*13.8 m*]	18.6′ l 7.2′		SAB(s)bc	easy
See March.				

Messier 88	NGC 4501	$12^h 32.0^m$	+14° 25′	March 30
9.6 m [*12.6 m*]	6.9′ l 3.7′		SA(rs)b	easy
See March.				

Caldwell 38	NGC 4565	$12^h 36.3^m$	+25° 59′	March 31
9.6 m [*13.5 m*]	15.5′ l 1.9′		SA(s)bsp	easy
See March.				

Messier 90	NGC 4569	$12^h 36.8^m$	+13° 10′	March 31
9.5 m [*13.6 m*]	9.5′l 4.4′		SAB(rs)ab	easy
See March.				

Caldwell 45	NGC 5248	$13^h 37.5^m$	+08° 53′	April 16
10.3 m [*13.8 m*]	6.8′l 5.0′		SAB(rs)bc	moderate
See April.				

Messier 101/2	NGC 5457	$14^h 03.2^m$	+54° 21′	April 22
7.9 m [*14.8 m*]	28.8′l 26.9′		SAB(rs)cd	moderate ©
See April.				

Messier 98	NGC 4192	$12^h 13.8^m$	+14° 54′	March 25
10.1 m [*13.2 m*]	9.8′l 2.8′		SAB(s)ab	moderate
See March.				

Messier 99	NGC 4254	$12^h 18.8^m$	+14° 25′	March 27
9.9 m [*13.0 m*]	5.4′l 4.7′		SA(s)c	moderate
See March.				

Messier 61	NGC 4303	$12^h 21.9^m$	+04° 28′	March 28
9.6 m [*13.4 m*]	6.5′l 5.8′		SAB(rs)bc	moderate
See March.				

Messier 100	NGC 4321	$12^h 22.9^m$	+15° 49′	March 28
9.3 m [*13.3 m*]	7.4′l 6.3′		SAB(s)bc	moderate
See March.				

Caldwell 36	NGC 4559	$12^h 35.9^m$	+27° 57′	March 31
9.8 m [*13.9 m*]	13.0′l 5.2′		SAB(rs)cd	moderate
See March.				

Caldwell 12	NGC 6946	$20^h 35.9^m$	+60° 09′	July 31
8.8 m [*13.8 m*]	11.0′l 9.8′		SAB(rs)cd	moderate ©
See July.				

Caldwell 40	NGC 3626	$11^h 20.1^m$	+18° 21′	March 12
11.0 m [*12.8 m*]	2.8′l 2.0′		(R)SA(rs)	difficult
See March.				

Caldwell 26	NGC 4244	$12^h 17.5^m$	+37° 48′	March 26
310.4 m [*14.0 m*]	18.5′l 2.3′		SA(s)cd:sp	difficult
See March.				

June

Messier 58	NGC 4579	$12^h 37.7^m$	+11° 49′	April 1
9.6 m [*13.0 m*]	5.9′ l 4.7′		SAB(rs)b	easy

Messier 104	NGC 4594	$12^h 40.0^m$	–11° 37′	April 1
8.0 m [*11.6 m*]	8.7′ l 3.5′		SA(s)asp	easy
See April.				

Messier 94	NGC 4736	$12^h 50.9^m$	+41° 07′	April 4
8.2 m [*13.0 m*]	11.2′ l 9.1′		(R)SA(r)ab	easy
See April.				

Messier 64	NGC 4826	$12^h 56.7^m$	+21° 41′	April 5
8.5 m [*12.4 m*]	9.3′ l 5.4′		(R)SA(rs)ab	easy
See April.				

Caldwell 29	NGC 5005	$13^h 10.9^m$	+37° 03′	April 9
9.8 m [*12.6 m*]	6.3′ l 3.0′		SAB(rs)bc	easy
See April.				

Messier 63	NGC 5055	$13^h 15.8^m$	+42° 02′	April 10
8.6 m [*13.6 m*]	12.6′ l 7.2′		SA(rs)bc	easy
See April.				

Messier 51	NGC 5194	$13^h 29.9^m$	+47° 12′	April 14
8.4 m [*13.1 m*]	11.2′ l 6.9′		SA(s)bcP	easy
See April.				

Messier 83	NGC 5236	$13^h 37.0^m$	–29° 52′	April 16
7.5 m [*13.2 m*]	12.9′ l 11.5′		SAB(s)c	easy
See April.				

Caldwell 30	NGC 7331	$22^h 37.1^m$	+34° 25′	August 30
9.5 m [*13.5 m*]	11.4′ l 4.0′		SA(s)bc	easy
See August.				

Caldwell 45	NGC 5248	$13^h 37.5^m$	+08° 53′	April 16
10.3 m [*13.8 m*]	6.8′ l 5.0′		SAB(rs)bc	moderate
See April.				

Messier 101/2	NGC 5457	$14^h 03.2^m$	+54° 21′	April 22
7.9 m [*14.8 m*]	28.8′ l 26.9′		SAB(rs)cd	moderate ©
See April.				

July

Messier 58	NGC 4579	$12^h 37.7^m$	+11° 49′	April 1
9.6 m [*13.0 m*]	5.9′ × 4.7′		SAB(rs)b	easy

See April.

Messier 104	NGC 4594	$12^h 40.0^m$	−11° 37′	April 1
8.0 m [*11.6 m*]	8.7′ × 3.5′		SA(s)asp	easy

See April.

Messier 94	NGC 4736	$12^h 50.9^m$	+41° 07′	April 4
8.2 m [*13.0 m*]	11.2′ × 9.1′		(R)SA(r)ab	easy

See April.

Messier 64	NGC 4826	$12^h 56.7^m$	+21° 41′	April 5
8.5 m [*12.4 m*]	9.3′ × 5.4′		(R)SA(rs)ab	easy

See April.

Caldwell 29	NGC 5005	$13^h 10.9^m$	+37° 03′	April 9
9.8 m [*12.6 m*]	6.3′ × 3.0′		SAB(rs)bc	easy

See April.

Messier 63	NGC 5055	$13^h 15.8^m$	+42° 02′	April 10
8.6 m [*13.6 m*]	12.6′ × 7.2′		SA(rs)bc	easy

See April.

Messier 51	NGC 5194	$13^h 29.9^m$	+47° 12′	April 14
8.4 m [*13.1 m*]	11.2′ × 6.9′		SA(s)bcP	easy

See April.

Messier 83	NGC 5236	$13^h 37.0^m$	−29° 52′	April 16
7.5 m [*13.2 m*]	12.9′ × 11.5′		SAB(s)c	easy

See April.

Caldwell 30	NGC 7331	$22^h 37.1^m$	+34° 25′	August 30
9.5 m [*13.5 m*]	11.4′ × 4.0′		SA(s)bc	easy

See August.

Caldwell 12	NGC 6946	$20^h 35.9^m$	+60° 09′	July 31
8.8 m [*13.8 m*]	11.0′ × 9.8′		SAB(rs)cd	moderate ©

A challenging galaxy to locate using binoculars. It will appear, if at all, as a small round hazy patch of light with a barely perceptible increase at its centre. What makes this galaxy difficult to locate and observe, even though it is close to us, is that it lies in the part of the sky near the plane of the Milky Way. This results in the light from the galaxy being dimmed by the intervening dust and stars. With telescopes of 20 cm, and a dark location, the faint outer halo can be glimpsed. In order to observe further detail such as spiral arms, a very dark location along with large aperture is needed. Research indicates that there is a hectic period of star formation occurring in its inner nuclear region. Such an outburst is termed a *starburst*.

Caldwell 43	NGC 7814	$00^h\ 03.2^m$	+16° 08′	September 21
310.6 m [*13.3 m*]	6.3′ l 3.0′		SA(s)ab:sp	moderate
See September.				

Caldwell 45	NGC 5248	$13^h\ 37.5^m$	+08° 53′	April 16
10.3 m [*13.8 m*]	6.8′ l 5.0′		SAB(rs)bc	moderate
See April.				

Messier 101/2	NGC 5457	$14^h\ 03.2^m$	+54° 21′	April 22
7.9 m [*14.8 m*]	28.8′ l 26.9′		SAB(rs)cd	moderate ©
See April.				

August

Caldwell 30	NGC 7331	$22^h\ 37.1^m$	+34° 25′	August 30
9.5 m [*13.5 m*]	11.4′ l 4.0′		SA(s)bc	easy
This is the brightest galaxy in Pegasus, and with binoculars will appear a faint patch of light that has a brighter core. Easily visible in a telescope of 20 cm, which will show its structure in a little more detail. Apparently this galaxy is similar to M31, but lies much further from us at a distance of 50 million l.y. There is also some debate as to whether it is linked with the famous Stephen's Quintet (see the section on Groups and Clusters of Galaxies).				

Messier 31	NGC 224	$00^h\ 42.7^m$	+41° 16′	October 1
3.4 m [*13.6 m*]	3° l 1°		Sb	easy
See October.				

Caldwell 65	NGC 253	$00^h\ 47.6^m$	–25° 17′	October 3
7.2 m [*12.6 m*]	25.0′ l 7.0′		SAB(s)c	easy
See October.				

Caldwell 70	NGC 300	$00^h\ 54.9^m$	–37° 40′	October 4
8.1 m [*14.7 m*]	20.0′ l 15.0′		SA(s)d	easy
See October.				

Messier 33	NGC 598	$01^h\ 33.9^m$	+30° 39′	October 14
5.7 m [*14.2 m*]	71′ l 42′		SA(s)cd	easy
See October.				

Caldwell 12	NGC 6946	$20^h\ 35.9^m$	+60° 09′	July 31
8.8 m [*13.8 m*]	11.0′ l 9.8′		SAB(rs)cd	moderate ©
See July.				

Caldwell 43	NGC 7814	$00^h\ 03.2^m$	+16° 08′	September 21
10.6 m [*13.3 m*]	6.3′ l 3.0′		SA(s)ab:sp	moderate
See September.				

Messier 74	NGC 628	$01^h 36.7^m$	+15° 47′	October 15
9.2 m [*14.4 m*]	10.0′ l 9.6′		SA(s)c	moderate
See October.				

Caldwell 23	NGC 891	$02^h 22.6^m$	+42° 20′	October 27
9.9 m [*13.8 m*]	14.0′ l 3.0′		SA(s)sp	moderate
See October.				

Caldwell 62	NGC 247	$00^h 47.1^m$	–20° 45′	October 2
9.1 m [*14.1 m*]	20.0′ l 7.0′		SAB(s)	moderate
See October.				

September

Caldwell 30	NGC 7331	$22^h 37.1^m$	+34° 25′	August 30
9.5 m [*13.5 m*]	11.4′ l 4.0′		SA(s)bc	easy
See August.				

Messier 31	NGC 224	$00^h 42.7^m$	+41° 16′	October 1
3.4 m [*13.6 m*]	3° l 1°		Sb	easy
See October.				

Caldwell 65	NGC 253	$00^h 47.6^m$	–25° 17′	October 3
7.2 m [*12.6 m*]	25.0′ l 7.0′		SAB(s)c	easy
See October.				

Caldwell 70	NGC 300	$00^h 54.9^m$	–37° 40′	October 4
8.1 m [*14.7 m*]	20.0′ l 15.0′		SA(s)d	easy
See October.				

Messier 33	NGC 598	$01^h 33.9^m$	+30° 39′	October 14
35.7 m [*14.2 m*]	71′ l 42′		SA(s)cd	easy
See October.				

Messier 77	NGC 1068	$02^h 42.7^m$	–00° 01′	November 1
8.9 m [*13.2 m*]	7.1′ l 6.0′		(R)SA(rs)b	easy
See November.				

Caldwell 43	NGC 7814	$00^h 03.2^m$	+16° 08′	September 21
10.6 m [*13.3 m*]	6.3′ l 3.0′		SA(s)ab:sp	moderate
This is no binocular object, but nevertheless a splendid sight, especially with larger aperture telescopes. It is a fine example of an edge-on galaxy, and bears many similarities to its better-known cousin M104. Easily seen in a telescope of aperture 20 cm, it does however provoke debate among amateurs as to whether its dust lane can be seen with small telescopes. Some profess to have seen it with 20 cm, while others claim that at least 40 cm is needed. Try observing with as high a power as it can take, as this may help you to resolve this dilemma.				

Caldwell 12	NGC 6946	20h 35.9m	+60° 09′	July 31
8.8 m [*13.8 m*]	11.0′ l 9.8′		SAB(rs)cd	moderate ©
See July.				

Messier 74	NGC 628	01h 36.7m	+15° 47′	October 15
9.2 m [*14.4 m*]	10.0′ l 9.6′		SA(s)c	moderate
See October.				

Caldwell 23	NGC 891	02h 22.6m	+42° 20′	October 27
9.9 m [*13.8 m*]	14.0′ l 3.0′		SA(s)sp	moderate
See October.				

Caldwell 62	NGC 247	00h 47.1m	–20° 45′	October 2
9.1 m [*14.1 m*]	20.0′ l 7.0′		SAB(s)	moderate
See October.				

Caldwell 5	IC 342	03h 46.8m	+68° 05′	November 17
9.2 m [*15.2 m*]	17.8′ l 17.4′		SAB(rs)cd	difficult ©
See November.				

October

Messier 31	NGC 224	00h 42.7m	+41° 16′	October 1
3.4 m [*13.6 m*]	3° l 1°		Sb	easy

Also known as the Andromeda Galaxy. The most famous galaxy in the sky is probably also the most often visited one, and is always a first observing object for the beginner. It is visible to the naked eye, even on those nights when the conditions are far from perfect. Many naked-eye observers claim to have seen the galaxy spread over at least 2¹/₂° of sky, but this depends on the transparency. In binoculars it presents a splendid view, with the galactic halo easily seen along with the bright nucleus. Large binoculars may even show one or two dust lanes. Using averted vision and with a very dark sky, several amateurs report that the galaxy can be traced to about 3° of sky in telescopes of aperture 10 cm. In larger telescopes a wealth of detail becomes visible. With an aperture of about 20 cm, a star-like nucleus is apparent, cocooned within several elliptical haloes. Another striking feature are the dust lanes, especially the one running along its north-western edge. Many observers are often disappointed with what they see when observing M31, as the photographs seen in books actually belie what is seen at the eyepiece. M31 is so big that any telescope cannot really encompass all there is to show. Patience when observing this wonderful galaxy will reward you with a lot of surprises. Spend several nights observing, and try to chose a dark night in a country location. This really is a spectacular galaxy. It contains about 300 million stars with a diameter of 130,000 l.y. and is among the largest galaxies known. It is the largest member of the Local Group. In older texts it is often referred to as the Great Nebula in Andromeda.

Caldwell 65	NGC 253	00^h 47.6^m	–25° 17′	October 3
7.2 m [*12.6 m*]	25.0′ l 7.0′		SAB(s)c	easy

This is a wonderful object and is often referred to as the southern hemisphere's answer to the Andromeda Galaxy. Easily seen in binoculars as a long, spindle-shaped glow with a bright nucleus. Under absolutely perfect conditions, some structure can even be glimpsed with large binoculars. Its size makes it impressive, as it is about as long as the Moon is wide and around a third as thick. Any telescope will suffice to see this object, even as small as 6 cm. Larger apertures will reveal more detail, and with averted vision the spiral arms can be glimpsed. There are several reports that considerable mottling can be seen within the galaxy with a 15 cm telescope. It has the dubious honour of being one of the dustiest galaxies known, as well as one that is undergoing a period of frenetic star formation in its nuclear regions.

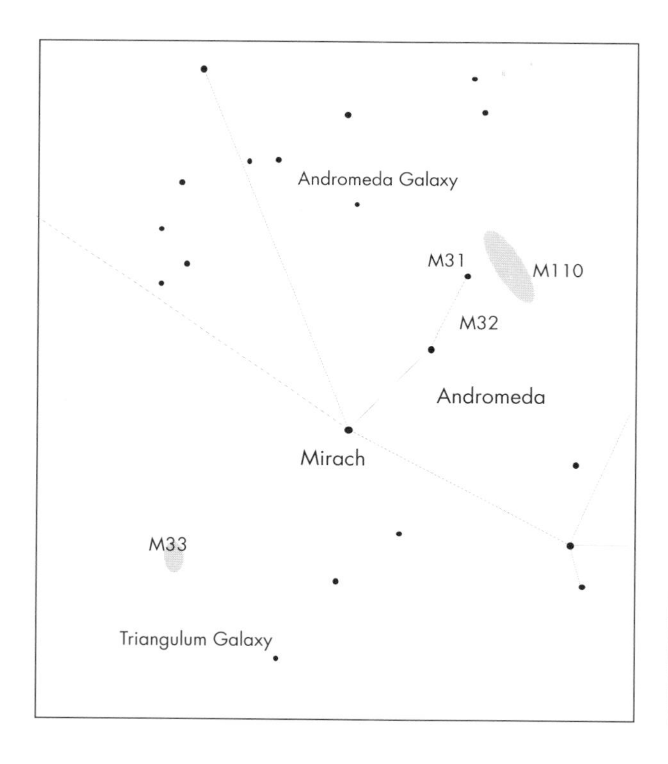

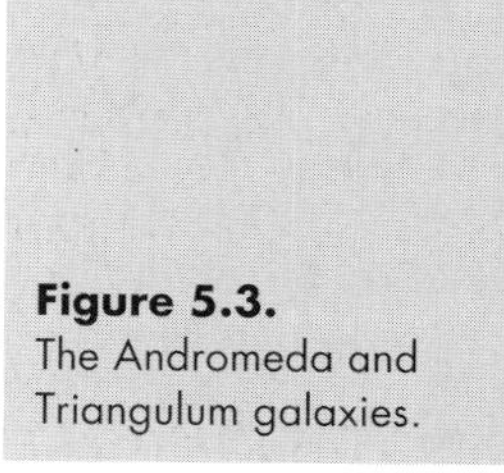

Figure 5.3. The Andromeda and Triangulum galaxies.

Caldwell 70	NGC 300	00^h 54.9^m	–37° 40′	October 4
8.1 m [*14.7 m*]	20.0′ l 15.0′		SA(s)d	easy

A difficult object to locate owing to its very low surface brightness, it will also present a considerable challenge to binocular observers. Nevertheless, once found it can be an impressive object. With a 20 cm aperture telescope, the nucleus is readily seen engulfed in the unresolved haze of the spiral arms. Larger telescopes will begin to resolve some further detail. It lies at the close distance of about 7 million l.y.

Messier 33	NGC 598	$01^h 33.9^m$	+30° 39′	October 14
5.7 m [*14.2 m*]	71′ l 42′		SA(s)cd	easy

Also known as the Pinwheel Galaxy, is famous for several reasons. It is without doubt one of the most impressive examples of a face-on spiral. But at the same time it has a reputation as one of the most difficult galaxies to find. Many amateurs have never seen it, while others have no trouble locating it. The problem arises from its having such a large surface area. Although it has an integrated magnitude of 5.7, it spreads out the light to such an extent that it is very faint. As a result, the galaxy may be all but invisible in telescopes, whereas it will be easily found with binoculars. It will look like a large, very faint cloud, with a slight brightening at its centre. In addition, there are several reports of it being visible to the naked eye, and I can testify to this, as it was strikingly visible from a totally dark sight under perfect conditions, when it was impossible not to see it! In a telescope of aperture 10 cm, several spiral arms can be glimpsed arcing from the very small nucleus. With large telescopes a plethora of detail becomes visible, such as star clusters, stellar associations and nebulae, all located within the galaxy.[14] This truly is a spectacular galaxy.

Caldwell 30	NGC 7331	$22^h 37.1^m$	+34° 25′	August 30
9.5 m [*13.5 m*]	11.4′ l 4.0′		SA(s)bc	easy

See August.

Messier 77	NGC 1068	$02^h 42.7^m$	–00° 01′	November 1
8.9 m [*13.2 m*]	7.1′ l 6.0′		(R)SA(rs)b	easy

See November.

Caldwell 62	NGC 247	$00^h 47.1^m$	–20° 45′	October 2
9.1 m [*14.1 m*]	20.0′ l 7.0′		SAB(s)	moderate

This galaxy is barely discernible in large binoculars, where it is seen as an elongated hazy patch of light with a brighter nucleus. However, it is low down in the skies for UK observers, so is often neglected. In larger-aperture telescopes, its mottled appearance becomes visible, along with the brighter, southern part of the galaxy. The northern aspect is much fainter and will require averted vision, along with clear skies. It was thought that the galaxy was a member of the Sculptor Group of galaxies, but recently doubts have arisen, as the most recent estimates of its distance put it at about 13.5 million l.y., which is twice the distance of the cluster.

Caldwell 43	NGC 7814	$00^h 03.2^m$	+16° 08′	September 21
10.6 m [*13.3 m*]	6.3′ l 3.0′		SA(s)ab:sp	moderate

See September.

[14]A large HII region, NGC 604, is visible. See the entry under Emission Nebulae in Chapter 4.

Messier 74	NGC 628	01h 36.7m	+15° 47′	October 15
9.2 m [*14.4 m*]	10.0′ l 9.6′		SA(s)c	moderate

This is another object that proves frustrating to amateurs. Again, as in M33 above, the reason is the low surface brightness. It can be glimpsed in binoculars but only under excellent conditions. It is also a paradox that the galaxy is often better glimpsed with small telescopes, even those of 5 cm aperture rather than, say, 10 cm. One way to find this elusive object is to locate the star Eta (η) Piscium and put it at the southern edge of your field of view. Then let the stars drift across, and within a few minutes M74 will enter the field. Of course, averted vision and excellent sky transparency will all help.

Caldwell 23	NGC 891	02h 22.6m	+42° 20′	October 27
9.9 m [*13.8 m*]	14.0′ l 3.0′		SA(s)sp	moderate

This is a fine example of an edge-on galaxy, and is thought by many to be the finest. Just visible in binoculars as a hazy, but distinct elongated smudge. With a telescope of aperture 20 cm, its spindle shape is very apparent, and with a larger aperture the distinctive dust lane will be resolved.

Caldwell 5	IC 342	03h 46.8m	+68° 05′	November 17
9.2 m [*15.2 m*]	17.8′ l 17.4′		SAB(rs)cd	difficult ©

See November.

November

Messier 77	NGC 1068	02h 42.7m	–00° 01′	November 1
8.9 m [*13.2 m*]	7.1′ l 6.0′		(R)SA(rs)b	easy

A famous galaxy for several reasons. In binoculars it is visible just as a hazy patch of light, and under excellent seeing a faint star-like nucleus may be glimpsed. In telescopes of about 10 cm and greater, and providing that dark skies are available, then the spiral arms can be glimpsed. But what makes this galaxy so special is that it is the archetypal active galaxy of a class known as *Seyferts*.[15] It was first discovered in the middle of the twentieth century by Carl Seyfert, who noticed that it had very prominent emission lines. These are due to the high velocity of gas close to the nucleus of the galaxy. The high speed of the gas, in the order of 350 kilometres per second, is believed to be due to the influence of a massive black hole. Seyferts are distant cousins of the famous quasars. It is one of the brightest active galaxies visible to the amateur astronomer.

Messier 31	NGC 224	00h 42.7m	+41° 16′	October 1
3.4 m [*13.6 m*]	3° l 1°		Sb	easy

See October.

Caldwell 65	NGC 253	00h 47.6m	–25° 17′	October 3
7.2 m [*12.6 m*]	25.0′ l 7.0′		SAB(s)c	easy

See October.

[15]M77 is in fact classified as a Seyfert II galaxy which indicate that it has only narrow emission lines. A Seyfert I galaxy had both broad and narrow emission lines. The width of the line is a measure of the velocity of the gas that produced the emission line.

Caldwell 70	NGC 300	00h 54.9m	–37° 40′	October 4
8.1 m [*14.7 m*]	20.0′ l 15.0′		SA(s)d	easy
See October.				

Messier 33	NGC 598	01h 33.9m	+30° 39′	October 14
5.7 m [*14.2 m*]	71′ l 42′		SA(s)cd	easy
See October.				

Caldwell 43	NGC 7814	00h 03.2m	+16° 08′	September 21
10.6 m [*13.3 m*]	6.3′ l 3.0′		SA(s)ab:sp	moderate
See September.				

Messier 74	NGC 628	01h 36.7m	+15° 47′	October 15
9.2 m [*14.4 m*]	10.0′ l 9.6′		SA(s)c	moderate
See October.				

Caldwell 23	NGC 891	02h 22.6m	+42° 20′	October 27
9.9 m [*13.8 m*]	14.0′ l 3.0′		SA(s)sp	moderate
See October.				

Caldwell 62	NGC 247	00h 47.1m	–20° 45′	October 2
9.1 m [*14.1 m*]	20.0′ l 7.0′		SAB(s)	moderate
See October.				

Caldwell 5	IC 342	03h 46.8m	+68° 05′	November 17
9.2 m]*15.2 m*]	17.8′ l 17.4′		SAB(rs)cd	difficult ©

A very difficult galaxy to observe for the same reason that plagues so many others: its large size causes the surface brightness to be low. Reportedly impossible to see in a telescope of 20 cm unless conditions are perfect, and even then large apertures (40 cm) only show barely discernible structure. It suffers from being located in an area of the sky where the interstellar dust and gas dim its light appreciably. We can only wonder what it would look like were it at some other location. It is an active galaxy of the type known as a *starburst* galaxy. This means that it is undergoing a period of vigorous star formation. I often wonder what activity started this period of star forming – was it an encounter with another unseen galaxy, or perhaps something else?

December

Messier 77	NGC 1068	02h 42.7m	–00° 01′	November 1
8.9 m [*13.2 m*]	7.1′ l 6.0′		(R)SA(rs)b	easy
See November.				

Caldwell 7	NGC 2403	$07^h 36.9^m$	+65° 35′	January 14
8.5 m [*13.9 m*]	17.8′ × 10.0′		SAB(s)cd	easy ©

See January.

Messier 31	NGC 224	$00^h 42.7^m$	+41° 16′	October 1
3.4 m [*13.6 m*]	3° × 1°		Sb	easy

See October.

Caldwell 65	NGC 253	$00^h 47.6^m$	–25° 17′	October 3
7.2 m [*12.6 m*]	25.0′ × 7.0′		SAB(s)c	easy

See October.

Caldwell 70	NGC 300	$00^h 54.9^m$	–37° 40′	October 4
8.1 m [*14.7 m*]	20.0′ × 15.0′		SA(s)d	easy

See October.

Messier 33	NGC 598	$01^h 33.9^m$	+30° 39′	October 14
5.7 m [*14.2 m*]	71′ × 42′		SA(s)cd	easy

See October.

Caldwell 43	NGC 7814	$00^h 03.2^m$	+16° 08′	September 21
10.6 m [*13.3 m*]	6.3′ × 3.0′		SA(s)ab:sp	moderate

See September.

Messier 74	NGC 628	$01^h 36.7^m$	+15° 47′	October 15
9.2 m [*14.4 m*]	10.0′ × 9.6′		SA(s)c	moderate

See October.

Caldwell 23	NGC 891	$02^h 22.6^m$	+42° 20′	October 27
9.9 m [*13.8 m*]	14.0′ × 3.0′		SA(s)sp	moderate

See October.

Caldwell 62	NGC 247	$00^h 47.1^m$	–20° 45′	October 2
9.1 m [*14.1 m*]	20.0′ × 7.0′		SAB(s)	moderate

See October.

Caldwell 5	IC 342	$03^h 46.8^m$	+68° 05′	November 17
9.2 m [*15.2 m*]	17.8′ × 17.4′		SAB(rs)cd	difficult ©

See November.

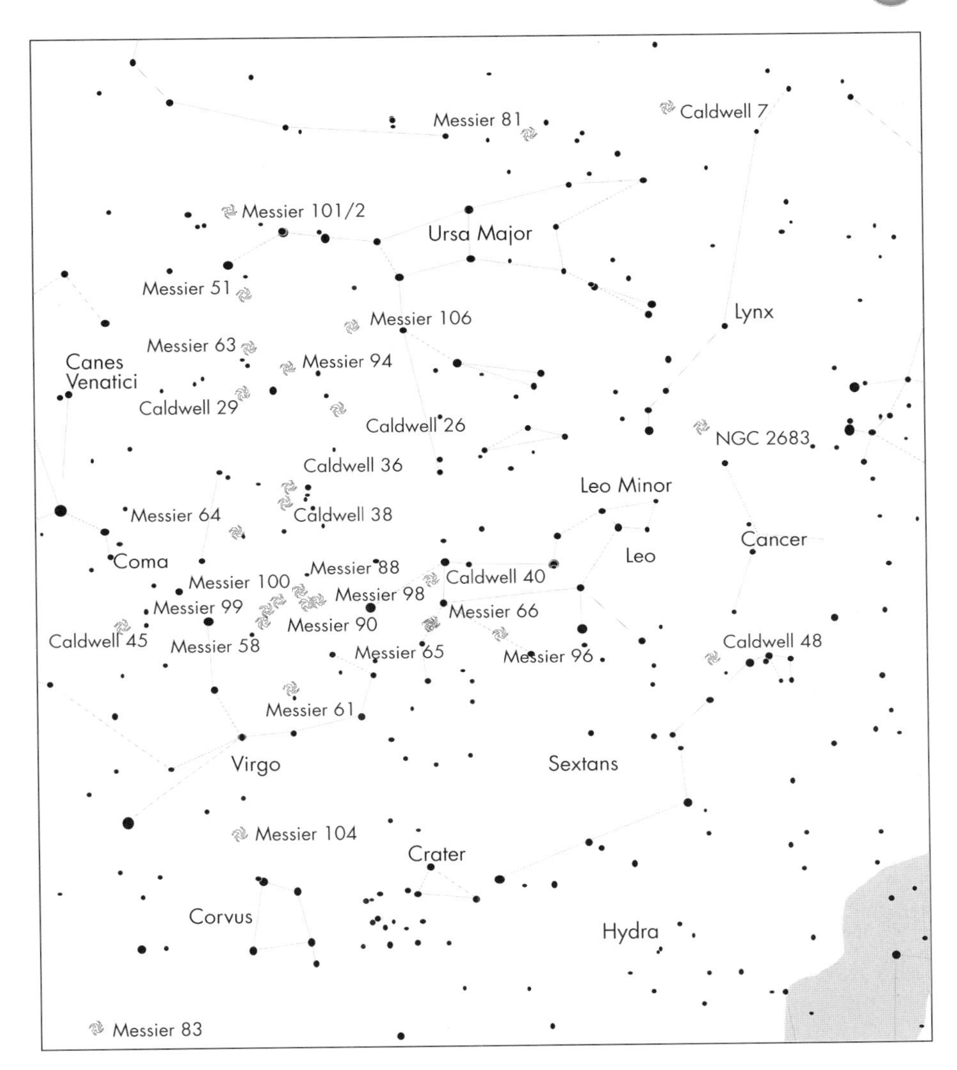

Figure 5.4. Spiral Galaxies 1.

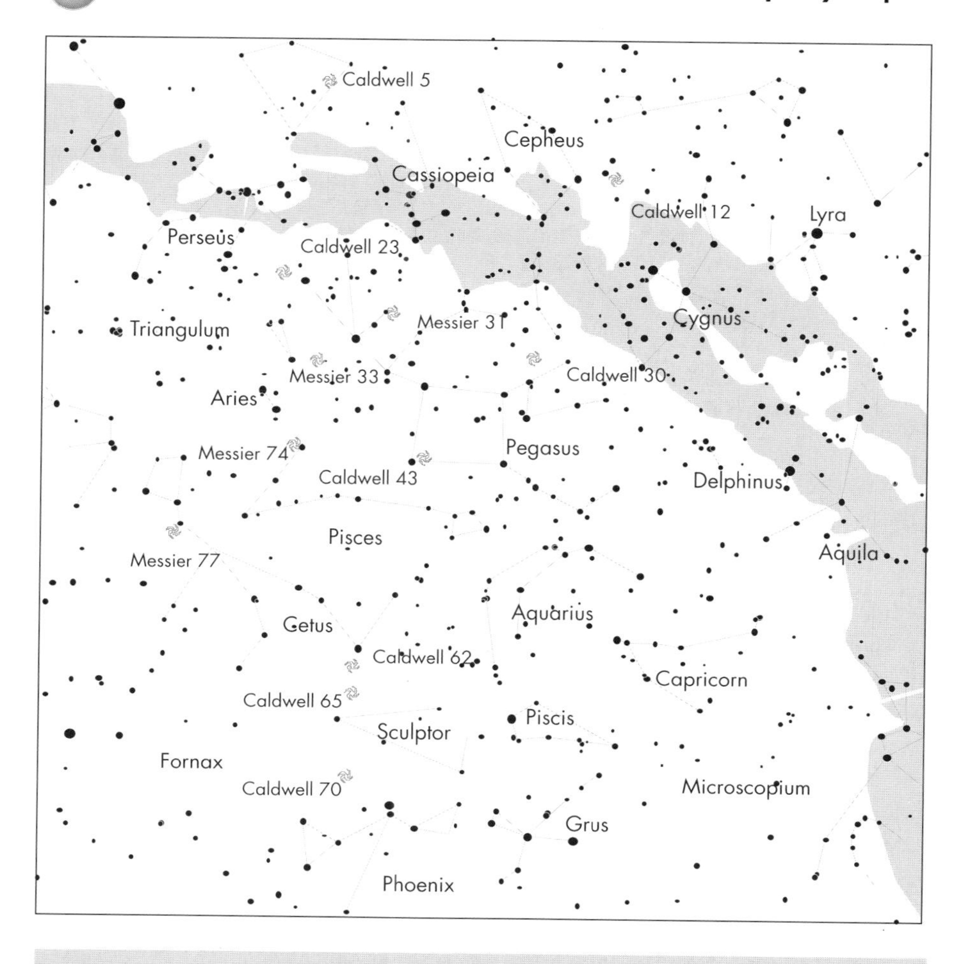

Figure 5.5. Spiral galaxies 2.

5.6 Barred Spiral Galaxies

January–June

Caldwell 67	NGC 1097	02h 46.3m	–30° 16′	November 2
9.5 m [*13.6 m*]	9.3′ l 6.6′		SB(s)b	easy
See July–December.				

–	NGC 1365	03h 33.6m	–36° 08′	November 14
9.5 m [*13.7 m*]	9.8′ l 5.5′		(R′)SB(s)b	moderate
See July–December				

Messier 95	NGC 3351	10h 44.0m	+11° 42′	March 3
9.7 m [*13.5 m*]	7.4′ l 5.0′		SB(R)b	moderate

This is a faint galaxy, that shows little if any detail in binoculars. It will just appear as a hazy patch, but it will be in the same field of view as M96. With a telescope of at least 15 cm, some structure can be glimpsed, with larger apertures showing the distinctive bar feature. There is some debate as to the real magnitude of the galaxy, with some observers putting it at 9.2 m. Does it seem too bright to you?

Messier 108	NGC 3556	11h 11.5m	+55° 40′	March 10
10.0 m [*13.0 m*]	8.7′ l 2.2′		SB(s)cdsp	moderate ©

This galaxy is visible in binoculars as a very faint streak of light. The central condensation has been reported visible in an 8 cm telescope. Larger aperture shows a surprising amount of detail with considerable mottling and structure. This is a very small galaxy, only one-twentieth the mass of the Milky Way, and seems to lack a central bulge. Although it is a recent addition to Messier's list, it is known that he was aware of the galaxy's existence; for some reason he just didn't include it.

Messier 109	NGC 3992	11h 57.6m	+53° 23′	March 21
9.8 m [*13.5 m*]	7.6′ l 4.7′		SB(rs)bc	moderate ©

Another recent addition to the Messier catalogue, this galaxy can be glimpsed in binoculars providing the conditions are right. With a low power and small aperture, it is evident that you are looking at a galaxy, but no further detail can be seen. High power and larger aperture does begin to show some structure such as the core and halo regions. The central bar also becomes prominent with apertures around 25 cm. The penultimate Messier object.

Caldwell 3	NGC 4236	12h 16.7m	+69° 27′	March 26
9.6 m [*14.7 m*]	23.0′ l 8.0′		SB(s)dm	difficult ©

Although this is a large galaxy, it is also very faint, and so difficult to locate. In addition, as it is edge-on to us, it presents a very slim view and so spiral arm features are absent. With apertures around 20 cm, its distinctive spindle shape is conspicuous. A nice galaxy for those who like to test the limits of a small telescope, as well as their observing skill. It lies at a distance of about 10 million l.y.

Messier 91	NGC 4548	12h 35.4m	+14° 30′	March 31
10.1 m [*13.3 m*]	5.4′ l 4.3′		SB(rs)b	difficult

A mystery. If one tries to locate M91 from Messier's original notes, they will make an interesting discovery. There is nothing there! Most observers agree that he made a mistake and that in fact the galaxy NGC 4548 what he originally observed but incorrectly plotted. Whatever the reason, the galaxy is a faint object, and telescopes of medium aperture will be needed for you to see any detail, although it is visible in apertures of 10 cm as a faint, hazy circular patch.

Caldwell 32	NGC 4631	$12^h 42.1^m$	+32° 32′	April 2
9.2 m [*13.3 m*]	17.0′ l 3.5′		SB(s)dsp	moderate

An often neglected galaxy, which is surprising as it has a lot to offer. Visible in binoculars as a faint elongated object, it really needs a telescope to be appreciated. It is a very big galaxy, which owing to its appearance has led to it being unofficially nicknamed the Whale Galaxy. Its eastern end is appreciably thicker than its western, hence the name. This aspect can be seen with an aperture of 20 cm, and larger telescopes will show further details such as patches of light and dark, along with two prominent knots. On the northern side of the galaxy is a faint 12th-magnitude star which, providing the seeing is good, will act as a pointer to a faint companion galaxy. Several theories have arisen as to the origin of its strange and disturbed appearance. The most probable is tidal interactions with several nearby galaxies.

July–December

Caldwell 67	NGC 1097	$02^h 46.3^m$	–30° 16′	November 2
9.5 m [*13.6 m*]	9.3′ l 6.6′		SB(s)b	easy

This is a nice galaxy, and its bar can be seen easily. With a 20 cm aperture telescope, the core is resolved is easy to see, as well as a faint elongated glow, which in fact is the bar. Larger apertures will resolve this feature quite well, along with the spiral arms which emanate form the bar's end. It is an active galaxy, and classified as a Seyfert galaxy of type I. This means that gas close to the nucleus is moving at extremely fast speeds, maybe in excess of 1000 kilometres per second. The most likely cause of this motion is the influence of a black hole.

Caldwell 72	NGC 55	$00^h 15.1^m$	–39° 13′	September 24
7.9 m [*13.5 m*]	25.0′ l 4.1′		SB(s)m:sp	easy

Although this galaxy lies so far south as to make it invisible from the northern hemisphere, it still warrants inclusion. In binoculars it appears a faint spindle-shaped object, and large binoculars hint at some delicate structure. Telescopes show even more detail, and it is one of the few galaxies where an H-alpha filter will highlight its HII regions.

Caldwell 44	NGC 7479	$23^h 04.9^m$	+12° 19′	September 7
10.9 m [*13.6 m*]	4.4′ l 3.4′		SB(s)c	moderate

This faint galaxy can be glimpsed in a small telescope of aperture 8 cm, but do not expect any detail – it will just appear as a smudge. With aperture around 20 cm the central bar will be seen, along with a suggestion of some structure. The spiral arms at the end of the bar will need at least a 30 cm telescope, although some observers claim they can be seen with 25 cm aperture under perfect conditions. The nucleus is easily resolved however. It has the honour, among some amateurs, of being the finest barred spiral on offer for the northern hemisphere. It lies at about 100 million l.y. from us. Quite a distance!

Messier 95	NGC 3351	$10^h 44.0^m$	+11° 42′	March 3
9.7 m [*13.5 m*]	7.4′ l 5.0′		SB(R)b	moderate

See January–June

Messier 108	NGC 3556	$11^h 11.5^m$	+55° 40′	March 10
10.0 m [*13.0 m*]	8.7′ l 2.2′		SB(s)cdsp	moderate ©
See January–June.				

Messier 109	NGC 3992	$11^h 57.6^m$	+53° 23′	March 21
9.8 m [*13.5 m*]	7.6′ l 4.7′		SB(rs)bc	moderate ©
See January–June.				

Caldwell 32	NGC 4631	$12^h 42.1^m$	+32° 32′	April 2
9.2 m [*13.3 m*]	17.0′ l 3.5′		SB(s)dsp	moderate
See January–June.				

–	NGC 1365	$03^h 33.6^m$	–36° 08′	November 14
9.5 m [*13.7 m*]	9.8′ l 5.5′		(R′)SB(s)b	moderate
A very impressive galaxy, easily visible in binoculars as an elongated hazy object with a brighter centre. In a telescope with an aperture as small as 8 cm, its galaxy origin is obvious, and larger apertures will show considerably more detail. Although not visible from the UK, it should be a nice observing target from the USA.				

Caldwell 3	NGC 4236	$12^h 16.7^m$	+69° 27′	March 26
9.6 m [*14.7 m*]	23.0′ l 8.0′		SB(s)dm	difficult ©
See January–June.				

Messier 91	NGC 4548	$12^h 35.4^m$	+14° 30′	March 31
10.1 m [*13.3 m*]	5.4′ l 4.3′		SB(rs)b	difficult
See January–June.				

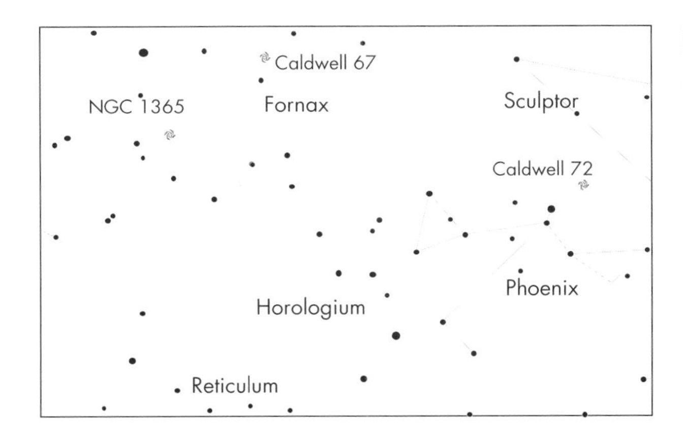

Figure 5.6. Barred spirals 1.

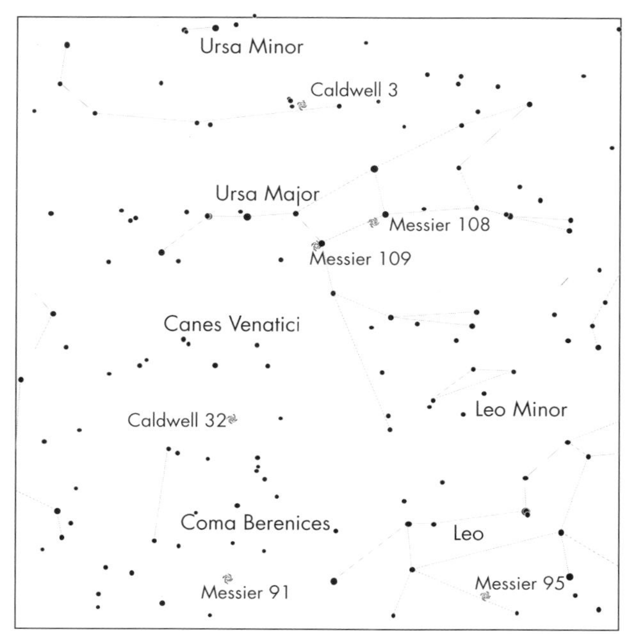

Figure 5.7.
Barred spirals 2.

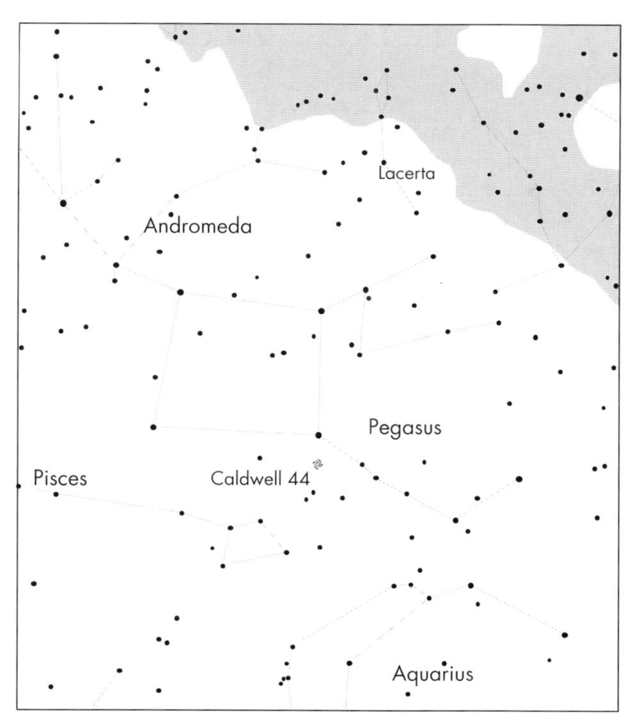

Figure 5.8.
Barred spirals 3.

5.7 Elliptical Galaxies

January–June

Messier 105	NGC 3379	10h 47.8m	+12° 35′	March 4
9.3 m [*12.1 m*]	5.4′ l 4.8′		E1	easy

This is the third galaxy in a nice triple system that can be seen in the same field of view. The other members are M95 and M96. The galaxy can be seen in binoculars but will appear only as a small, completely featureless, oval patch of light. With a telescope, not much more is seen. With a large aperture and medium to high magnification, a slight brightening of the core may be observed, but all in all, the galaxy remains a smooth-featured object. It lies at a distance of 26 million l.y.

Messier 84	NGC 4374	12h 25.1m	+12° 53′	March 28
9.1 m [*12.3 m*]	6.5′ l 5.6′		E1	easy

This nice elliptical galaxy forms a pair with M86. Located close to the Virgo Cluster of galaxies, it presents a small oval patch of light when seen through binoculars. The bright nuclei can be glimpsed weather when conditions permit. As with most ellipticals observed with amateur telescopes, there is never much detail seen in the galaxy; most remain smooth objects with little structure and perhaps only a brightening of the core is all that is ever resolved. Nevertheless they make worthwhile observing objects. There seems to be some debate as to whether M84 is in fact an E1 galaxy, or an S0, which is a galaxy between an elliptical and a spiral. It is located in the Virgo Cluster at about 55 million l.y. The area around M86 is full of very faint galaxies, and although only a handful will show any perceptible detail, it is nevertheless worthwhile sweeping the area for these most elusive objects.

Messier 86	NGC 4406	12h 26.6m	+12° 57′	March 29
8.9 m [*13.9 m*]	8.9′ l 5.8′		E3	easy

The companion to M84, this is the brighter of the two, and is very similar in appearance, with M86 being perhaps slightly brighter with a less condensed core. Visible in binoculars and telescopes of all sizes. As with the entry preceding this one, the area is full of galaxies, and with a dark sky, and patience, many more will be seen.

Messier 49	NGC 4472	12h 29.8m	+08° 00′	March 30
8.4 m [*12.9 m*]	10.2′ l 8.3′		E2	easy

This is the second-brightest galaxy in Virgo, and easily spotted in binoculars as a featureless, oval patch of light. Although most ellipticals are rather featureless, M49 stands up quite well with higher power and large aperture, when some resolution can be seen in the nuclear area. It seems to have a bright nucleus surrounded by a diffuse core region, which in turn is surrounded by a rather diffuse halo. Some observers report that the nucleus shows a mottled appearance under magnification. The galaxy is at the centre of a subcluster of galaxies called the Virgo Cloud, which in turn is part of the much larger Virgo Cluster. In addition it appears the elliptical galaxy is cocooned in an envelope of hot gas at a temperature of about 10,000,000 °K. At such a high temperature, X-rays are formed and it was with an X-ray telescope that this feature was detected.

Messier 87	NGC 4486	$12^h\ 30.8^m$	+12° 24′	March 30
8.6 m [*12.7 m*]	8.3′ l 6.6′		E0.5P	easy

A very special galaxy. Its is bright, and easily seen in binoculars, but with telescopes little else is resolved. But this rather bland appearance is deceiving. This is a monster of a galaxy, with a mass estimated to be that of 800 billion Suns. This makes it one of the most massive galaxies known in the entire universe. But that isn't all. It is an active galaxy, and lurking at its core is a black hole with a mass of 3 billion Suns. Another feature which some observers report seeing with telescopes of aperture 50 cm is the famous "jet" that streams out from M87. It would be a challenge indeed, and a triumph, were this ever to be observed from the light-polluted skies of the UK. The jet is a stream of plasma (hot ionised gas) several thousand light years in length that is believed to be due to some sort of interaction between the black hole and its surroundings. It is, however, very easy to photograph and image with a CCD camera.

M87 lies at the heart of the Virgo Cluster and most of the surrounding galaxies are influenced by its tremendous gravitational attraction. The cluster has about 300 large galaxies and perhaps as many as two thousand smaller ones. It is the closest large cluster, lying at a distance of around 55 million l.y. It spans over 100 square degrees in both Virgo and Coma Berenices. Such is its influence that the Milky Way is actually gravitationally attracted to it.

Messier 89	NGC 4552	$12^h\ 35.7^m$	+12° 33′	March 31
9.7 m [*12.3 m*]	5.1′ l 4.7′		E0	moderate

A difficult galaxy to find in binoculars, especially if the seeing conditions are far from ideal. Even if it is located it will appear as just a small hazy spot of light. With a telescope and medium magnification a bright and well-defined nucleus is seen enveloped by the mistiness of the halo. With large aperture and magnification some mottling has been reported on the halo, but again, the atmospheric conditions will limit observability. In the same field of view as M89 is the spiral galaxy, M90. Both are members of the Virgo Cluster.

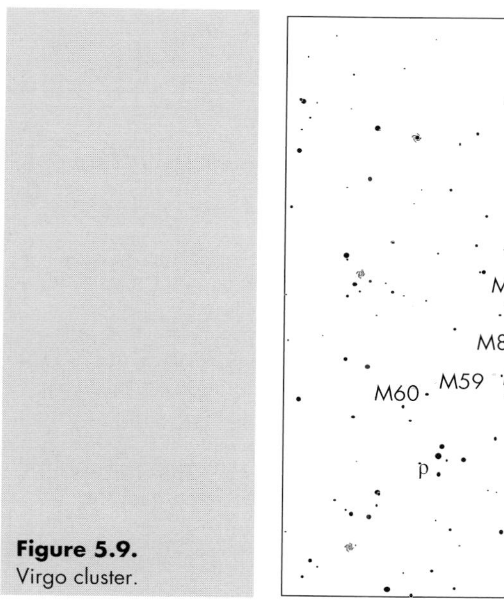
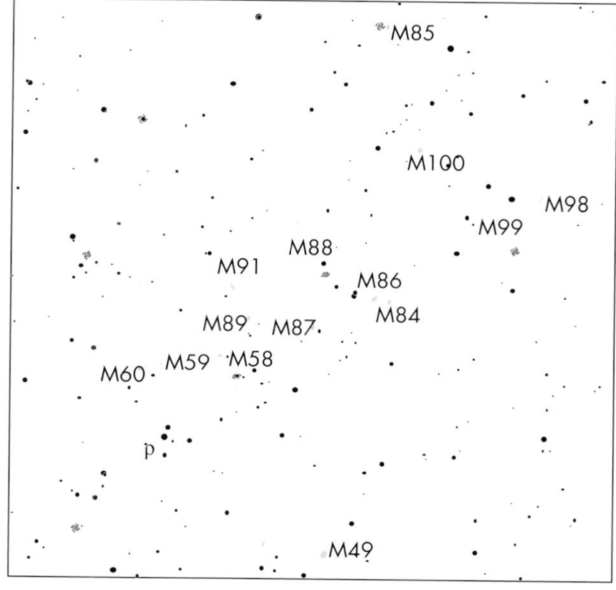

Figure 5.9.
Virgo cluster.

Messier 60	NGC 4649	12h 43.7m	+11° 33′	April 2
8.8 m [*12.8 m*]	7.2′ 6.0′		E2	easy

This is one of the biggest ellipticals at 118,000 l.y. across, and like its close companion M59 it to has a star-like nucleus surrounded by a faint halo. It is easier to see with binoculars, and the nucleus may be glimpsed with averted vision. Although there are no reports of any features visible telescopically, there are two aspects of interest. A close and tiny companion galaxy of M60 is located only 2.5′ to the north-west. This little spiral galaxy can only really be seen with averted vision. Also, M60 has been reported having a nice pale yellow tint. What do you see?

Messier 59	NGC 4621	12h 42.0m	+11° 39′	April 2
9.6 m [*12.5 m*]	5.4′ 3.7′		E5	moderate

Although visible in binoculars, this galaxy will pose a challenge to most observers. It will probably need the use of averted vision to be spied, and dark adaption will undoubtedly be needed. However, telescopically, M59 is rather nice as it is one of the few ellipticals that seems to show detail. It has a star-like nucleus, and some observers report a faint mottled appearance, although it must be mentioned that this could be an effect of foreground stars seen against the oval of the galaxy. It would be interesting to find out if this is correct. Try observing it under excellent conditions to see if you can detect any features. Also in the same field of view is the elliptical galaxy M60 (see above).

Caldwell 52	NGC 4697	12h 48.6m	–05° 48′	April 3
9.2 m [*12.7 m*]	6.0′ 3.8′		E6	moderate

This is a nice galaxy that is often ignored and left out of most observing schedules. It is rather bland in appearance, but stands out well against the background star field. Not really a binocular object, telescopically it is featureless, although at large aperture some increased brightness can be seen at its core. It is the dominant member of a small cluster of galaxies and lies at a distance of about 60 million l.y.

Caldwell 18	NGC 185	00h 38.9m	+48° 20′	September 30
9.2 m [*14.3 m*]	12′ 10′		dE0	moderate

See July–December.

Caldwell 35	NGC 4889	13h 00.1m	+27° 58′	April 6
11.5 m [*13.4 m*]	2.8′ 2.0′		E4	difficult

This is well worth seeking out as it is a very distant galaxy, at a distance of about 350 million l.y. Excellent seeing conditions are needed to glimpse this tiny object, with a telescope of at least a minimum 20 cm aperture. It has a bright core, surrounded by the usual faint halo. It is a dominant member of the Coma Galaxy Cluster, which contains about 1000 galaxies (several of which can be seen in large-aperture telescopes of at least 40 cm). The cluster itself is made of many elliptical galaxies and S0-type galaxies. Apparently it is the result of a merger of two older clusters. Observing any of these galaxies is a feat indeed, but well worth the effort.

Caldwell 17	NGC 147	00h 33.1m	+48° 30′	September 29
9.5 m [*14.5 m*]	13′ 8.1′		dE4	difficult

See July–December.

July–December

Messier 105	NGC 3379	10h 47.8m	+12° 35′	March 4
9.3 m [*12.1 m*]	5.4′ l 4.8′		E1	easy
See January–June.				

Messier 84	NGC 4374	12h 25.1m	+12° 53′	March 28
9.1 m [*12.3 m*]	6.5′ l 5.6′		E1	easy
See January–June.				

Messier 86	NGC 4406	12h 26.6m	+12° 57′	March 29
8.9 m [*13.9 m*]	8.9′ l 5.8′		E3	easy
See January–June.				

Messier 49	NGC 4472	12h 29.8m	+08° 00′	March 30
8.4 m [*12.9 m*]	10.2′ l 8.3′		E2	easy
See January–June.				

Messier 87	NGC 4486	12h 30.8m	+12° 24′	March 30
8.6 m [*12.7 m*]	8.3′ l 6.6′		E0.5P	easy
See January–June.				

Caldwell 18	NGC 185	00h 38.9m	+48° 20′	September 30
9.2 m [*14.3 m*]	12′ l 10′		dE0	moderate
This is another companion galaxy to M31, as mentioned in an earlier entry. However, this is much easier to locate and observe. In a telescope of 10 cm, it will just be detected, whereas in 20 cm it is easily seen. It remains featureless even at larger apertures (40 cm), but with a perceptibly brighter core. Several reports mention that with the very large aperture of 75 cm some resolution of the galaxy becomes apparent. Another dwarf elliptical galaxy.				

Messier 89	NGC 4552	12h 35.7m	+12° 33′	March 31
9.7 m [*12.3 m*]	5.1′ l 4.7′		E0	moderate
See January–June.				

Messier 59	NGC 4621	12h 42.0m	+11° 39′	April 2
9.6 m [*12.5 m*]	5.4′ l 3.7′		E5	moderate
See January–June.				

Caldwell 17	NGC 147	$00^h\ 33.1^m$	+48° 30′	September 29
9.5 m [*14.5 m*]	13′ l 8.1′		dE4	difficult

Located in Cassiopeia, this is classified as a dwarf elliptical galaxy. Although some distance from M31, the Andromeda Galaxy, it is in fact a companion to that famous galaxy. It is difficult to locate and observe, however, so dark skies are a prerequisite. It has been said that a minimum of 20 cm aperture is needed to see this galaxy, but I have recently had reports that under excellent conditions a 10 cm telescope is sufficient, though averted vision was needed. The moral of this story is that dark skies are essential to see faint objects. Increased aperture will help as well as higher magnification, when its nuclear region then becomes visible. A member of the Local Group, it is one of over 30 galaxies which are believed to be companions to either M31 or the Milky Way.

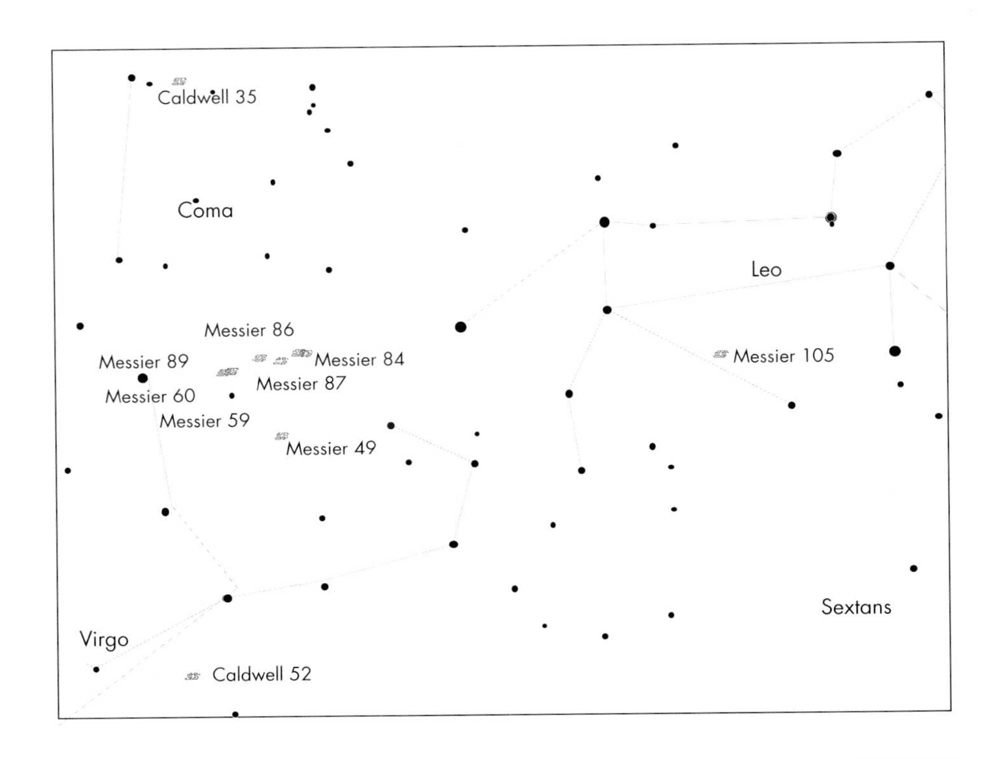

Figure 5.10. Elliptical Galaxies 1.

Messier 110	NGC 205	00^h 40.4^m	+41° 41′	October 1
8.0 m [*13.9 m*]	21.9′ l 11.0′		E5P	easy

The final entry in the Messier catalogue, and added to the original list in 1967. It is the second satellite galaxy of M31, and although it has a brighter magnitude than M32, the first satellite, it has a much lower surface brightness. Consequently it is much harder to see. It is visible in large binoculars, but will only appear as a very faint, featureless glow, north-west of M31. In a telescope it shows a surprising amount of detail, and high magnification will bring out its mottled nucleus. In addition, it shows detail that is peculiar for an elliptical galaxy; furthermore, these details are visible to the amateur. Of course, exceptionally dark skies and perfect seeing and transparency will be needed, but in a telescope of even modest aperture, say 10 cm, with a high magnification they are readily seen. Look for dark patches near a bright centre. Strangely enough, they are reminiscent of features normally found in a spiral galaxy. A definite observing challenge!

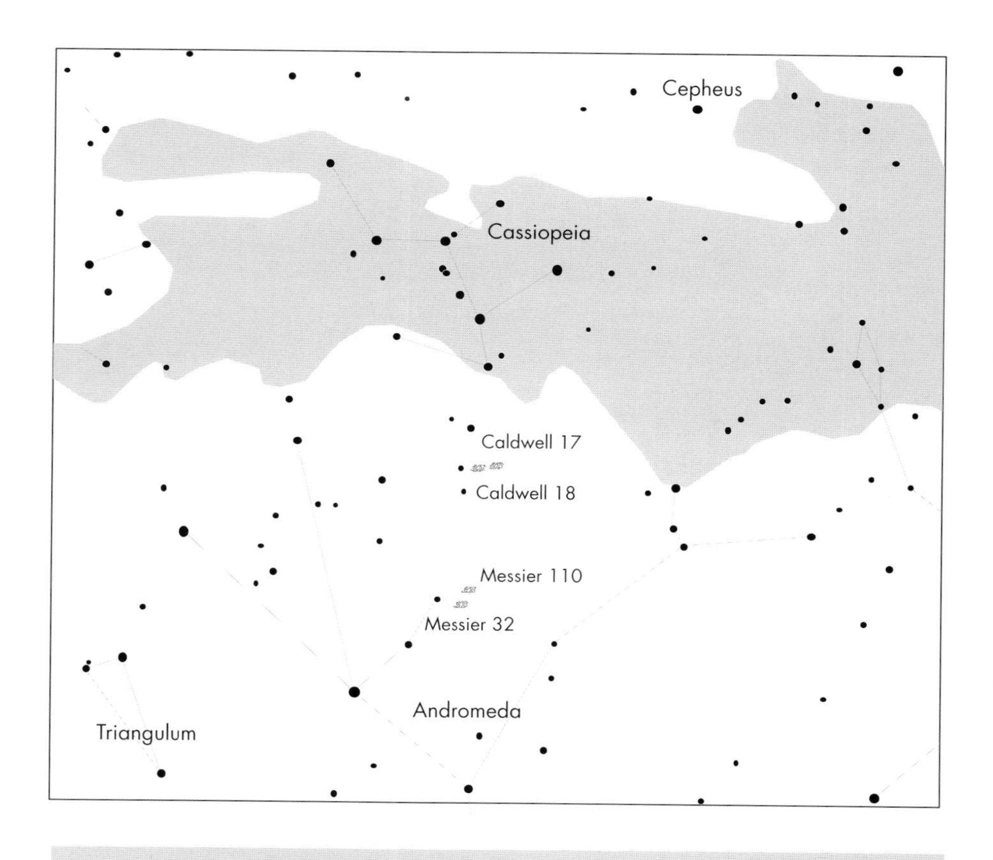

Figure 5.11. Elliptical galaxies 2.

Messier 32	NGC 221	00h 42.7m	+40° 52′	October 1
8.2 m [12.7 m]	8.7′ l 6.5′		E2	easy

The first satellite galaxy to M31 (see preceding entry) is visible in binoculars as a tiny round patch of light. However, it can often be mistaken for a star, so large binoculars are recommended. With a telescope, however, its elliptical shape becomes obvious, and under medium magnification a star-like nucleus can be glimpsed. Perhaps the most intriguing aspect of this little galaxy is the fact that it is believed to have a black hole at its heart. Just like its big brother, M31, although ten times less massive.

5.8 Lenticular, Peculiar and Irregular Galaxies

January–February

Messier 82	NGC 3034	09h 55.8m	+69° 41′	February 19
8.4 m [12.8 m]	11.2′ l 4.3′		IOsp	easy ©

A very strange galaxy, which is apparent when it is seen through a telescope. It can be glimpsed with binoculars, where it will appear as a elongated pale glow. Large binoculars will begin to hint at some detail, and with averted vision the dark dust lane may be seen. In even a small telescope of 10 cm aperture, it is evident that something strange has happened to M82. The western part is obviously brighter than the eastern. The core region appears jagged and angular. Throughout the length of the galaxy, starlight appears to stream through the gaps in the dark dust lanes. It is a galaxy that will repay long and detailed study, especially at large aperture and high magnification. The galaxy is an active galaxy, of the starburst type, and is undergoing an immense amount of star formation. This may have been caused by the close passage of its companion M81. During that time, which was about 40 million years ago, the gravitational effect of M81 caused the interstellar material within M82 to collapse and form new stars. Subsequently, material that was dragged from M82 is now believed to be falling back onto it, which gives rise to both its appearance and the new era of star formation. Both M81 and M82 can be seen in the same field of view and are a stunning sight.

Caldwell 53	NGC 3115	10h 05.2m	–07° 43′	February 21
8.9 m [12.6 m]	8.3′ l 3.2′		SOsp	easy

Also known as the Spindle Galaxy. This galaxy is often overlooked, which is a shame because it is a fine example of its type, as well as being quite bright. In binoculars it will appear as a small, faint elongated cloud, while in large binoculars it displays the characteristic lens shape. It is easily located in telescopes because of its high surface brightness. In telescopes of aperture 20 cm, it will appear as a featureless oval cloud, with perhaps a slight brightening toward its centre. As it is classed as an SO-type galaxy, it will not show any further detail even with larger aperture. It is a very big galaxy, some five times larger than the Milky Way. It is also one of the most favoured objects that is purported to have a black hole at its centre.

Caldwell 60/61	NGC 4038/9	12h 01.6m	–18° 51′	March 22
10.3/10.6 m [*14.4 m*]	7.6′ l 4.9′		Sp S(B)p	moderate
See March–April.				

Messier 85	NGC 4382	12h 25.4m	+18° 11′	March 28
9.1 m [*13.0 m*]	7.1′ l 5.5′		SA(s)OP	moderate
See March.				

Caldwell 21	NGC 4449	12h 28.1m	+44° 05′	March 29
9.6 m [*12.5 m*]	6.0′ l 4.5′		IBm	moderate
See March–April.				

Caldwell 24	NGC 1275	03h 19.8m	+41° 30′	November 10
11.9 m [*13.2 m*]	3.5′ l 2.5′		P	difficult
See November–December.				

March–April

Caldwell 77	NGC 5128	13h 25.5m	–43° 01′	April 13
6.8 m [*12.9 m*]	18.2′ l 14.5′		SOpec	easy

Also known as Centaurus A. Although this galaxy is too far south for UK observers, it nevertheless warrants inclusion because it is so spectacular. Photographs show it as a nearly circular object bisected by a very prominent dark dust lane. Visible in binoculars as a hazy star; with larger binoculars the famous dark lane can just be glimpsed. In small telescopes, aperture 15 cm, the dark lane is easily seen. Larger aperture will of course give a more detailed view, with the dark lane showing some structure. The well-known writer and astronomer Iain Nicolson says about the galaxy, "Centaurus A (NGC 5128) is a magical object, one of those rare extragalactic objects that, when it swims into the field of view, looks just like the photographs that grace the pages of astronomy books. The first time I saw it, it seemed almost to fill the field of view. It's a beautiful object in its own right: a near-spherical elliptical galaxy with a pronounced dark lane right through the middle. It is especially intriguing because of its status as the nearest active galaxy. To know that this object has a compact core that probably houses a supermassive black hole, even although it cannot be seen directly by eye, makes Centaurus, for me, one of the most exciting objects in the sky".

Messier 82	NGC 3034	09h 55.8m	+69° 41′	February 19
8.4 m [*12.8 m*]	11.2′ l 4.3′		IOsp	easy ©
See January–February.				

Caldwell 53	NGC 3115	10h 05.2m	–07° 43′	February 21
8.9 m [*12.6 m*]	8.3′ l 3.2′		SOsp	easy
See January–February.				

Caldwell 60/61	NGC 4038/9	12h 01.6m	−18° 51′	March 22
10.3/10.6 m [*14.4 m*]	*7.6′ l 4.9′*		Sp S(B)p	moderate

Also known as the Antennae or Ring-Tail Galaxies. Together, these probably make one of the most famous objects in the entire sky, but few amateurs ever observe it, believing it to be too faint. A telescopic object, it will appear as an asymmetrical blur in apertures of about 20 cm. Large apertures will begin to hint at its detailed structure, and at 25 cm aperture it will begin to resemble the famous apostrophe shape. Using apertures of about 30 cm, along with medium to high magnification, will show you that there are two objects involved, and it would be a worthwhile project to see just how much detail is resolved with a different group of telescopes and observers. It is one of those celestial objects that is so familiar from photographs that your perception of what is seen will be tainted by what you expect to see. Nevertheless, it is a wonderful object. Sadly, it is very low down for UK observers, so perfect observing conditions will be necessary. The marvellous shape of the Antennae is the result of spiral galaxies passing close by each other, so that tidal interaction causes material to be dispersed. Witness the amazing long tails that can be seen on deep images of these galaxies. Furthermore, recent work has shown that the interaction has led to a vast bout of star formation. Also, it has encouraged astronomers to put forward the idea that spiral galaxies evolve into elliptical galaxies after such an encounter.

Messier 85	NGC 4382	12h 25.4m	+18° 11′	March 28
9.1 m [*13.0 m*]	*7.1′ l 5.5′*		SA(s)OP	moderate

This is a bright galaxy that can be glimpsed in binoculars on clear nights. With large binoculars it is even easier to see, where it will show a star-like nucleus surrounded by the faint glow of the halo. Using a telescope will just magnify the rather featureless aspect, though a few observers report that at high magnification some faint detail can be glimpsed to the south of the nucleus which may be a trace of some spiral structure. Also, there is an indication of a faint blue tint to the galaxy.

Caldwell 21	NGC 4449	12h 28.1m	+44° 05′	March 29
9.6 m [*12.5 m*]	6.0′ l 4.5′		IBm	moderate

A member of the Canes Venaticorum Group of galaxies, this is a faint and frequently ignored object. Its irregular shape is often mistaken for a comet. Under good skies, a telescope of 20 cm aperture will easily discern its fan-shaped morphology, along with its faint nucleus. Larger telescopes will of course resolve the galaxy with a considerable amount of detail. An interesting point is that several HII regions are visible, especially one at the northern corner of the open fan shape. It is apparently a site of much ongoing star formation, and is similar in many ways to the Large Magellanic Cloud.

May–June

Caldwell 77	NGC 5128	13h 25.5m	−43° 01′	April 13
6.8 m [*12.9 m*]	18.2′ l 14.5′		SOpec	easy

See March–April.

Messier 82	NGC 3034	09h 55.8m	+69° 41′	February 19
8.4 m [*12.8 m*]	11.2′ l 4.3′		IOsp	easy ©

See January–February.

Caldwell 60/61	NGC 4038/9	$12^h 01.6^m$	–18° 51′	March 22
10.3/10.6 m [*14.4 m*]	7.6′ l 4.9′		Sp S(B)p	moderate
See March–April.				

Messier 85	NGC 4382	$12^h 25.4^m$	+18° 11′	March 28
9.1 m [*13.0 m*]	7.1′ l 5.5′		SA(s)OP	moderate
See March–April.				

Caldwell 21	NGC 4449	$12^h 28.1^m$	+44° 05′	March 29
9.6 m [*12.5 m*]	6.0′ l 4.5′		IBm	moderate
See March–April.				

Caldwell 57	NGC 6822	$19^h 44.9^m$	–14° 48′	July 18
8.8 m [*14.2 m*]	20.0′ l 10.0′		IB(s)m	moderate
See July–August.				

July–August

Caldwell 77	NGC 5128	$13^h 25.5^m$	–43° 01′	April 13
6.8 m [*12.9 m*]	18.2′ l 14.5′		SOpec	easy
See March–April.				

Messier 82	NGC 3034	$09^h 55.8^m$	+69° 41′	February 19
8.4 m [*12.8 m*]	11.2′ l 4.3′		IOsp	easy ©
See January–February.				

Caldwell 57	NGC 6822	$19^h 44.9^m$	–14° 48′	July 18
8.8 m [*14.2 m*]	20.0′ l 10.0′		IB(s)m	moderate

Also known as Barnard's Galaxy. This is a challenge for binoculars astronomers. Even though it is fairly bright, it has a low surface brightness and so is difficult to locate. Once found, however, it will just appear as a hazy indistinct glow running east–west. This is in fact the bar of the galaxy. Strangely enough, it is one of those objects that seems to be easier to find using small aperture, say 10 cm, rather than large. Nevertheless, dark skies are essential to locate this galaxy.

September–October

Messier 82	NGC 3034	$09^h 55.8^m$	+69° 41′	February 19
8.4 m [*12.8 m*]	11.2′ l 4.3′		IOsp	easy ©
See January–February.				

Caldwell 51	IC 1613	01ʰ 04.8ᵐ	+02° 07′	October 7
9.2 m [–]	11.0′ l 9.0′		dIA	moderate

A very difficult galaxy to observe. A few reports state that it is visible in large binoculars as a very faint hazy glow, while others claim that a minimum of 20 cm aperture is needed. Whatever you choose, one thing is paramount, namely that a dark sky will be needed. A member of the Local Group, it is similar in many respects to Caldwell 57 (see previous entry). It is an old galaxy that is still forming stars.

Caldwell 57	NGC 6822	19ʰ 44.9ᵐ	–14° 48′	July 18
8.8 m [*14.2 m*]	20.0′ l 10.0′		IB(s)m	moderate

See July–August.

Caldwell 24	NGC 1275	03ʰ 19.8ᵐ	+41° 30′	November 10
11.9 m [*13.2 m*]	3.5′ l 2.5′		P	difficult

See November–December.

November–December

Messier 82	NGC 3034	09ʰ 55.8ᵐ	+69° 41′	February 19
8.4 m [*12.8 m*]	11.2′ l 4.3′		IOsp	easy ©

See January–February.

Caldwell 53	NGC 3115	10ʰ 05.2ᵐ	–07° 43′	February 21
8.9 m [*12.6 m*]	8.3′ l 3.2′		SOsp	easy

See January–February.

Caldwell 24	NGC 1275	03ʰ 19.8ᵐ	+41° 30′	November 10
11.9 m [*13.2 m*]	3.5′ l 2.5′		P	difficult

The final galaxy in our list is a very important galaxy, even though visually it is not impressive. It nevertheless should be observed for several reasons. It is the main member of the Perseus Galaxy Cluster, also known as Abell 426. Several amateurs have stated that it can be seen in telescopes as small as 15 cm aperture, while larger apertures will make it easier to locate. If you can have access to a telescope of aperture 40 cm or larger, use the opportunity to look at this galaxy as it will be surrounded by several fainter ones which are all part of the cluster. In many respects it is the most concentrated field of galaxies in the winter sky for the northern observer. Caldwell 24 is a strong radio galaxy and is believed to be the remnant of a merger between two older galaxies. It is one of over 500 members of the Perseus Cluster. The cluster itself forms part of an even larger super cluster, the Pisces–Perseus Supercluster.

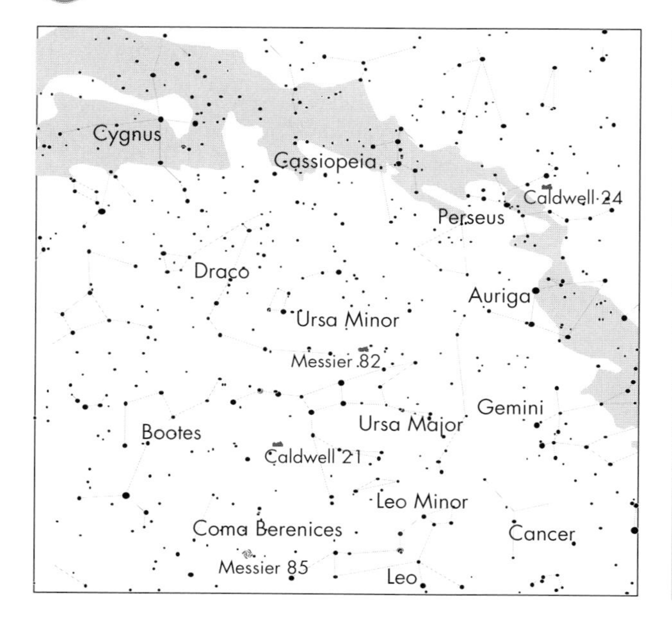

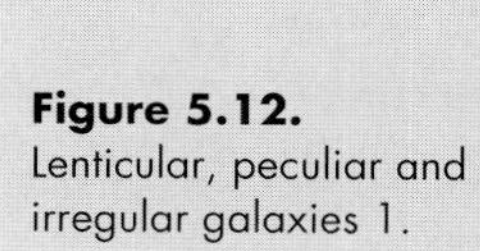

Figure 5.12. Lenticular, peculiar and irregular galaxies 1.

Figure 5.13. Lenticular, peculiar and irregular galaxies 2.

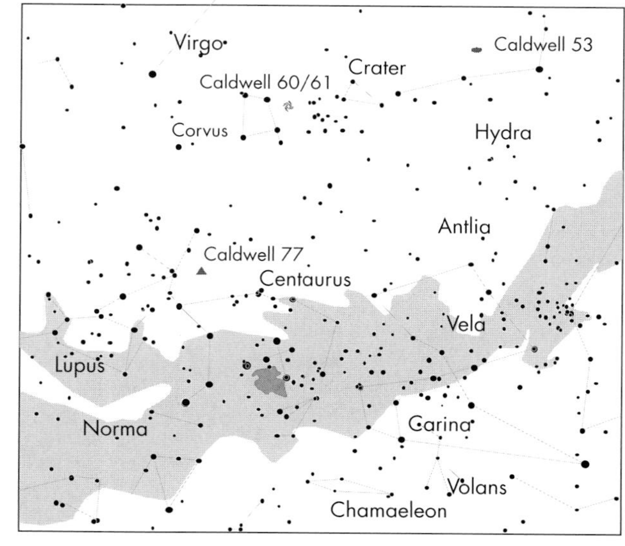

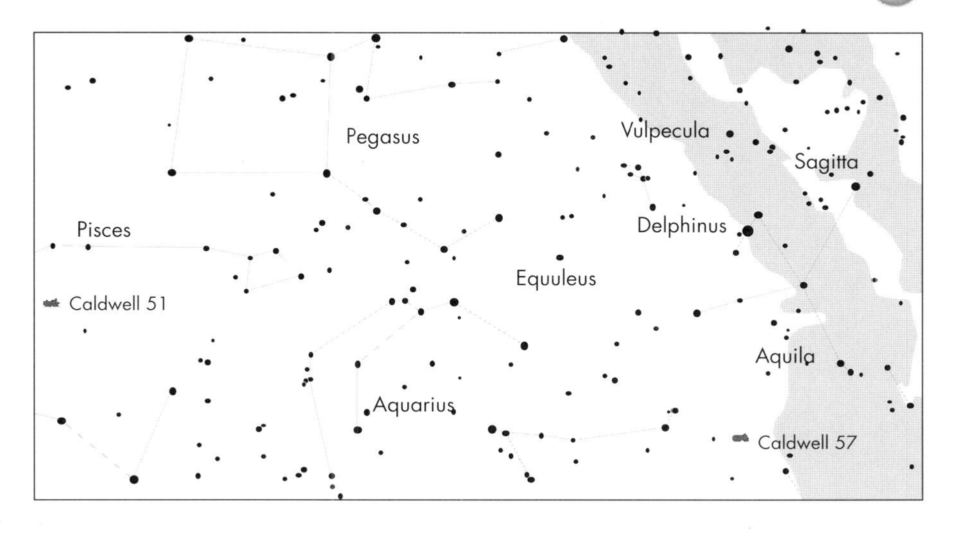

Figure 5.14. Lenticular, peculiar and irregular galaxies 3.

5.9 Groups and Clusters of Galaxies

January–June

Hickson Group 68	NGC 5353	$13^h\ 53.4^m$	+40° 47′	April 20
11.1 m	→11.2′←	5		moderate

This is a very nice group of stars for amateur instruments. The brightest member can be seen in a telescope as small as 6 cm, and with a 15 cm aperture it will show a slight brightening of its centre. The other galaxies will appear as faint patches of light, and to see the faintest member would most certainly require an aperture of about 25 cm.

Copeland's Septet	NGC 3753	$11^h\ 37.9^m$	+21° 59′	March 16
13.4 m	→7.0′←	7		difficult

Also known as the Hickson Galaxy Group 57. This is a very small group of galaxies. Situated in the constellation Leo, all within about 7 arc seconds. Telescopes of at least 25 cm aperture will be needed, and even then you may not spot the fainter members of the group but just the four brighter galaxies. Larger apertures should of course allow you to spot them. Nevertheless, seeing conditions will determine what you observe, regardless of aperture. The group is a mix of barred spirals, ordinary spirals and lenticular galaxies.

Coma Cluster	NGC 4889	$12^h\ 57.7^m$	+28° 15′	April 6
11.4 m	→120+′←	10+		difficult

This is a large cluster of galaxies. Many of its members are within reach of amateur telescopes of aperture 25 cm and larger. It is fairly well spread out, and so the field of view will be dotted with many indistinct faint patches of light. It contains many elliptical, spiral, barred spiral and lenticular galaxies.

Stephen's Quintet	NGC 7320	$22^h\ 36.1^m$	+33° 57′	April 6
12.6 m	→4′←	5		difficult

A very famous group of galaxies located in Pegasus, but one that has in the past proved strangely difficult for amateurs. Under perfect seeing conditions, the group is visible in a 20 cm telescope. However, I stress the word *perfect*! The largest member of the quintet is only 2.2 × 1.2 arc seconds in size, so it is very small, but it is the brightest. To actually see the group as a distinct unit and not a faint smudge of light will require a telescope of aperture 25 cm. This will show at least four of the group, but the fifth requires an aperture of at least 30 cm. Under high magnification and large aperture, structure can be seen within the brighter members. It is believed that four of the group are interacting with each other, and there is debate as to whether the fifth is in fact a line-of-sight galaxy. A challenge to the urban astronomer.

Seyfert's Sextet	NGC 6027	$15^h\ 59.2^m$	+20° 46′	May 22
13.3 m	→1.5′←	6		very difficult

This is a real challenge! In all but the largest telescopes it is questionable if you will see anything at all, and even in apertures around 40 cm the galaxies will barely be resolved. Nevertheless, it would be interesting to find out what would be the smallest aperture required to spot these faint galaxies.

Virgo Cluster

See Messier 87 in Section 5.7.

Perseus Cluster

See Caldwell 24 in Section 5.7.

July–December

Hickson Group 68	NGC 5353	$13^h\ 53.4^m$	+40° 47′	April 20
11.1 m	→11.2′←	5		moderate

See January–June.

Fornax Cluster	NGC 1316	$03^h\ 20.9^m$	−37° 17′	November 10
11.4 m	→12+′←	10+		difficult

This is another large cluster of galaxies. Amateur telescopes should be able to pick out the brightest members with no difficulty. What makes this cluster so spectacular, however, is that with a modest aperture, say 25 cm, and clear and dark skies, there are so many galaxies that identification is very difficult. The brightest member is visible even in an 8 cm telescope! A galaxy of note in the cluster is NGC 1365, which is a nice barred spiral 8 arc seconds in length and visible in an 8 cm aperture telescope as a faint blur. The cluster also contains a galaxy known as the Fornax System. It is a dwarf spheroidal galaxy, a very small and faint class of galaxy. To my knowledge this galaxy has never been observed with amateur instruments.

Copelands Septet	NGC 3753	$11^h\ 37.9^m$	+21° 59′	March 16
13.4 m	→7.0′←	7		difficult

See January–June.

Seyfert's Sextet	NGC 6027	$15^h\ 59.2^m$	+20° 46′	May 22
13.3 m	→1.5′←	6		very difficult

See January–June.

Perseus Cluster

See Caldwell 24 in Section 5.7.

The massive cluster of galaxies, Abell 2218, acts like a giant zoom lens in space. The gravitational field of the cluster magnifies the light of more distant galaxies far behind it, providing a deep probe of the very distant Universe! The cluster was imaged in full colour, giving astronomers a spectacular and unique new view of the early Universe. The colour of a distant source is preserved by gravitational lensing. By matching images of the same colour, families of multiple images produced by the lensing process can be identified.

Astronomers are particularly fascinated by the unusual red feature in this field. This extraordinary object has colours which indicate it is either a rare, extremely cool, dwarf star in our own galaxy, or alternatively one of the most distant objects ever viewed by Hubble, lensed into visibility by the mass of the cluster. Further observations may confirm the identity of this unusual object.

NASA/A. Fruchter/ERO Team (STScI)

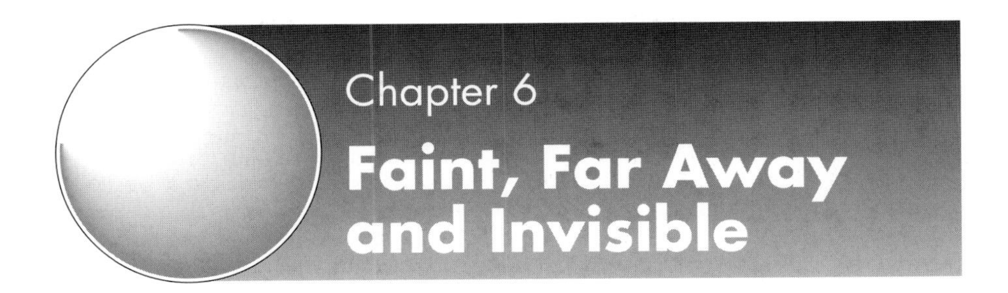

Chapter 6
Faint, Far Away and Invisible

6.1 Introduction

You're probably wondering what possible objects there are that can be listed as faint, far away and invisible, and *still* be of interest to an amateur astronomer!

There are a few, if you think about it. So far I have just covered those celestial objects which are familiar to nearly every amateur – stars, nebulae and galaxies – but in fact you're familiar with quite a few other objects (which I am sure you have read about in other books or astronomy magazines).

Quasars, for example. But have you ever observed one? You may have read about a *lensed quasar*, but thought that there couldn't possibly be any that are visible with amateur instruments. Could there? Well, there is one – faint, tiny, but visible under perfect conditions.

What about the centre of our Galaxy, or even a star which we know has planets orbiting it? (No, it isn't the Sun!) These are the types of objects I will briefly cover. If you are anything like me, then even knowing the position of these objects is interesting and worthwhile in itself. It has always interested me to know that a particular star, which may look very ordinary, is in fact, very extra-ordinary, even though, when seen through as telescope, it appears much like every other star. In a similar vein, knowing that a rather featureless patch of sky actually harbours what is believed to be a *black hole* has always held a fascination. Many of the objects mentioned in this chapter will, to all intents and purposes, be invisible to amateur telescopes or show no detail, but that shouldn't deter you from seeking out the area of sky in which they lurk.

As usual, I'll begin with some background astronomy about the objects.

6.2 The Centre of Our Galaxy

The Milky Way, our home galaxy, is as you know a spiral galaxy, similar to many that have been mentioned in Chapter 5. It is about 1000 l.y. thick and 100,000 l.y. in diameter, and contains over 100 billion stars. It is classified as a SAB(rs)bc-type galaxy. The Solar System is located about halfway from the centre of the Galaxy to the edge of the disc, and around 70 l.y. above the plane of the Galaxy. It is near the inner edge of a short spiral arm called the Orion Arm, which is around 15,000 l.y. long and contains the Cygnus Rift and the Orion Nebula (M42).

April–September

Galactic Centre	Sgr A*	$17^h 45.6^m$	–29° 00′	June 18

The centre of the Galaxy has always posed several problems to astronomers. Unfortunately, owing to the vast amount of gas and dust that lies between us, it has been impossible to see the centre using visible light. However, infrared and radio waves can escape from the centre, and so a picture can be built up of the inaccessible region. What has been learnt is impressive. There is a radio source located very near or even at the exact centre, called Sagittarius A*,[1] and this was the first cosmic radio source discovered. Measurements of the radio source indicate that it is no bigger than the diameter of Mars's orbit. One of the surprising results is that Sagittarius A* is stationary, which would indicate that it is very massive.[2] The current estimate for the mass is about a thousand solar masses. All this information suggests that at the centre of our Galaxy lies a black hole! This, of course, would mean that the actual centre of our Galaxy must be invisible. Although, if you look through a telescope at the region, the field of view will be full of numerous star fields, these actually lie much closer to us than to the centre. And when you are observing this region you may like to contemplate the almost certain fact that even though you cannot see the centre, there in your eyepiece, ever invisible, lies a supermassive black hole, around which you and I, the Solar System and the Galaxy rotate.

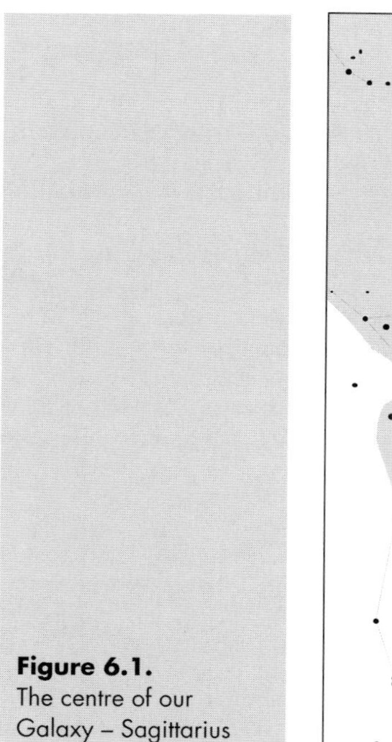

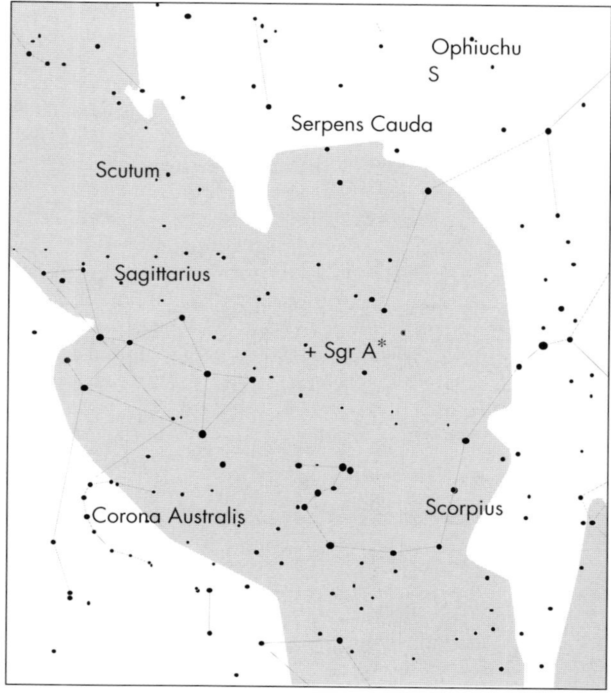

Figure 6.1. The centre of our Galaxy – Sagittarius A*.

[1] Sagittarius A* is now believed to be made of two components, SgrA East and SgrA West. The former is a supernova remnant and the latter is an ultra-compact, non-thermal source, i.e., a black hole.
[2] Recent analysis suggests that the density around the centre of the Galaxy is about a million times greater than any known star cluster. It is probably made up of stars, dead stars, gas and dust, and of course a black hole.

6.3 Stellar Black Holes

Although black holes have been mentioned several times in preceding chapters, they have mainly been associated with galaxies. Thus they have several million times the mass of our sun, and are massive beasts. There exist however (or should I correctly say, are believed to exist), black holes which have smaller mass, and are associated with the evolution of stars. The current scenario[3] which would allow black holes to form is for a star to evolve from the main sequence, becoming a red giant star. If the star has a mass from about 1.4 to 3 times that of the Sun, it would eventually form a neutron star. But if the mass is greater than 3 times the Sun's, then not even the neutron degeneracy present in a neutron star will stop collapse, and the star will shrink down to become a black hole.

The problem with detecting possible black holes is that because of their immense gravitational influence they do not emit radiation, and so cannot be detected directly. How does one begin to search for such an elusive and invisible object?

The answer is to search for the effect a black hole may have on a companion star. First search for a star whose motion, determined by measuring the Doppler shift in its spectrum is a member of a binary stars. (The Doppler shift is the change in the appearance of light from an object that is moving away from or towards an observer.) If it proves possible to see both stars, then give up on that object. The search is for a binary system where one companion is invisible, no matter how powerful the telescope used. However, just because it is unseen it is not a black hole candidate. It may be just too faint to be seen, or the glare from the companion may swamp out its light. It could even be a neutron star. Thus further evidence is needed to see if the invisible companion has a mass greater than that allowable for a neutron star. Kepler's laws are used at this point to determine whether the star, or rather the invisible object, has a mass greater than three solar masses. If this is so, then the unseen companion may be a black hole. Further information is still needed, however, and this may appear in the form of X-rays, which can arise either from material flowing from one star into the black hole or from an accretion disc which has formed around the black hole. Either way, the presence of X-ray emission is a good indicator that a black hole may be the unseen companion object. Of course, the measurements as stated above are a bit more complicated than this. For instance, it is known that neutron stars can emit X-rays and have an accretion disc. So careful analysis of the data is necessary. However, a few candidates are known, and one is even visible to the amateur astronomer. Or perhaps I should say that the companion star to the black hole is visible!

May–October

Cygnus X-1	HDE 226868	$19^h 58.4^m$	35° 12′	July 21
This is one of the strongest X-ray sources in the entire sky, and is possibly the most convincing candidate for a black hole. Its position is coincident with the star HDE 226868, which is a B0Ib supergiant of magnitude 9. It lies about 0.5° ENE of Eta Cygni. It is a very hot star, of around 30,000 °K, and analysis shows it is a binary with a period of around 5.6 days. Observations by satellite have detected variations in the X-ray emission on a time scale of less than 50 milliseconds. The estimated mass of the unseen black hole companion is in the range of 6 to 15 solar masses. This would mean that it has a diameter less of about 45 kilometres.				

[3]Please refer to the books listed in Appendix 2 for a full and proper account of current theory on stellar evolution.

Other black hole binary systems are:

Star	Type	Orbital period (days)	Black hole mass estimates (M_{Sun})
LMC X-3	B main sequence	1.7	4 to 11
V616 Mon	K main sequence	7.8	4 to 15
V404 Cyg	K main sequence	6.5	> 6
Nova Sco 1994	F main sequence	2.4	4 to 15
Nova Velorum	M dwarf	0.29	4 to 8

In Fig. 6., Cygnus X-1 is the companion to the bottom star in the close double system.

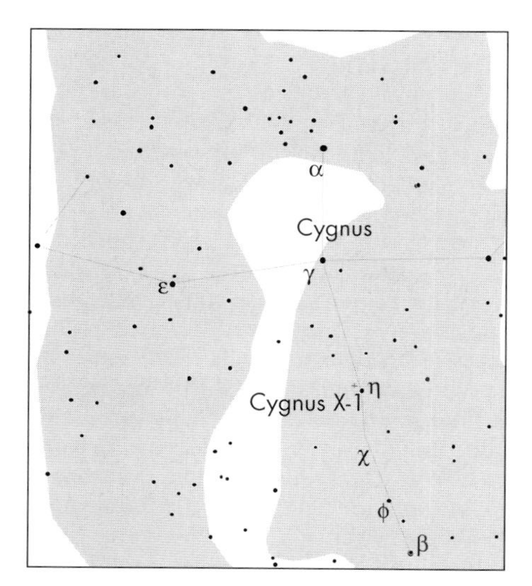

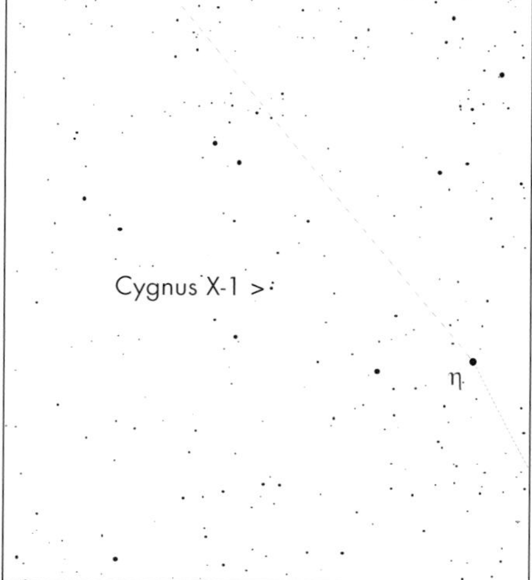

Figure 6.2. Cygnus X-1.

6.4 Quasars

I doubt whether there are any amateur astronomers who haven't heard of the term *quasar*. It actually stands for quasi-stellar radio source. In 1963 the astronomer Maarten Schmidt was looking at some spectra of what was then believed to be radio stars, when he noticed that one object (which looked to all appearances like a star) was in fact something very different. On analysing the spectra he discovered that the object was in fact *very* luminous and *very* far away. He measured the red shift of the object, and discovered that it was moving away from us at 45,000 kilometres per second, or about 15 per cent of the speed of light. Obviously some thing strange and spectacular was going on. The object in question was known as 3C 273.[4]

[4] It is entry number 273 entry in the 3rd Cambridge catalogue of radio sources.

We now know that *quasars*, and their radio-quiet relatives, QSOs (quasi-stellar objects), are in fact the ultra-luminous central regions of very young galaxies. They appear star-like owing to their vast distances, and are very bright because it is believed that supermassive black holes lie at their hearts, creating immense amounts of energy. Some very luminous quasars are thought to have black holes with masses of between 2 million and 2 billion solar masses. Because they are so far away, the galaxy (and thus the stars) in which they reside are too faint to be seen, so you only see the energy that is produced from the region around the black hole. Quasars are classed as *active galaxies*, and are an important aspect of a galaxy's evolution. The active galaxies mentioned earlier in the book such as Seyfert galaxies can be thought of as having mini-quasars at their cores. Many more quasars have been found, and they are among the furthest and, subsequently, the youngest objects in the Universe.

Visually, a quasar will just appear like a faint star. After all, this is what astronomers believed them to be up until recently. However, the brightest and most famous, 3C 273, is easily within reach of amateur telescopes. Needless to say, dark skies and excellent seeing are necessary. Also, a good star atlas is a must, as there will be countless background stars visible.

January–June

3C 273		12^h 29.1^m	02° 03′	March 29
12.8 m	Redshift (z) 0.158		2,000,000,000 l.y.	

This quasar is the brightest in the sky, and within reach of medium-to-large-aperture telescopes. There are reports that it has been glimpsed in telescopes of 20 cm and thus is well within reach of most amateurs. Averted vision will also help locate this distant object. To locate it, the following directions should help.

It is situated about 3.5 north-east of Eta Virginis.

Find the galaxy NGC 4536 (magnitude 10.6, surface brightness 13.2, at position R.A.12^h 34.5^m Dec. 02° 11′). At about 1.25° east of the galaxy is the quasar.

In the immediate vicinity is a double star, arranged east–west with 3 arc seconds separation. The double has magnitudes 12.8 and 13, and the quasar is a bright, blue-tinted stellar object east of the double system.

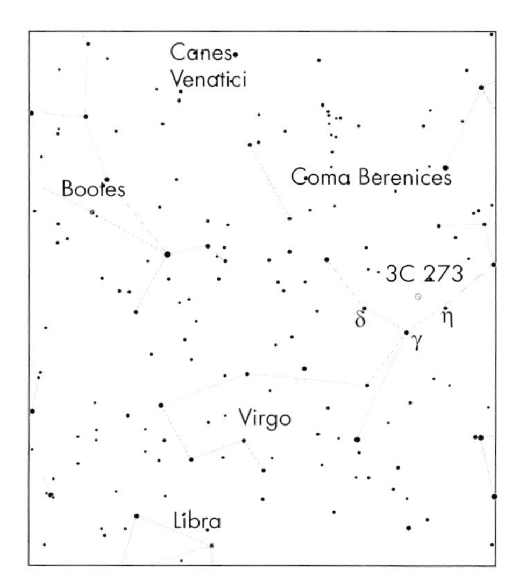

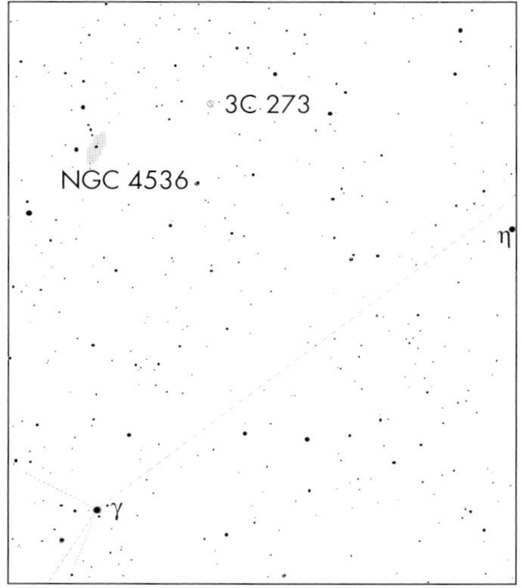

Figure 6.3. Quasar 3C 273.

PKS 405–123	MSH 04–12	04^h 07.8^m	–12° 11′	November 22
14.8 m	Redshift (z) 0.57		6,000,000,000 l.y.	

See July–December.

July–December

PKS 405–123	MSH 04–12	04^h 07.8^m	–12° 11′	November 22
14.8 m	Redshift (z) 0.57		6,000,000,000 l.y.	

This is another quasar that should be within the reach of amateur astronomers, and has been glimpsed in telescopes of 20 cm. It is located in the constellation Eridanus. The quasar lies about 3° to the north-east of Zaurak (Gamma Eri). When seen through an eyepiece, you may spot a tiny green dot to the left . This is the planetary nebula NGC 1535 (Cleopatra's Eye). If you do manage to see the quasar, and you will need detailed star maps to confirm the observation, then you will be a member of a very small and elite group of observers. It is also incredible to note that the light that enters your eye from this quasar started its journey some 1.5 billion years *before* the Solar System was formed!

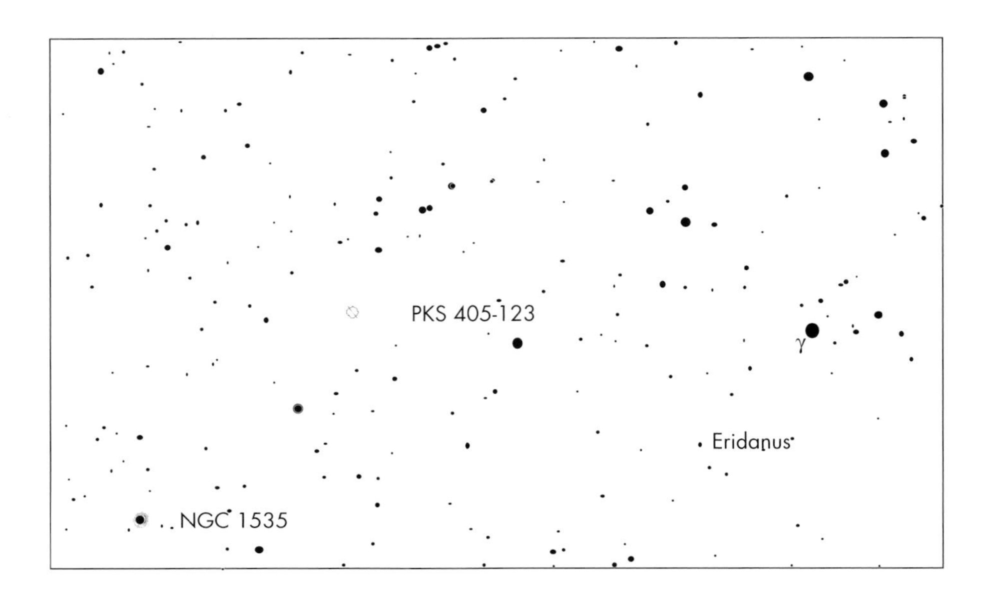

Figure 6.4. Quasar PKS 405–123.

3C 273		12^h 29.1^m	02° 03′	March 29
12.8 m	Redshift (z) 0.158		2,000,000,000 l.y.	

See January–June.

6.5 Gravitational Lensing

Before I leave the topic of quasars for the final time, there is one more amazing aspect of them – gravitational lensing – and it can be seen by amateurs. It is possible, under the right conditions, to see one of the most fascinating consequences of Einstein's general theory of relativity.

The general theory of relativity is far beyond the scope of this book, but a simple outline of what happens according to it is fairly straightforward. Gravity has the ability to "bend" light, if the gravitational force is strong enough. The first experimental justification of Einstein's theory was in fact a measure of this light bending, when on 29 May 1919, the British astronomer Arthur S. Eddington measured the amount that starlight was deflected by the Sun. He used a total eclipse of the Sun so that any faint stars would not be rendered invisible by the glare of the Sun. The accuracy of the measurements was about 20 per cent, but it was enough to vindicate the theory. Subsequent measurements using radio waves have managed to confirm the predictions made by Einstein to within 1 per cent.

The Sun is not the only thing that can bend a ray of light. Any object that has sufficient mass can deflect light waves. Calculations show that when light rays from a distant object pass close to a compact but massive galaxy, the bending of the light can result in the appearance of multiple or twisted images. It is as if the galaxy were acting like a lens, and so any object emitting light from *behind* the galaxy has its light bent as it passes close to the galaxy. This bizarre effect is called *gravitational lensing*.

In 1979, Astronomers noticed that a pair of quasars known as QSO 0957+561 had identical spectra and redshifts, and it was suggested that these two quasars may in fact be one, and that the two images were produced by an intervening object. This was subsequently proved to be the true explanation, and the light from the distant quasars was being lensed by a cluster of galaxies.

You may have seen images of such objects in various books and magazines, and always thought that it would be nearly impossible to see these through a telescope. The examples always given are usually of quasars so distant that the Hubble Space Telescope or at least the world's largest ground-based telescopes are needed to image them.

This is (only) more or less true, but there are one or two quasars which can be and have been seen by amateurs. It isn't easy. I must stress that good seeing conditions are essential to observe these faint objects, and that a detailed star atlas is required to confirm the observation.

January–December

Twin quasar	Q0957+0561A/B	$10^h\ 01^m$	55° 53′	February 2
16.8 m (17.1 17.4 A/B)	Separation 6″		8,000,000,000 l.y.	

The quasar is in the constellation Ursa Major, and so is a fine target for northern hemisphere observers. The starting point for the quasar is the bright edge-on galaxy NGC 3079 (8.1′ × 1.4′, magnitude 11.5, within reach of a 20 cm telescope; several fainter galaxies lie nearby). The galaxy points to the quasar to the south-east, about two galaxy lengths away near a parallelogram of 13th- and 14th-magnitude stars. The quasar lies off the south-east corner.

The two components are 17.1 and 17.4 magnitude, separated by 6″. Observers with very large instruments of aperture 50 cm have reported seeing the two objects cleanly split. Like most quasars, Q0957+0561 is slightly variable in brightness. With small telescopes, the two images will appear as one, but slightly elongated. In this case, the lensing is done by a cluster of galaxies, which lie 3.5 billion l.y. away, and is splitting the light of the more distant Q0957+0561 into multiple images. Two of these images are much brighter than the others, and this is what is observed. It may be wise to try as high a magnification as possible. This is a good observing challenge for CCD owners. The quasar lies at a distance of at almost 8 billion l.y., and may well be the most distant object visible to the amateur astronomer.

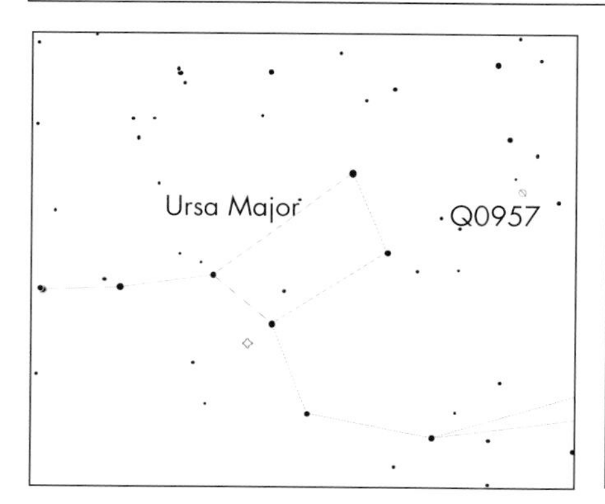

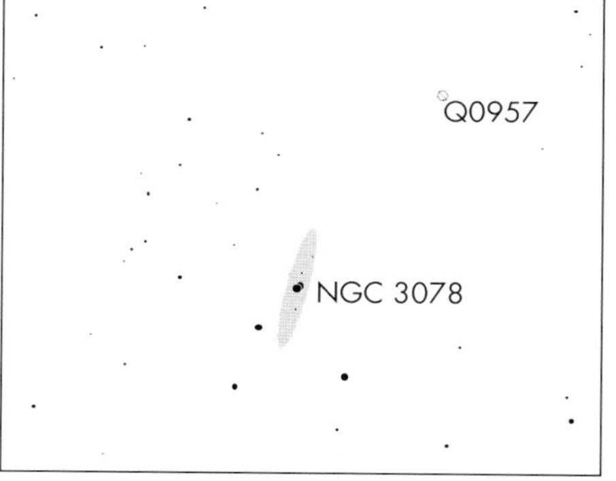

Figure 6.5. Twin quasar.

Leo Double Quasar	QSO 1120+019	$11^h\ 23.3^m$	01° 37′	March 13
15.7 20.1 m (A/B)	Redshift (z) 1.477			

An extremely difficult quasars to resolve. The brighter A component is easily seen in large telescopes, but the fainter B component is very difficult.

Cloverleaf Quasar	H 1413+117	$14^h\ 15.8^m$	11° 29′	April 25
17 m (A/B/C/D)	Redshift (z) 2.558			

An exceedingly difficult object to observe, and only with perfect conditions and very large-aperture telescopes will it be seen. The greatest separation among the four is about 1.36″. To my knowledge it has never visually been observed from the UK, but US observers report seeing just an asymmetric, faint hazy and tiny blob of light, although it has been imaged by CCD.

6.6 Extra-Solar Planets

It is sobering to think that in a period of about ten years our view of the universe has changed so much that we no longer believe that our solar system is the only one there is, but in fact is just one of many that exist in the Milky Way, and that this belief is based on firm evidence.

As I write these words I have just heard that yet another star has been discovered which has planets orbiting it. It is as if there are new solar systems being discovered every month! Soon, with the advent of larger telescopes and improved image-processing techniques it will only be a matter of time before the first image of a non-Solar-System planet is obtained. That will be a very special day. However, until then any indication as to what these new planets will look like will have to come from the minds of artists and scientists (not forgetting amateur astronomers).

Surprisingly many of the stars which have been reported as having planets are quite bright, and so easily within reach of small telescopes. Even though any sign of planets will be absent if these stars are observed, it is still a sobering and also wonderful thing to be able to view them and to think that circling these bright stellar objects are new worlds. And what else besides, one wonders ...

The technique used for detecting these planets makes use of the Doppler effect, which is the change in the appearance of light from an object that is moving away from or towards an observer. The gravitational pull of a large planet orbiting a star causes the star to wobble slightly. When a star wobbles towards Earth, the star's light appears from Earth to be shifted towards the blue part of the visible light. When the star wobbles away from Earth, the opposite effect occurs.

The Doppler shift is proportional to the speed with which the star approaches or recedes from an observer on Earth. Unfortunately the Doppler shift caused by the wobbling of stars with companion planets is very small; the wavelength of the star's light changes by only about one part in 10 million under the influence of a large, Jupiter-sized planet. For example, the Sun's "wobble speed" is only about 12.5 m/s. To detect planets around other stars, the errors in the measured speed of stars must be smaller than about 10 m/s, or about 36 km/h. This is just within the range of today's instruments.

Astronomers noticed that the Doppler shift of the star 51 Pegasi in the constellation Pegasus varied in a periodic way, first shifting towards the blue and then toward the red. The timing of the Doppler shift suggested that the star was wobbling owing to a closely orbiting planet that completes a full revolution around its star in only 4.2 days. This means that the planet is orbiting its star at a speed of about 134 km/sec or 482,000 km/h (299,000 miles per hour), more than four times faster than Earth's velocity around the Sun. Similar techniques were applied to several other stars and many more extra solar planets were discovered.

Another possible planet, orbiting the star HD114762, has a large mass (more than 10 times that of Jupiter), not unlike the planet around 70 Virginis, which has a mass more than 6.5 times that of Jupiter. Both HD114762 and 70 Virginis are so large that most astronomers are not sure whether they are large planets or low-mass *brown dwarfs*. A brown dwarf is an object intermediate in mass between a planet and a star.

The wobble motion of a star with a planetary companion can provide a great deal of information about the star's companion planet, including an estimate of its mass and the size and frequency of its orbit. The *orbital period* of a planet (the time it takes the planet to complete one full revolution around its star) is equal to the time it takes the star to finish one wobble cycle. The size of the star's wobble is also proportional to the size of the planet's orbit, and by using Kepler's third law of planetary motion, which states that the cube of the average distance (a^3) between two orbiting bodies equals the orbital period squared (p^2), the distance between the star and its companion can be determined. Knowing that the orbital period is the same as the period of the star's wobble, one can calculate the average distance between a star and its companion.

The following is a complete (as of 25 July 2000) list of extra-solar planets and possible companion objects. No doubt that by the time this book is published, several more will have been discovered. Those objects marked by * will be covered in more detail. For an amateur astronomer, what is really remarkable about these stars is that so many of them are visible with the naked eye!

Confirmed Planets/Brown Dwarfs Around Main Sequence Stars

1. HD 16141
2. HD168746
3. HD 46375
4. HD108147
5. HD83443
6. HD 75289
7. 51 Peg*
8. BD-103166
9. HD 187123
10. HD 209458
11. upsilon And*
12. HD 192263
13. HD38529
14. 55 Cnc*
15. HD 37124
16. HD 130322
17. rho CrB*
18. HD52265
19. HD 177830
20. HD 217107
21. HD 210277
22. 16 Cyg B*
23. HD 134987
24. Gliese 876
25. HD92788
26. HD82943
27. HR810
28. 47 Uma*
29. HD 12661
30. HD169830
31. 14 Her*
32. GJ 3021
33. HD 195019
34. Gl 86
35. tau Boo*
36. HD 168443
37. HD 222582
38. HD 10697
39. 70 Vir*
40. HD 89744
41. HD114762

– 13 Jupiter mass limit – possible brown dwarfs

42. HD162020
43. HD 110833
44. BD –04 782
45. HD 112758
46. HD 98230
47. HD 18445
48. HD 29587
49. HD 140913
50. HD 283750
51. HD 89707
52. HD 217580
53. Gl 229

Confirmed Planets Around Pulsars

54. PSR 1257+12
55. PSR B1620–26

Discs (Potentially Protoplanetary or Associated to Planets) (Sorted by Distance to the Sun)

56. Beta Pictoris
57. L 1551
58. BD +31 o 643

The usual nomenclature is given in the details below, with the addition of the mass of the suspected planet, given in Jupiter's mass [J] or Earth's mass [E], and the period of its revolution about the star in days [d] or years [y].

January

55 Cancri	HD 75732	08^h 52.6^m	+28° 20′	February 2	
5.96 m	5.47 M	G8V	0.84 [J]	14.7 [d]	41 l.y.
			> 5 [J]	8 [y]	

47 Ursae Majoris	HD 95128	10^h 59.4^m	+40° 25′	March 7	
5.03 m	4.29 M	G0V	2.41 [J]	3.0 [y]	46 l.y.
See March.					

υ Andromedae	HD 982	01^h 36.8^m	+41° 24′	October 15	
4.1 m	3.45 M	F8V	0.71 (J)	4.62 [d]	44 l.y.
			2.11 (J)	241.2 [d]	
			4.61 (J)	1266.6 [d]	
See October.					

February

55 Cancri	HD 75732	08^h 52.6^m	+28° 20′	February 2	
5.96 m	5.47 M	G8V	0.84 [J]	14.7 [d]	41 l.y.
			> 5 [J]	8 [y]	
Also known as r1 Cancri. This is one of the so-called "51 Peg" planets (see 51 Pegasi). There is recent evidence that a possible second companion may exist. In addition, the star is surrounded by a dust disc extending at least a 40 AU, with an inclination ≈25°. Furthermore, there may by a hole with a radius of ~~10 AU in the disc. The star has a companion r2 Cancri, about 1150 AU away.					

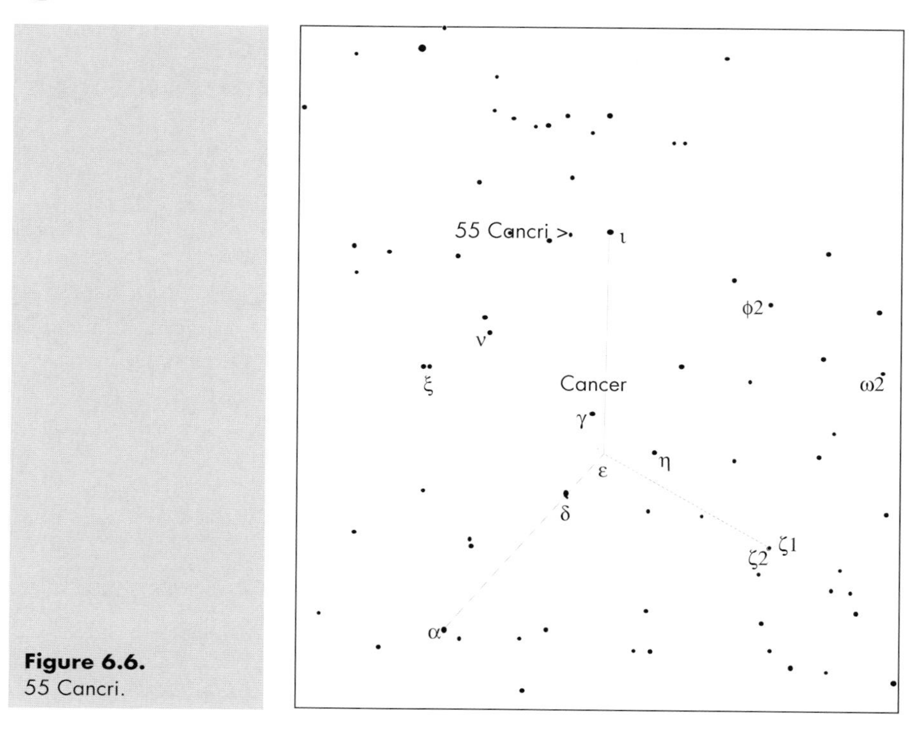

Figure 6.6.
55 Cancri.

47 Ursae Majoris	HD 95128	$10^h 59.4^m$	+40° 25′	March 7	
5.03 m	4.29 M	G0V	2.41 [J]	3.0 [y]	46 l.y.
See March.					

υ Andromedae	HD 982	$01^h 36.8^m$	+41° 24′	October 15	
4.1 m	3.45 M	F8V	0.71 (J)	4.62 [d]	44 l.y.
			2.11 (J)	241.2 [d]	
			4.61 (J)	1266.6 [d]	
See October.					

March

47 Ursae Majoris	HD 95128	$10^h 59.4^m$	+40° 25′	March 7	
5.03 m	4.29 M	G0V	2.41 [J]	3.0 [y]	46 l.y.
This planet is one of the few (perhaps the only one) that appears to fit all the current models and theories about planetary formation. Its orbit, which is circular, lies at about 2.1 AU, which in our Solar system would place it between Mars and Jupiter, , and it is only slightly more massive than Jupiter. If placed in our Solar System, it would fit rather nicely as Jupiter's big brother. Recent analysis of the Hipparchos data suggest an upper mass limit of 7 Jupiters for the planet.					

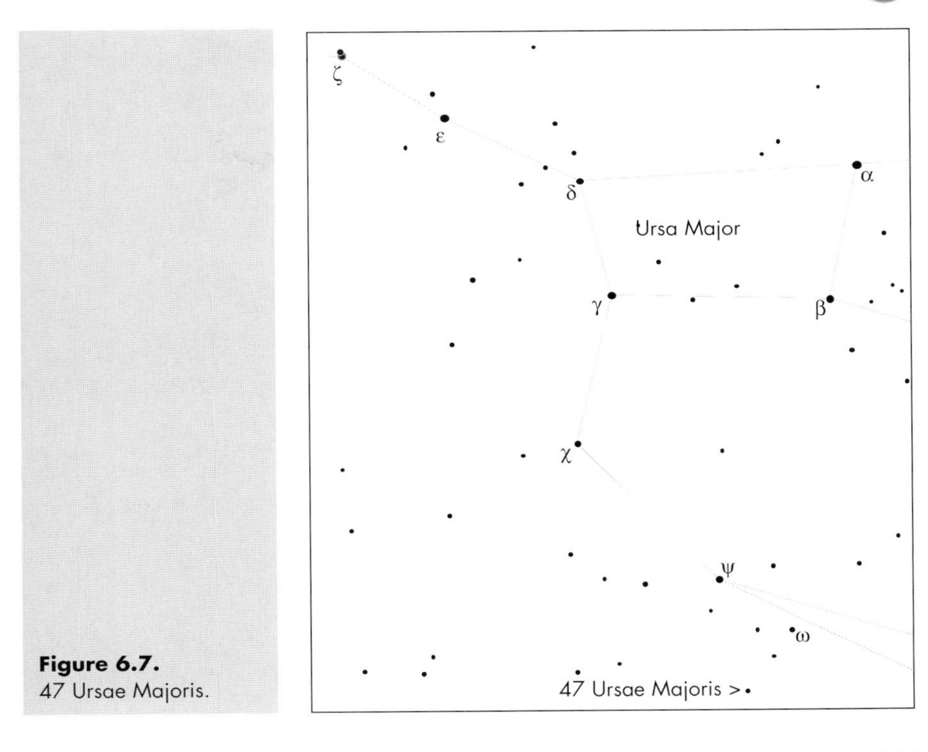

Figure 6.7.
47 Ursae Majoris.

55 Cancri	HD 75732	08^h 52.6^m	+28° 20′	February 2	
5.96 m	5.47 M	G8V	0.84 [J]	14.7 [d]	41 l.y.
			> 5 [J]	8 [y]	
See February.					

70 Virginis	HD 117179	13^h 28.3^m	+13° 46′	April 13	
4.97 m	3.68 M	G5V	6.6 [J]	116.6 [d]	59 l.y.
See April.					

τ Boötis	HD 120136	13^h 47.2^m	+17° 27′	April 18	
4.50 m	3.53 M	F6IV	3.87 [J]	3.3 [d]	51 l.y.
See April.					

April

70 Virginis	HD 117179	13^h 28.3^m	+13° 46′	April 13	
4.97 m	3.68 M	G5V	6.6 [J]	116.6 [d]	59 l.y.

This planet is so large that it may be reclassified as a brown dwarf. This is an object which in intermediate between a planet and a star. The planet also has a very eccentric orbit, with an eccentricity of 0.4 . A value of 0 would be a perfect circle, while a value of 1.0 is a long flattened oval. Mercury and Pluto have the largest eccentricities in our Solar System, with values around 0.2.

τ Boötis	HD 120136	13^h 47.2^m	+17° 27′	April 18	
4.50 m	3.53 M	F6IV	3.87 [J]	3.3 [d]	51 l.y.

This is another of the 51 Peg planets. It is the only system (so far) that has had a probable detection of the starlight reflected by the planet. This, the albedo, is claimed to be detected only in the wavelength range from 456 to 524 nm. The star has a companion (GJ 527B) about 240 AU away.

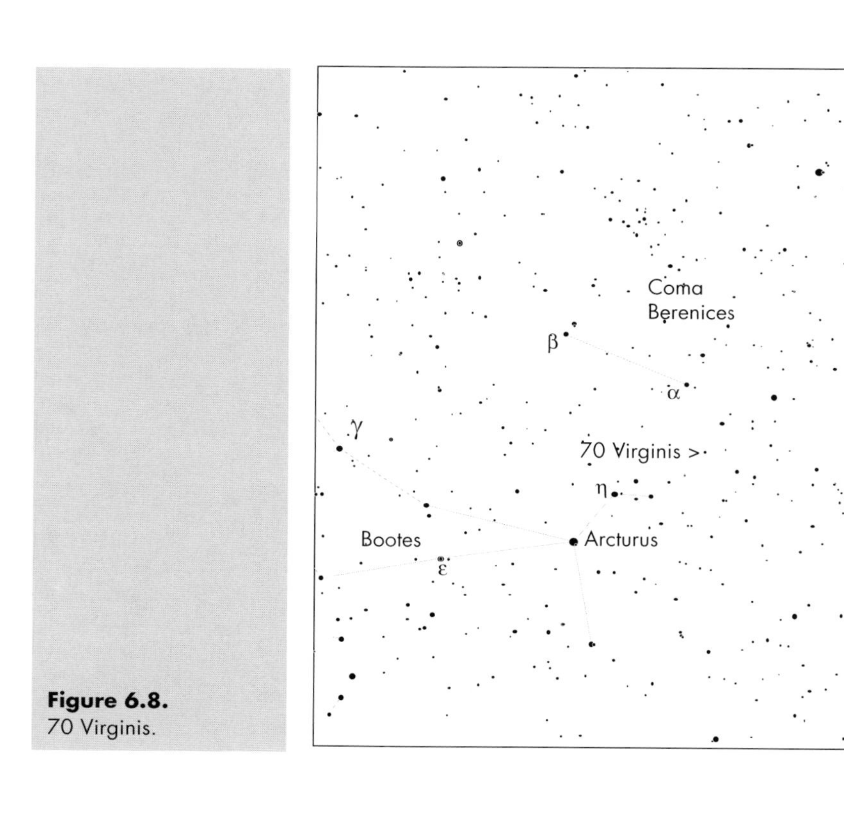

Figure 6.8.
70 Virginis.

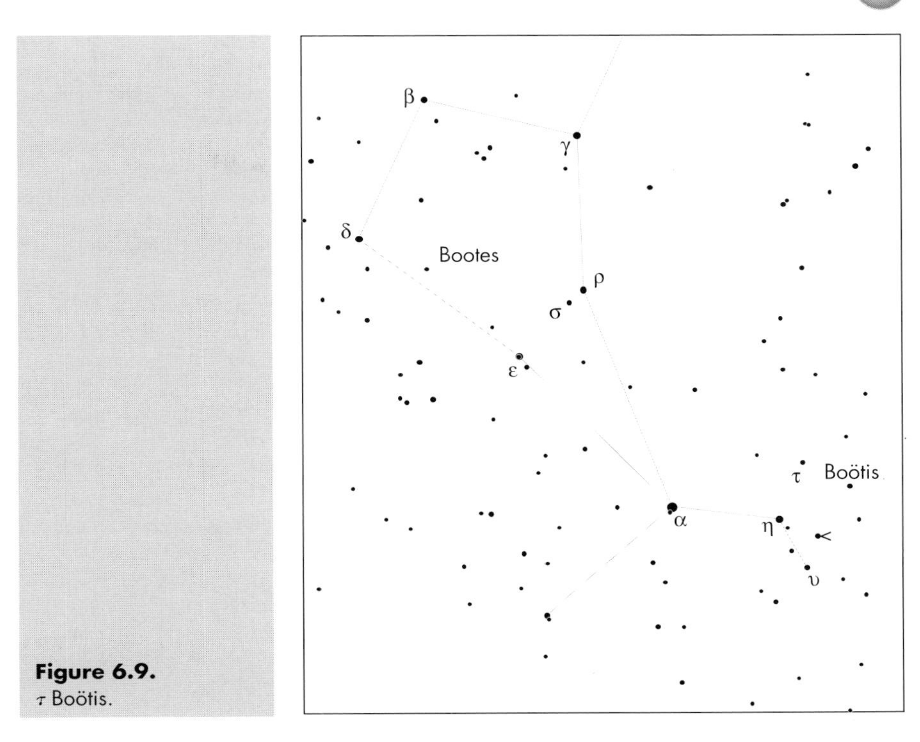

Figure 6.9.
τ Boötis.

3 47 Ursae Majoris	HD 95128	10^h 59.4^m	+40° 25′	March 7	
5.03 m	4.29 M	G0V	2.41 [J]	3.0 [y]	46 l.y.
See March.					

ρ Coronae Borealis	HD 143761	16^h 01.1^m	+33° 18′	May 22	
5.39 m	4.18 M	G2V	1.1 [J]	39.6 [d]	57 l.y.
See May.					

14 Herculis	HD 145675	16^h 10.4^m	+43° 49′	May 24	
6.61 m	5.32 M	K0V	3.3 [J]	1619 [d]	59 l.y.
See May.					

May

ρ **Coronae Borealis**	HD 143761	$16^h\ 01.1^m$	+33° 18′	May 22	
5.39 m	4.18 M	G2V	1.1 [J]	39.6 [d]	57 l.y.

Recent observations using infrared techniques have led astronomers to believe that there is a circumstellar disc of gas and dust around the star. From the disc inclination (46°) a planet of mass of 1.5 Jupiters can be inferred; however, this value differs from other results. The orbital period and amplitude imply a mass of around 1.1 Jupiter masses, and a semi-major axis (which is half the distance across the long axis of an ellipse – is usually referred to as the *average* distance of an orbiting object) of around .23 AU, or roughly half the distance between the Sun and Mercury. In situ formation of such a planet is thought to be unlikely. A more plausible scenario is that the planet formed at several AU from the parent star by means of gas accretion onto a rocky core, and then migrated inward. This could have happened by interactions with another giant gas planet which was ejected in the process, through interactions with the protoplanetary gas disc, or by interactions with planetesimals – the building blocks of planets, formed by accretion in the solar nebula.

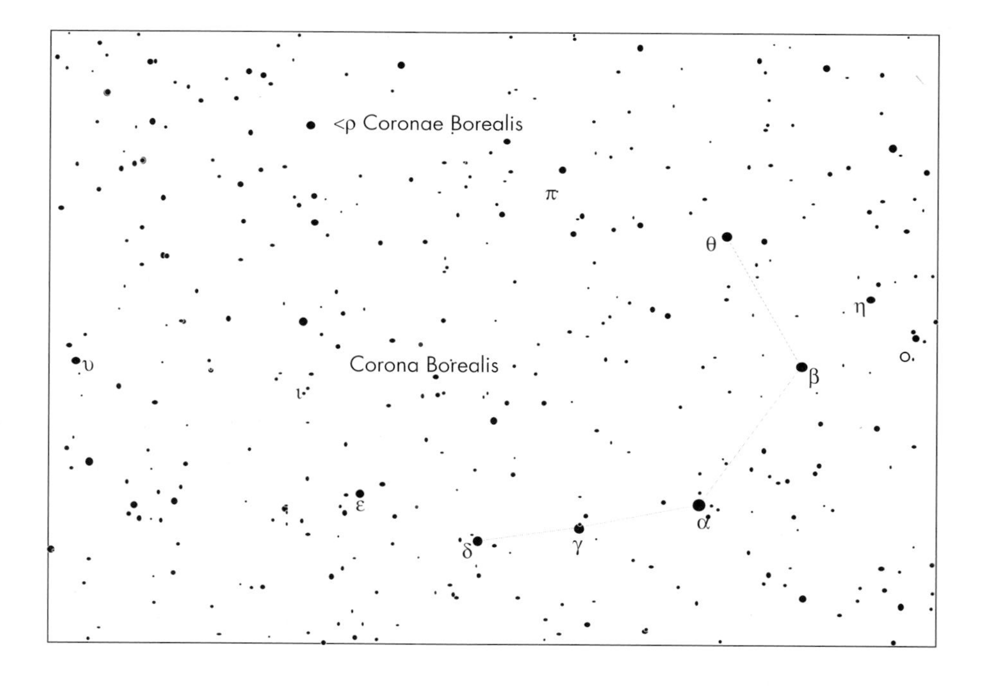

Figure 6.10. *ρ* Coronae Borealis.

14 Herculis	HD 145675	$16^h 10.4^m$	+43° 49′	May 24	
6.61 m	5.32 M	K0V	3.3 [J]	1619 [d]	59 l.y.

14 Herculis (Gliese 614) is a star somewhat less massive than the Sun (80 per cent), and its sole planet has a slightly elongated orbit of 4.4 years. Its mass is about 3.3 times that of Jupiter and it is at a distance of 2.5 AU from 14 Her. This giant planet is twice as close to 14 Her as Jupiter is to our Sun. The content in heavy chemical elements of 14 Her is rather large compared with that of the Sun, a discovery which reinforces the suggestion that giant planets are more frequently observed around metal-rich stars. Heavy chemical elements are needed to form dust or ice particles, and then, by agglomeration, planetesimals and the cores of giant planets. If the quantity of dust is large enough, this is certainly a factor in favour of the formation of giant planets.

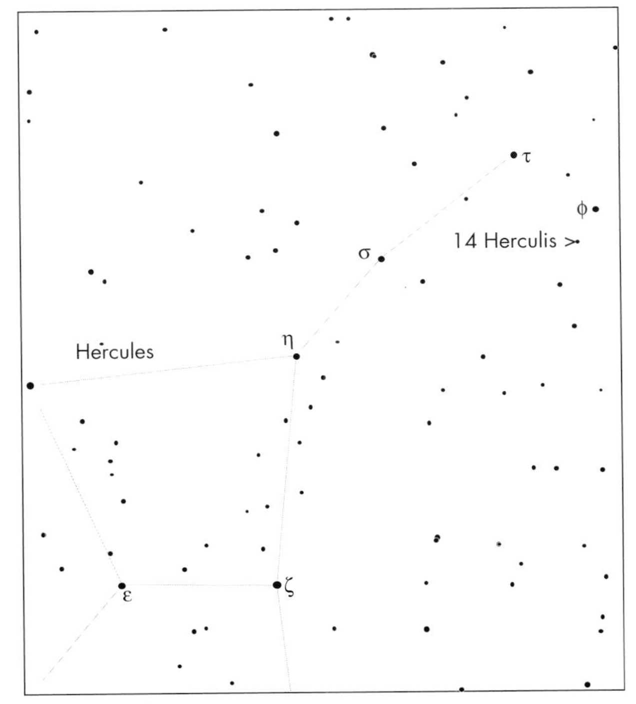

Figure 6.11.
14 Herculis.

70 Virginis	HD 117179	$13^h 28.3^m$	+13° 46′	April 13	
4.97 m	3.68 M	G5V	6.6 [J]	116.6 [d]	59 l.y.

See April.

τ Boötis	HD 120136	$13^h 47.2^m$	+17° 27′	April 18	
4.50 m	3.53 M	F6IV	3.87 [J]	3.3 [d]	51 l.y.

See April.

June

ρ Coronae Borealis	HD 143761	$16^h 01.1^m$	+33° 18′	May 22	
5.39 m	4.18 M	G2V	1.1 [J]	39.6 [d]	57 l.y.

See May.

14 Herculis	HD 145675	$16^h 10.4^m$	+43° 49′	May 24	
6.61 m	5.32 M	K0V	3.3 [J]	1619 [d]	59 l.y.

See May.

70 Virginis	HD 117179	$13^h 28.3^m$	+13° 46′	April 13	
4.97 m	3.68 M	G5V	6.6 [J]	116.6 [d]	59 l.y.

See April.

τ Boötis	HD 120136	$13^h 47.2^m$	+17° 27′	April 18	
4.50 m	3.53 M	F6IV	3.87 [J]	3.3 [d]	51 l.y.

See April.

16 Cygni B	HD 186427	$19^h 41.8^m$	+50° 31′	July 17	
6.25 m	3.40 M	G5V	1.5 [J]	804 [d]	70 l.y.

See July.

July

16 Cygni B	HD 186427	$19^h 41.8^m$	+50° 31′	July 17	
6.25 m	3.40 M	G5V	1.5 [J]	804 [d]	70 l.y.

The star is a visual binary, and the companion, 16 Cyg A, is about 700 AU away. The planet also has a very large eccentricity, value 0.6, which is causing some concern among astronomers, as they cannot explain it!

ρ Coronae Borealis	HD 143761	$16^h 01.1^m$	+33° 18′	May 22	
5.39 m	4.18 M	G2V	1.1 [J]	39.6 [d]	57 l.y.

See May.

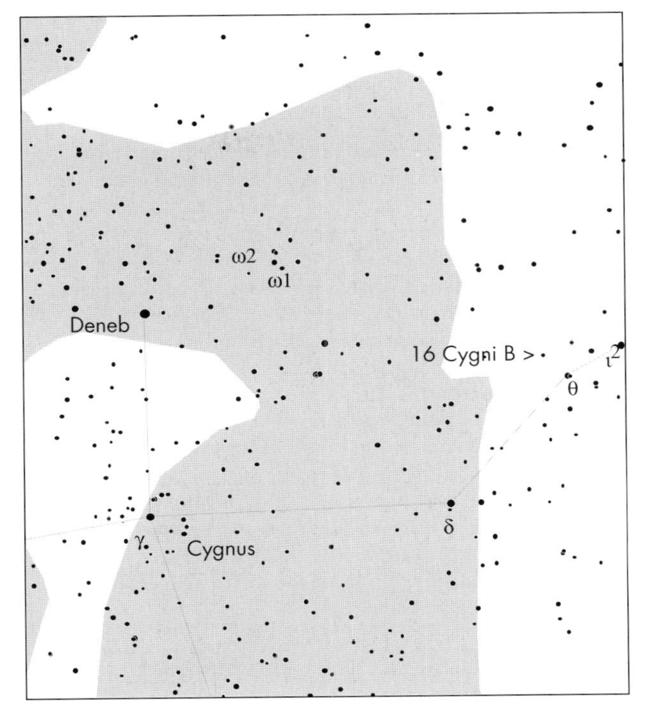

Figure 6.12.
16 Cygni B.

14 Herculis	HD 145675	$16^h\ 10.4^m$	+43° 49′	May 24	
6.61 m	5.32 M	K0V	3.3 [J]	1619 [d]	59 l.y.
See May.					

August

16 Cygni B	HD 186427	$19^h\ 41.8^m$	+50° 31′	July 17	
6.25 m	3.40 M	G5V	1.5 [J]	804 [d]	70 l.y.
See July.					

51 Pegasi	HD 217014	$22^h\ 57.4^m$	+20° 46′	September 5	
5.45 m	4.52 M	G5V	0.47 [J]	4.23 [d]	50 l.y.
See September.					

September

51 Pegasi	HD 217014	$22^h\ 57.4^m$	+20° 46′	September 5	
5.45 m	4.52 M	G5V	0.47 [J]	4.23 [d]	50 l.y.

This is a peculiar type of system characterised by orbital periods shorter than 15 days. The orbits are small, with radii less than 0.11 AU, which is about a tenth the distance between the Earth and the Sun. Such an orbit is in fact much smaller than that of Mercury's (radius 0.38 AU, period 88 days). However, these planets are similar in mass to that of Jupiter and in some cases even larger. Research indicates that the planets have circular orbits. The 51 Peg planets are a problem because they do not fit current planet formation theory. This predicts that giant planets like those in our Solar System (Jupiter, Saturn, Neptune and Uranus) should be formed in the colder, more distant parts of a proto-planetary disc, some 5 AU from a star. It seems that a possible solution to this problem may be the rotation of a young star. Such a star spins very rapidly, perhaps making one complete revolution in about 5 to 10 days. As a planet approaches the star, tides will be raised on the star. However, the star is spinning quicker than the planet orbits around it, and so the bulge caused by the planet on the star will move in front of the motion of the planet, rather than staying on a straight line directly in line with the planet. The net result will be that the gravitational pull from the tide on the star pulls the planet closer. A circumstellar disc was searched for using the UKIRT and KECK telescopes, but no evidence for such an object was detected. The temperature of the planet is calculated to be in the range 1200 to 1400 °K.

υ Andromedae	HD 982	$01^h\ 36.8^m$	+41° 24′	October 15	
4.1 m	3.45 M	F8V	0.71 (J)	4.62 [d]	44 l.y.
			2.11 (J)	241.2 [d]	
			4.61 (J)	1266.6 [d]	

See October.

16 Cygni B	HD 186427	$19^h\ 41.8^m$	+50° 31′	July 17	
6.25 m	3.40 M	G5V	1.5 [J]	804 [d]	70 l.y.

See July.

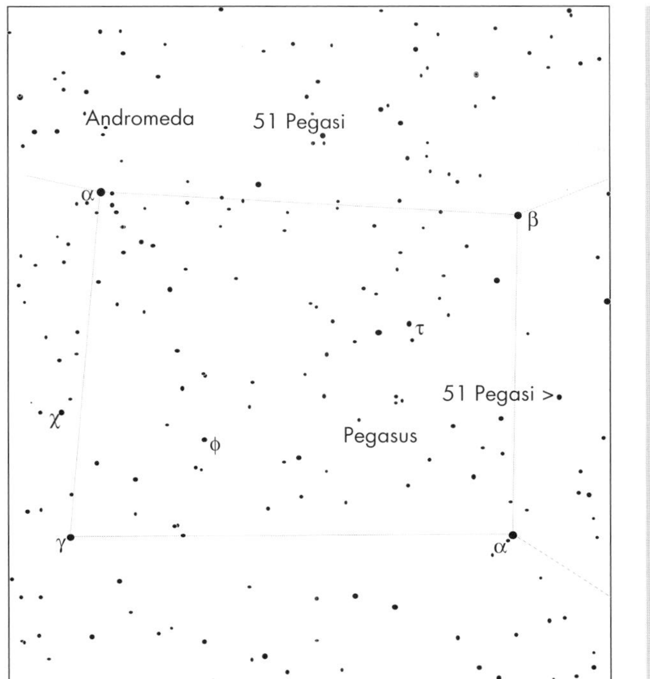

Figure 6.13.
51 Pegasi.

October

υ **Andromedae**	HD 982	$01^h 36.8^m$	+41° 24′	October 15	
4.1 m	3.45 M	F8V	0.71 (J)	4.62 [d]	44 l.y.
			2.11 (J)	241.2 [d]	
			4.61 (J)	1266.6 [d]	

This is another of the 51 Peg-type planets, and was the first multiple planet system discovered. The innermost (and first discovered) planet contains at least three-quarters of the mass of Jupiter and orbits only 0.06 AU from the star. The middle planet resides approximately 0.83 AU from the star, similar to the orbital distance of Venus. The outermost planet has a mass of at least four Jupiters and completes one orbit every 3.5 to 4 years, placing it 2.5 AU from the star. The two outer planets are both recent discoveries and have elliptical orbits. The formation of these planets has posed several problems. The usual picture is that gas giant planets form at least four AU away from a star, where temperatures are low enough for ice to condense and start the process of planet formation, but all three giant planets around Upsilon Andromedae now reside inside this theoretical ice boundary. A circumstellar disc was searched for using the UKIRT and KECK telescopes, but no evidence for such an object was detected.

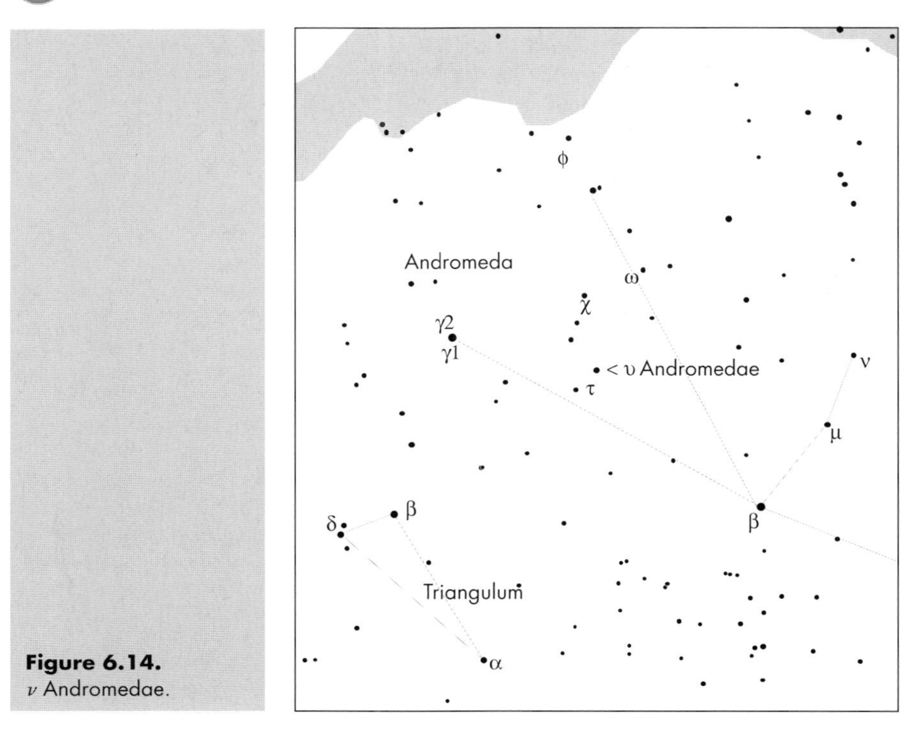

Figure 6.14.
υ Andromedae.

51 Pegasi	HD 217014	22^h 57.4^m	+20° 46′	September 5	
5.45 m	4.52 M	G5V	0.47 [J]	4.23 [d]	50 l.y.
See September.					

November

***υ* Andromedae**	HD 982	01^h 36.8^m	+41° 24′	October 15	
4.1 m	3.45 M	F8V	0.71 (J)	4.62 [d]	44 l.y.
	2.11 (J)	241.2 [d]			
	4.61 (J)	1266.6 [d]			
See October.					

51 Pegasi	HD 217014	22^h 57.4^m	+20° 46′	September 5	
5.45 m	4.52 M	G5V	0.47 [J]	4.23 [d]	50 l.y.
See September.					

December

υ Andromedae	HD 982	01h 36.8m	+41° 24′	October 15	
4.1 m	3.45 M	F8V	0.71 (J)	4.62 [d]	44 l.y.
	2.11 (J)	241.2 [d]			
	4.61 (J)	1266.6 [d]			
See November.					

55 Cancri	HD 75732	08h 52.6m	+28° 20′	February 2	
5.96 m	5.47 M	G8V	0.84 [J]	14.7 [d]	41 l.y.
		> 5 [J]	8 [y]		
See February.					

Appendix 1
Optical Filters

One of the most useful accessories an amateur can possess is one of the ubiquitous optical filters. Having been accessible previously only to the professional astronomer, they came onto the marker relatively recently, and have made a very big impact. They are useful, but don't think they're the whole answer! They can be a mixed blessing.

From reading some of the advertisements in astronomy magazines you would be correct in thinking that they will make hitherto faint and indistinct objects burst into vivid observability. They don't.

What the manufacturers do not mention is that regardless of the filter used, you will still need dark and transparent skies for the use of the filter to be worthwhile. Don't make the mistake of thinking that using a filter from an urban location will *always* make objects become clearer. The first and most immediately apparent item on the downside is that in all cases the use of a filter *reduces* the amount of light that reaches the eye, often quite substantially. The brightness of the field of view and the objects contained therein is reduced.

However, what the filter does do is select specific wavelengths of light emitted by an object, which may be swamped by other wavelengths. It does this by suppressing the unwanted wavelengths. This is particularly effective in observing extended objects such as emission nebulae and planetary nebulae.

In the former case, use a filter that transmits light around the wavelength of 653.2 nm, which is the spectral line of hydrogen alpha (Hα), and is the wavelength of light responsible for the spectacular red colour seen in photographs of emission nebulae. Some filters may transmit light through perhaps two wavebands: 486 nm for hydrogen beta[1] (Hβ) and 500.7 nm for oxygen-3 (OIII), two spectral lines which are very characteristic in planetary nebula. Use of such filters will enhance the faint and delicate structure within nebulae, and, from a dark site, they really do bring out previously invisible detail.

[1]This filter can be used to view dark nebulae that are overwhelmed by the proximity of emission nebulae. A case in point is the Horsehead Nebula, which is incredibly faint, and swamped by light from the surrounded emission nebulosity.

Don't forget (as the advertisers sometimes seem to) that "nebula" filters do not (usually) transmit the light from stars, and so when in use, the background will be pitch-black with only nebulosity visible. This makes them virtually useless for observing stars, star clusters and galaxies.

One kind of filter that does help in heavily light-polluted areas is the LPR (light pollution reduction) filter, which effectively blocks out the light emitted from sodium and mercury street lamps, at wavelengths 366, 404.6, 435.8, 546.1, 589.0 and 589.5 nm. Clearly, the filter will be effective only if the light from the object you want to see is significantly different from the light-polluting source: fortunately, this is usually the case. Light pollution reduction filters can be very effective visually and photographically, but remember that there is always some overall reduction in brightness of the object you are observing.

Whatever filters you decide on, it is worthwhile trying to use them before you make a purchase (they are expensive!), by borrowing them either from a fellow amateur or from a local astronomical society. This will show you whether the filter really makes any difference to *your* observing.

There is no doubt that modern filters can be an excellent purchase, but it may be that your location or other factors will prevent the filter from realising its full potential or value for money. Most commercially available filters are made for use at a telescope and not for binoculars, so unless you are mechanically minded and can make your own filter mounts (and are happy to pay – two LPR filters could easily cost more than the binoculars!), it's likely that only those observers with telescopes can benefit.

Appendix 2
Books, Magazines and Organisations

There are many fine astronomy and astrophysics books in print, and to choose between them is a difficult task. Nevertheless I have selected a few which I believe are among the best on offer. I do not expect you to buy, or even read them all, but it would be in your better interests to check at your local library to see if they have some of them.

Star Atlases and Observing Guides

Ridpath, I. (ed.) (1999) Norton's star atlas and reference handbook. Longmans, Harlow.
Tirion, W., Sinnott, R. (1999) Sky atlas 2000.0. Sky Publishing and Cambridge University Press, Cambridge, MA.
Sinnott, R., Perryman, M. (1999) Millennium star atlas. Sky Publishing, Cambridge, MA.
Luginbuhl, C., Skiff, B. (1990) Observing handbook and catalogue of deep-sky objects. Cambridge University Press, Cambridge, MA.
Kepple, G., Sanner, G. (1999) The night sky observer's guide, vols I and II. Willman-Bell, Richmond, VA.
O'Meara, S. (1999) Deep-sky companions: the Messier objects. Cambridge University Press, Cambridge, UK.
Ratledge, D. (2000) Observing the Caldwell objects. Springer, London.
Burnham, R. (1978) Burnham's celestial handbook. Dover Books, New York.

Astronomy and Astrophysics Books

Sidgwick, J. (1979) Amateur Astronomer's Handbook. Pelham Books, London.
Kitchin, C. (1998) Astrophysical Techniques. Institute of Physics, Bristol.
Kitchin, C. (1995) Telescopes and Techniques. Springer, London.
Bless, R. (1996) Discovering the cosmos. University Science Books, Sausilito, ??STATE??.
Kaufmann, W., Comins, N. (1996) Discovering the universe. Freeman, New York.
Bennett, J., Donahue, M., Schneider, N., Voit, M. (1999) The cosmic perspective. Addison-Wesley, Reading, MA.
Fraknoi, A., Morrison, D., Wolff, S. (2000) Voyages through the universe. Saunders, Philadelphia, PA.
Zeilik, M., Gregory, S., Smith, E. (1999) Introductory astronomy and astrophysics. Saunders, Philadelphia, PA.
Phillips, A. (1994) The physics of stars. Wiley, Chichester.
Kitchin, C. (1987) Stars, nebulae and the interstellar medium. Hilger, Bristol.
Elmegreen, D. (1998) Galaxies and galactic structure. Prentice-Hall, Englewood Cliffs, NJ.

Magazines

Astronomy Now (UK)
New Scientist (UK)
Sky & Telescope (USA)
Nature (UK)
Science (USA)
Scientific American (USA)

The first three magazines are aimed at a general audience and so are applicable to everyone; the last three are aimed at the well-informed layperson. In addition there are many research-level journals which can be found in university libraries and observatories.

Organisations

The Federation of Astronomical Societies, 10 Glan y Llyn, North Cornelly, Bridgend County Borough, CF33 4EF, UK.
[http://www.fedastro.demon.co.uk/]
Society for Popular Astronomy, The SPA Secretary, 36 Fairway, Keyworth, Nottingham, NG12 5DU. UK.
[http://www.popastro.com/]
The British Astronomical Association, Burlington House, Piccadilly, London, W1V 9AG, UK.
[http://www.ast.cam.ac.uk/~~baa/]
The Royal Astronomical Society, Burlington House, Piccadilly, London, W1V 0NL.
[http://www.ras.org.uk/membership.htm]
Campaign for Dark Skies, 38 The Vineries, Colehill, Wimborne, Dorset, BH21 2PX, UK.
[http://www.dark-skies.freeserve.co.uk/]

Appendix 3

The Greek Alphabet

The following is a quick reference guide to the Greek letters, which are used in the Bayer classification system. Each entry shows the uppercase letter, the lowercase letter, and the pronunciation.

A α Alpha	H η Eta	N ν Nu	T τ Tau
B β Beta	Θ θ Theta	Ξ ξ Xi	Υ υ Upsilon
Γ γ Gamma	I ι Iota	O o Omicron	Φ ϕ Phi
Δ δ Delta	K κ Kappa	Π π Pi	X χ Chi
E ϵ Epsilon	Λ λ Lambda	P ρ Rho	Ψ ψ Psi
Z ζ Zeta	M μ Mu	Σ σ Sigma	Ω ω Omega

Index

The entry for an astronomical object refers to its most familiar name and/or its main entry in the book.